MUSCLE, MATRIX, AND BLADDER FUNCTION

ADVANCES IN EXPERIMENTAL MEDICINE AND BIOLOGY

Recent Volumes in this Series

Volume 379
SUBTILISIN ENZYMES: Practical Protein Engineering
Edited by Richard Bott and Christian Betzel

Volume 380
CORONA- AND RELATED VIRUSES: Current Concepts in Molecular Biology and Pathogenesis
Edited by Pierre J. Talbot and Gary A. Levy

Volume 381
CONTROL OF THE CARDIOVASCULAR AND RESPIRATORY SYSTEMS IN HEALTH AND DISEASE
Edited by C. Tissa Kappagoda and Marc P. Kaufman

Volume 382
MOLECULAR AND SUBCELLULAR CARDIOLOGY: Effects of Structure and Function
Edited by Samuel Sideman and Rafael Beyar

Volume 383
IMMUNOBIOLOGY OF PROTEINS AND PEPTIDES VIII: Manipulation or Modulation of the Immune Response
Edited by M. Zouhair Atassi and Garvin S. Bixler, Jr.

Volume 384
FATIGUE: Neural and Muscular Mechanisms
Edited by Simon C. Gandevia, Roger M. Enoka, Alan J. McComas, Douglas G. Stuart, and Christine K. Thomas

Volume 385
MUSCLE, MATRIX, AND BLADDER FUNCTION
Edited by Stephen A. Zderic

Volume 386
INHIBITORS TO COAGULATION FACTORS
Edited by Louis M. Aledort, Leon W. Hoyer, Jeanne M. Lusher, Howard M. Reisner, and Gilbert C. White III

Volume 387
BIOLOGICAL REACTIVE INTERMEDIATES V: Basic Mechanistic Research in Toxicology and Human Risk Assessment
Edited by Robert Snyder, I. Glenn Sipes, George F. Kalf, David J. Jollow, Helmut Greim, James J. Kocsis, Terrence J. Monks, and Charlotte M. Witmer

---

A Continuation Order Plan is available for this series. A continuation order will bring delivery of each new volume immediately upon publication. Volumes are billed only upon actual shipment. For further information please contact the publisher.

MUSCLE, MATRIX, AND BLADDER FUNCTION

Edited by

Stephen A. Zderic

University of Pennsylvania
School of Medicine
Children's Hospital of Philadelphia
Philadelphia, Pennsylvania

PLENUM PRESS • NEW YORK AND LONDON

Library of Congress Cataloging-in-Publication Data

On file

Proceedings of a symposium on Muscle, Matrix, and Bladder Function,
held March 18–19, 1994, in Philadelphia, Pennsylvania

ISBN 0-306-45193-X

A Division of Plenum Publishing Corporation
233 Spring Street, New York, N. Y. 10013

10 9 8 7 6 5 4 3 2 1

Printed in the United States of America

PREFACE

This conference and monograph were the result of many collective efforts. The whole concept was formulated one early Wednesday morning at our weekly research meeting at Children's Hospital in our division of urology. We have been most fortunate to have a close collaboration with Bob Levin, Ed Macarak, and Pam Howard who have helped steer the course of our division's growing interest in basic science. At our weekly meetings our laboratory fellow will summarize their current work. Other ongoing areas of investigation in our labs and elsewhere are discussed. We have always made an effort to try and understand what other groups are doing who are working in the area of bladder smooth muscle research. It occurred to us that the best way to really know what everyone working in this field was doing would be to sponsor a 2-day meeting where we could all gather to discuss our ongoing work. A major limitation of the annual meeting of the American Urologic Association or the urology section of the American Academy of Pediatrics is that the scientfic sessions are limited as these are meant to be primarily clinical meetings (as they should be). For this reason the idea of a meeting devoted solely to research about the urinary bladder had great appeal. In addition to allowing for longer presentations than the standard 5 to 7 minutes, every effort would be made to encourage a dialogue amongst the presenters and the audience. We were pleased that so many people took their time to come to Philadelphia, and participate.

No editor should ever fail to acknowledge the many assistants without whom such a meeting and book could be completed. It all started with Olga Korytko from our Office of Medical Education at Children's who tended to the details of putting on such a meeting. The administration of this effort was made so much easier thanks to her experience and know how. My office staff in the division of urology fielded endless phone calls and last minute requests, and over the course of the past two years have handled countless changes in scheduling and other last minute requests. To Marolyn Buchanan I extend special thanks for her assistance with the pesky details associated with the preparation of a camera-ready manuscript. At Plenum Publishing Eileen Bermingham and Robert Wheeler provided valuable guidance in steering this project to completion. A listing of our corporate and NIH sponsors appears on the next page and I am grateful to them for the interest they have shown.

No project like this ever gets completed without a major boost from one's family and I wish to thank my wife Kate (who has the stamina to direct a busy department of pediatric emergency medicine and organize our home), and our children, Olivia and Colin, for all they have given me. Kate and I wish to acknowledge her parents Joan and Charles Cronan, who always seem to magically appear when our schedules are on a collision course. They have all tolerated the endless piles of manuscripts, late hours, and occasional lapses of memory that often pulled me away physically and mentally, at times when I was especially needed. I am forever grateful for their inspiration, love, and support.

Stephen A. Zderic MD

ACKNOWLEDGMENTS

Support from the following organizations is gratefully acknowledged, for without this funding, the meeting and this book would not have been possible.

The National Institutes of Health
Conference Grant R-13 DK DK47242

Smith Kline Beecham Pharmaceuticals

Sandoz Pharmaceuticals

Glaxo Research Institute

Ortho McNeil Pharmaceuticals

Pfizer Research PLC

C.R. Bard Inc.

CONTENTS

THE CLINICAL RELEVANCE OF BASIC SCIENCE: A SURGEON'S PERSPECTIVE

John W. Duckett, M.D.

Division of Urology
Children's Hospital of Philadelphia
Philadelphia, PA, 19104

Pediatric urology has come a long way since 1970 when I began my career at The Children's Hospital of Philadelphia. For the first twenty years, efforts were focused on establishing the clinical subspecialty, gaining the respect first of the Pediatric Surgical discipline followed by our urology colleagues. Our Clinical Fellowship Program began in 1976, and we have seeded 25 outstanding products of the division across the country; all but two in academic centers. Although many of our fellows have made significant research contributions, it was apparent that the "new generation" of pediatric urologists should have a solid background in the basic sciences. This credibility will afford them a more effective edge as they move into a major research university so they may better focus a meaningful research enterprise. Our pediatric urology subspecialty must expand the third leg of the tripod: that is, success in basic sciences research - to complement the other two legs - education and clinical practice.

"The great Aim and End of all Learning should be to enable one to serve Humanity" (Benjamin Franklin, founder of the University of Pennsylvania).

With this new challenge for pediatric urology, we began a two year Fellowship Program in 1989, providing our Fellows with one full year of uninterrupted solid basic science. Stephen Zderic was the first, homegrown from the University of Pennsylvania Urology Training Program. It is Steve who has convened this NIH sponsored workshop on Muscle, Matrix and Bladder Function. David Ewalt followed and is now well established at Southwestern Medical Center in Dallas; Larry Baskin returned to UCSF; Douglas Coplen will be returning to Washington University in St. Louis, Harry Koo from Columbia is just finishing his laboratory effort and is being recruited now. Gregory Dean begins in July and has just been awarded the Valentine prize for his research grant. We fell that pediatric urology is enticing the most outstanding young urology residents in the country to our recognized subspecialty. The C.V.'s of the 20 applicants interviewed for this year's slot are so impressive that they may well be considered for Associate Professor slots at a teaching institution.

The question of priorities in research has now become the clinician's great concern. To avoid a scattering of small disjointed animal experiments, we are fortunate to enjoy collaboration with our own PhD basic urologists which is mutually beneficial. Not only are we able to deliver eager young scientists with great backgrounds able to embark on individual projects, but along with it comes the insight of the clinical needs which the scientists require. We have enjoyed an open door with Bob Levin in Alan Wein's department for many years. Ed Macarak and Pam Howard from the Connective Tissue Institute are now situated in the Penn Dental School with our clinical affiliations. Also contributing to this growing quorum are Sam Chacko is from the Veterinary School, and Joel Rosenbloom, an elastic fiber expert from the Dental School. Within our own division, Stephen Zderic continues his work on intracellular cancer handling. This fine group of scientists are collaborating in a Program Project Grant which has been submitted to NIH.

A classic example of basic research and its applicability is illustrated by the visit of Prime Minister Gladstone to the laboratory of Michael Faraday in the late 19th Century. Gladstone asked Faraday whether he thought this esoteric substance called electricity would ever have any practical uses. Faraday's reply was "One day, sir, you will tax it." Many similar examples sustain the fundamental goal of scientists, that good basic research almost invariably becomes of great value to mankind. Chick. Koop, our famous Surgeon General, whose medical career was centered here at Penn as Chief of Surgery at Children's Hospital, promoted the warning recently as a spokesman for the Clinton Healthcare Program:

"Warning: Insufficient medical research can be hazardous to your health"

Government funding through the National Institutes of Health has taken on a more 'hands on' primary responsibility of Congress whose job it is to identify the needs of its constituents in broad strategic goals. As Dan Koshland, Editor of Science says in a recent editorial, "It would be foolish of scientists to claim they have exclusive rights to priority-setting and folly of Congress to believe they can decide scientific feasibility without the advice of scientists". In Washington, the special interest lobbying groups for a number of health causes have increased from several hundred to nearly a thousand over the last several years. We clinicians have to show the relevance and need for our basic science and we must be able to effectively communicate this with our congressional representatives - an enormous challenge.

The outgrowth of our basic science focus at Penn was first stimulated by our desire to understand the dilated urinary tract in the fetus. This obviously has many ramifications as to the morphology and molecular biology of bladder and ureteral smooth muscle function, the pathophysiology of loss of compliance, and the development of the normal fetal urinary tract, for which we have chosen the bovine model. What may be the ultimate clinical relevance that impresses the grant givers? If we could alter the aging process which leads to loss of bladder compliance, billions of dollars could be saved keeping our elderly citizens out of nursing homes because of incontinence. We must not forget that urinary incontinence is the leading reason for which the elderly are institutionalized. Could an outpatient needle biopsy of the bladder with laboratory measurements of the changes in bladder extracellular matrix gene expression and protein levels predict the potential degree of bladder dysfunction? Can we alter with gene therapy and/or new pharmaco-therapy the effects of obstructive uropathy? In a world in which science will play crucial roles in almost every aspect of life, both the Gladstones and Faradays will be needed. A clinician cannot only create better avenues of communication, but should better focus the relevance of the science.

Perhaps a current analogy is apropos of what could happen without congenial collaboration of clinicians and scientists. Jerry Jones, the owner of the Dallas Cowboys had a good thing going with Jimmy Johnson, his enormously successful coach who is now replaced by Barry Switzer from Oklahoma. All three of them were teammates on the 1964 National Championship team at Arkansas. Jerry Jones went off to make millions in oil and bought the Cowboys, bringing his buddy Jimmy Johnson back from Miami to coach them to two Superbowl Championships within five years, an impressive accomplishment. Jerry Jones, however, failing to receive the recognition afforded Jimmy Johnson, has destroyed the whole system, hardly a happy symbiotic relationship. The clinicians can get their kicks from having their fingers in the research pies and tasting the excitement of success, while the kudos go to the real scientists who put together the touchdowns and eventually the Super bowl champions.

This symposium is your chance to give us the basic understanding of muscle, matrix, and bladder function while we ponder the excitement of what your revelations will make us alter in the operating rooms, clinics and academic conferences. Our rewards come from the clinical application of your science. Working together offers the mutual reward that Ben Franklin challenged "to serve Humanity."

BLADDER OUTLET OBSTRUCTION - AN OVERVIEW

Alan J. Wein M.D.

Division of Urology
Hospital of The University of Pennsylvania
University of Pennsylvania School of Medicine
Philadelphia, Pennsylvania, 19104

The first session of this meeting is dedicated to an in depth discussion about current themes in bladder outlet obstruction, and Steve had asked me to do a moderator's overview of this subject. I looked up what overview meant and it means it kind of a short summary. Consequently, I thought it would be kind of fun to put together a talk that is perhaps a bit short on facts but long on speculation. I would simply like to give you some things to think about during the first session of the conference, and also throughout the rest of it. I also want to stress what information we as clinical surgeons need in order to get the best outcomes for our patients, and where we feel that information is most lacking.

How can we begin to approach a patient who presents with the symptoms associated with bladder outlet obstruction. Basically, the diagnostic possibilities are anatomic obstruction, detrusor instability, or the loss of detrusor contractility. These last two possibilities may actually be secondary to the outlet obstruction. The diagnosis and characterization of bladder outlet obstruction when present is on the basis of symptoms, urodynamic studies (flowing, filling and voiding cystometrograms), and post-study processing of plots relating to detrusor pressure and flow, and maybe cystoscopy. Basically, what you can come out with is up to seven combinations including: 1)obstruction by itself, 2)instability by itself, 3)impaired detrusor contractility by itself, and various combinations of these three basic categories. Now unfortunately, all we are really able to do is make the diagnosis of any secondary bladder changes which occur. If we do that, then we can make a therapeutic choice. We can either not treat the patient (elect to wait at this point), we can treat the outlet obstruction by itself, we can treat only the secondary bladder changes. We can do that either by trying to institute corrective treatment or in some cases, trying to prevent them by prophylactic treatment. A good example of this is trying to manage myelomeningocele patients by maintaining their bladder storage pressures below 40 cm water pressure. This may be done medically with the use of anti-cholinergic agents with or without the use of clean intermittent catheterization. Alternatively we can treat the "outlet obstruction" (physiologic obstruction secondary to dyssinergia) by the use of sphincterotomy, balloon dilation, or even vesicostomy in the young patient. In some cases combinations of this kind of therapy may be utilized.

The problem is that urodynamics are the only tools we have right now can only make the diagnosis of outlet obstruction. It can't predict if the outlet is reversible or irreversible, or whether it is going to be progressive or non progressive. So urodynamics alone, which is the only clinical tool we have right now, can't decide whether treatment is necessary and it can't decide on the likelihood of success. There are other unresolved issues. 1)Can we prevent secondary bladder changes, what indicates that these bladder changes will become progressive? Are secondary bladder changes reversible? What is the optimal management of obstruction and any associated secondary bladder changes? What types of management should be utilized in these settings, and what should the timing of therapy be?

Basically, these are all the secondary bladder changes that can occur in response to outlet obstruction. The first area is really the only area we know a lot about. All of these other areas, are areas you are going to hear about changes in over the next couple of days. I feel that

Muscle, Matrix, and Bladder Function,
Edited by S. Zderic, Plenum Press, New York, 1995

these areas are where the advances are really going to come. We know that urodynamic activity in response to outlet obstruction and secondary bladder changes only exist basically in four different varieties: decreased compliance, increased compliance, instability which is involuntary bladder contractions, and impaired detrusor contractility. The macroscopic appearance of the bladder that is obstructed usually is that of increased weight and trabeculation. Light microscopy shows hypertrophy, fibrosis with collagen deposition, and decreased neural density. When one looks at the ultrastructure, one can see various changes that are associated with bladder outlet obstruction. This has been really only recent, but we haven't yet been put in the position where we use them in the diagnosis of progressive detrusor damage. However, recently, in the Journal of Urology, Ahmed El Badawi, Neal Resnick, and Subarao Yalla published a group of excellent papers (1). Anyone working in this area who has not seen these contributions must take the time to look them up. They correlated electron micrograph findings with various types of detrusor dysfunction. This is what they found in obstructive detrusor dysfunction with an almost 100% correlation in a group of patients evaluated and treated for voiding dysfunction. There were separations of enlarged smooth muscle cells with increased deposition of collagen between the muscle bundles resulting in diminished intermediate cell junctions. Superimposed on this were degeneration of smooth muscle cells and axons in the presence of impaired detrusor contractility. In the presence of bladder hyperactivity there were abundant axonal protrusions and ultra close abutments of axons to muscle fibers. They concluded that "Protrusion junctions and abutments probably mediate electrical coupling of muscle cells leading to involuntary contractions in the unstable detrusor associated with outlet obstruction. Excessive deposits of elastic fibers (hyperelastosis) between the muscle bundles and in the interstitium are probably the structural basis for increased bladder distensibility and chronic retention."

From a neurophysiologists perspective, Bill Steers and Chet DeGroat, will tell us about the nerve density changes that occur in response to outlet obstruction as well as changes that occur in reflex patterns. They have referred to these adaptations as neuroplasticity, and their contributions to this are have been most prolific. What we know most about at Penn is the neuropharmacology of bladder outlet obstruction that Bob Levin will be speaking about. A mild partial obstruction model in the rabbit would get increased bladder mass and increased frequency. The contractile response initially decreases and then goes back up to some 80% of its baseline in response to various forms of pharmacologic agents or electrical field stimulation. DNA content increases as does thymidine incorporation. One might say that in this stage there is secondary increase in smooth muscle mass (adaptive hypertrophy), with no secondary factors, and in this stage the bladder is said to be compensated because it actually works better in overcoming the obstruction. With an increase in obstruction , one begins to see functional impairment of contraction which is definitely associated with an increase of bladder mass. Possible explanations for this include a decrease in innervation with simultaneous production of various factors cells and cell production of key elements in force generation. This is referred to as the decompensation phase. It also had changes in the activation mechanisms of smooth muscle (excitation contraction coupling) and some of the speakers will talk about various changes in each area such as myogenic factors, or changes in neurotransmitters that trigger contraction.

Some of the speakers will talk about what happens to the mechanical properties of the bladder during the storage phase. Just what is this decrease in compliance due to? Is it due actually to collagen and what does that collagen formation do to the detrusor myocytes? During emptying what is important is the development of maximal active forces and shortening velocity which we will also be hearing about. These parameters, unfortunately, are not measured with our conventional present day urodynamic studies.

Jacek Mostwin will tell us about the various changes in electrophysiology that occur with bladder outlet obstruction. We will hear about changes in calcium fluxes, perhaps as calcium channels triggering bladder smooth muscle are altered in the face of obstruction. There are clearly changes in the contractile protein myosin, and these changes will be elaborated on by John McConnell and Sam Chacko. Finally, there are changes in metabolism, manifested by a decrease in oxygenated metabolism and an increase in lactate generation. Bengt Uvelius will comment for us on the changes in biochemical energetics that accompany bladder outlet obstruction. Hopefully some of our speakers will begin to tell us about the molecular changes that turn on some of these changes within the smooth muscle cell, and more importantly, what perhaps we should do in the way of turning them off. Other modifiers such as aging, vascular insufficiency, neurologic damage or disease, and diabetes will change these various responses in outlet obstruction. We have all these various things to look at in the way of secondary bladder changes in response to outlet obstruction.

As a clinician, what does this all mean? I am pleasantly surprised to hear John mention the concept of a detrusor biopsy, which is something that I had been thinking about over the last couple of weeks. I think that in addition to urodynamics, this procedure in the future will hopefully become more widely accepted beyond the research world to become something that everyone does. This is where our research should be focused on. In other words, can we predict those outcomes (progression) that we can't predict now, on the basis of various components of detrusor biopsy. Let me close this brief overview by telling you what my perception is; in the year 2010, a patient will present with the symptoms of bladder outlet obstruction and the first thing that will be done is not a set of urodynamics studies and a cystoscopy, but a bladder biopsy. We will come out with a scoring system that looks something like this. The patient has changes consistent with an El Badawi type 2, a Levin type 4, a McConnell type 3, a Uvelius type 3, a Steers and DeGroat type 1, a Mostwin type 2, and a Chacko type 3. By summing up these scores, a predictive index will be reached. This will tell us, a) whether these findings are progressive or not; b) whether the changes are reversible; c) whether the changes ought to be treated; and d) when they should be treated.

Finally, we go on, treatment will progress from the current invasive modes to medical or molecular treatment of the primary obstruction and the accompanying secondary bladder changes. Finally, in the year 2025, we won't have any patients because all of the various changes in outlet obstruction and the secondary bladder changes will have already been prevented with molecular and/or genetic treatment. Thank you.

REFERENCE

1) ElBadawi, A., Yalla, S.V., Resnick, N.M.: Structural basis of geriatric voiding dysfunction: bladder outlet obstruction. J. Urol. 150:1681,1993.

BLADDER FUNCTION IN EXPERIMENTAL OUTLET OBSTRUCTION: PHARMACOLOGIC RESPONSES TO ALTERATIONS IN INNERVATION, ENERGETICS, CALCIUM MOBILIZATION, AND GENETICS

Robert M. Levin, Ph.D.[1,2], Niels Haugaard, Ph.D.[1], Sheila S. Levin[1], Ralph Buttyan[3], Min-Wei Chen[2], Frederick C. Monson, Ph.D.[1], and Alan J. Wein, M.D.[1,2]

Division of Urology
University of Pennsylvania School of Medicine[1]
The Veteran's Administration Medical Center[2]
Philadelphia, PA 19104
Division of Urology,
Columbia School of Medicine[3]
New York, NY

INTRODUCTION:

Micturition is a complex neuromuscular process[1-3]. Although control mechanisms have been identified at several levels of the central nervous system, the final pathway for micturition control is the autonomic innervation of the urinary bladder and related structures. Based on current experimental studies, the bladder's progressive response to partial outlet obstruction can be divided into three distinct phases: **1)** An **initial response** to surgical induction of partial outlet obstruction. During this phase there are substantial alterations in bladder mass, pharmacology, and physiology. **2) Compensated bladder function** immediately follows the "initial phase" and lasts an indefinite, variable period of time. This period is characterized by relatively stable bladder mass, and stable contractile responses (which may be above control values) to field, bethanechol, and KCl stimulation. **3)** Finally, at some point, functional ability to empty degenerates and the bladder becomes "decompensated". The **decompensated phase** is characterized by progressive deterioration in contractile and functional status, a rapid increase in mass, and a progressive decrease in the volume fraction of smooth muscle elements within the bladder[4,5]. The end result is either an organ with a thick fibrous wall, low capacity, poor compliance, and little or no contractile function; or, a dilated bladder with a thin fibrous wall, high capacity and little or no contractile function.

In general, any stress, whether physiological, pathophysiological, or experimental, that either increases urethral resistance or increases urine volume induces a progressive increase in bladder mass. The increase in bladder mass secondary to partial outlet obstruction is a protective response to the increased intravesical pressure (and resultant distension) required to empty the bladder; the result is a bladder characterized by decreased compliance, increased tensile strength, and impaired function (ability to generate pressure, sustain pressure, and/or empty).

EXPERIMENTAL MODELS OF PARTIAL OUTLET OBSTRUCTION:

Urinary bladder outlet obstruction is a common medical problem. Over 80% of males 50--60 years of age and older have varying degrees of bladder outlet obstruction secondary to benign prostatic hypertrophy (BPH)[6-8]. Although BPH is the most common cause of outlet obstruction, other etiologic factors include carcinoma, sclerosis or fibrosis of the bladder neck, urethral stricture disease, urethral valves, and smooth and striated sphincter dyssynergia [1-3,6-8].

In order to understand the effects of outlet obstruction on bladder morphology, physiology, and pharmacology, several animal models of obstruction have been developed using a variety of species, including rat, rabbit, guinea-pig, and cat. Although there are marked differences in bladder size, capacity, compliance, physiology and pharmacology among these species, responses to outlet obstruction have many common characteristics. The following review is not comprehensive; it provides a relevent summary of urinary bladder response to partial outlet obstruction. For a comprehensive discussion on the topic, please see the recent review articles listed in the references [4,5,9].

Experimental studies of outlet obstruction have been performed on: rats[10-12], rabbits[13-17], guinea pigs[18,19], cats[20,21], dogs[22], and pigs[23,24]. Each species has advantages and disadvantages, and it is unclear which, if any, is most like man. Outlet obstruction is generally produced by the surgical placement of a ligature[25], cuff[13,14], or ring[18] around the urethra. The animal is allowed to recover; response to obstruction is studied temporally following surgery.

One of the most striking features of the bladder's response to experimentally induced outlet obstruction is the speed and magnitude with which it increases in mass (tissue hypertrophy). In rabbits, within one week, bladder mass increased from 1.7gm to 11gm[25]. In a similar study in rats, bladder mass increased from 56.6mg (control) to 138 mg after 3 days, 201 mg after 10 days, and 727 mg after 6 weeks[26]. It appears that in cases of partial urethral ligation ("severe" model), it may be the bladder's initial response, acute overdistension, which signals the rapid and substantial increase in bladder mass. In "mild" models of outlet obstruction, in which a "cuff" or ring is placed around the urethra, the degree of tissue hypertrophy is significantly less than that described for the more severe, ligature, model. In our studies using the cat, rabbit and rat models, the magnitude of increased bladder mass is inversely proportional to the level of contractile and functional impairments[14,21,27]. In chronic studies on the effect of mild partial outlet obstruction in rabbits and cats, functional impairment (ability of the *in vitro* whole bladder to generate pressure and empty) is directly related to bladder mass and not to duration of obstruction[14,21]. As stated previously, the bladder can maintain compensated function for a variable, prolonged time period before progressing to a decompensated state. Thus, one cannot judge the level of decompensation by duration of obstruction. In man, BPH related bladder hypertrophy can be present for years before significant symptoms present themselves, and medical attention is sought.

Studies on the effect of partial outlet obstruction on bladder contractile function are not without controversy. Experiments on these aspects of bladder function utilize a variety of methodologies including isolated tissue stips[10,11,18], bladder rings[17], *in vitro* whole bladder[14,21,25], and *in vivo*[10,11] preparations. A major problem encountered when comparing contractile data published by different researchers is that the units employed may vary considerably. Contractile data generated from strip studies may be presented as measured units such as grams of tension developed or mN (mNewtons); they may be normalized to isolated strip mass (i.e. gm tension per 100mg wet weight) or cross sectional area (gm/mm^2); they may be normalized to a maximum response to KCl administration; or, to a control response (% of maximum). It is obvious that, when comparing the responses of control vs obstructed strips, conclusions are directly related to the parameters (units) that one measures and the manner in which the data are presented.

The *in vitro* whole organ preparation can be used advantageously to evaluate both the bladder's contractile response (increasing intravesical pressure) as an intact organ and its ability to empty. In the rabbit and cat models of short-duration outlet obstruction, the bladder's ability to empty is reduced to a substantially greater degree than its ability to generate pressure[4,5,25]. In addition, the response to field stimulation is reduced to a significantly greater degree than the response to bethanechol, similar to the results described by Mattiasson et al.[28] and Speakman et al.[23].

Alterations in Energetics

As presented above, the obstructed bladder's ability to empty is reduced to a significantly greater degree than its ability to generate pressure (or generate tension as isolated strips). The ability to empty is related directly to the bladder's ability to sustain increased pressure rather than to its maximal pressure response[29]. Energetically, the initial phasic response appears to be related directly to the intracellular concentration of cytosolic ATP, whereas, the ability to sustain a contraction and empty is linked to mitochondrial oxidative phosphorylation. In this regard, partial outlet obstruction results in a significant reduction in mitochondrial respiration (aerobic metabolism); there is an increase in lactate generation (anaerobic metabolism) and a signficant decrease in CO_2 generation (aerobic metabolism)[29]. Further support for the theory that the contractile dysfunction observed in obstruction is related to decreased mitochondrial function comes from the recent demonstration that the activity of mitochondrial enzymes malate dehydrogenase and citrate synthase are decreased by 50%, whereas the activity of several cytosolic enzymes including creatine kinase and myosine ATP'ase are not significantly altered[30-32].

Alterations in Innervation

Under normal conditions, in the rat and rabbit bladder, the maximum contractile response to field stimulation (which acts primarily via the release of neurotransmitters) is approximately the same as the response to direct cholinergic stimulation[25,28,33]. Partial outlet obstruction reduces the response to field stimulation to a significantly greater degree than the response to direct cholinergic stimulation[13,21,23,25,28]. This is consistent with the morphological demonstration that outlet obstruction induces degeneration of specific neuronal elements (partial denervation) within bladder smooth muscle[4,5,23,28,34]. The loss of functional synapses may decrease field-stimulated release of contractile neurotransmitters and thus reduce the magnitude of the contractile response. Tammela et al. demonstrated that the altered field stimulated response induced by partial outlet obstruction is mimicked upon incubating isolated strips of control bladders in submaximal concentrations of tetrodotoxin (TTX)[35]. Progressively inhibiting acetylcholine release causes reduced tonic response to field stimulation at signifincatly lower TTX concentrations than those required to inhibit either maximal response to field stimulation or maximal rate of tension generation[35]. These changes are similar to those exhibited in response to partial outlet obstruction and support the hypothesis that a key alteration induced by outlet obstruction is a partial denervation. Our study demonstrating that partial outlet obstruction causes significantly reduced bladder choline acetyltransferase activity is consistent with this concept[36].

Also consistent with the indirect evidence of end-organ denervation described above, is a direct demonstration by Steers et al.[37-39] of markedly altered afferent neuronal pathways induced by outlet obstruction in rats. In these nerve-tracing studies employing wheat germ agglutinin-HRP injection into the bladder wall, outlet obstruction induced significantly hypertrophied labelled L_6 and S_1 dorsal root cells; the mean number of labelled dorsal root ganglion cells was unchanged. Furthermore, there was a 60% increase in the labelled afferent terminal field area in the intermedio-lateral L_6-S_1 spinal cord region.

Calcium Regulation of Bladder Smooth Muscle Contraction

Similar to that of all smooth muscle, bladder and urethral smooth muscle contraction depends upon a rise in intracellular free calcium[40,41], resulting from both influx from extracellular spaces and release from intracellular stores (calcium-induced calcium release[CICR])[42,43]. In normal bladder, calcium influx is mediated primarily (if not exclusively) through L-type (dihydropyridine-sensitive) calcium channels; no N or T-type channels are believed to participate in bladder smooth muscle response to stimulation[40,41,44,45].

Both extracellular influx and calcium release from intracellular storage sites occur in response to receptor stimulation[46-50]. The relative participation of calcium influx and calcium release from intracellular sites in mediating contractile response to stimulation (in human urogenetal smooth muscle) is the subject of a study by Maggi et al.[45]. These authors show that whereas KCl stimulated-contraction is mediated primarily (if not solely) by extracellular influx through voltage-operated, dihydropyridine-sensitive channels, muscarinic stimulated contraction is mediated by combined extracellular calcium influx and release from intracellular sites. The concept that calcium released from intracellular sites participates in the contractile response to receptor but not KCl mediated stimulation is also supported by the following observations: The maximal contractile response to receptor-mediated stimulation (carbachol, field stimulated release of acetylcholine) is signficantly greater than maximal response to KCl (membrane depolarization)[51,52]. Significantly less ATP hydrolysis results from KCl- than muscarinic stimulated contraction[52].

Recent studies from our laboratory demonstrate that calcium released from intracellular ryanodine-sensitive sites increases during both bladder development and hypertrophy secondary to partial outlet obstruction[53-55]. We propose that as bladder smooth muscle cell volume increases (during development or stress-induced hypertrophy), the relative role of CICR increases. Supportively, the difference between KCl and muscarinic (bethanechol) stimulated bladder strip contractions are relatively less in newborn rabbits than in adults.

Results of our Current Studies

We have identified four major cellular alterations induced by partial outlet obstruction that are directly related to the observed contractile and functional alterations: **a.** Partial

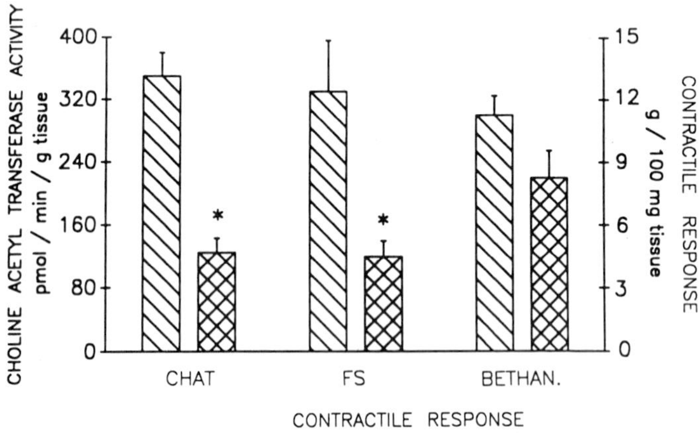

Figure 1. Effect of partial outlet obstruction on field stimulated contraction, bethanechol-stimulated contraction, and the activity of cholineacetyltransferase.

denervation. **b.** A shift from aerobic to anaerobic metabolism. **c.** A shift from extracellular to intracellular calcium utilization for contraction. **d.** Activation of DNA synthesis.

a. Partial oultet obstruction results in a greater reduction in the bladder's ability to respond to field stimulation (neurotransmitter release) than to bethanechol (direct muscarinic stimulation) (fig. 1). This decreased response to neurotransmitter release correlates proportionally with decreased choline acetyltransferase (CHAT) activity (CHAT is the enzyme responsible for acetylcholine synthesis in nerve terminals.) (fig. 1). With mild obstruction there is no decreased field stimulated response and no decreased CHAT activity[36]. There is a direct correlation between the increase in bladder mass and a selective decrease in field-stimulated contraction, as compared to direct muscarinic stimulated contraction. Figure 2 presents a comparison between the contractile response of isolated strips of bladder body to field stimulation and bethanechol as a function of bladder mass. There is a direct correlation between bladder mass and the decreased response to FS whereas the response to bethanechol did not begin to decrease until bladder mass increased to moderate--severe levels. This relationship was independent of duration of obstruction.

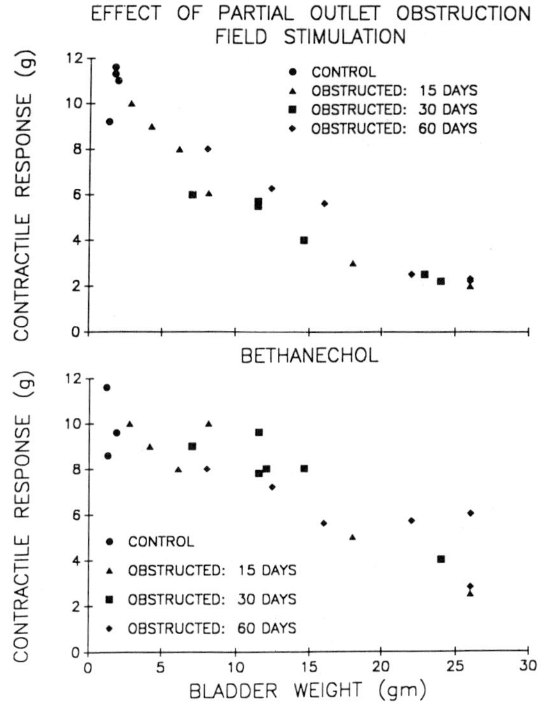

Figure 2. Correlation of the contractile response to FS and bethanechol as a function of bladder mass for rabbits partially obstructed for 15, 30, and 60 days.

b. Partial oultet obstruction results in a shift from aerobic to anaerobic metabolism as demonstrated by the shift in glucose metabolism from CO_2 to lactic acid generation (fig 3). Similarly, there is a marked decrease in the metabolism of pyruvate (mitochondrial substrate) to CO_2 (decreased V_{max}, no change in K_d). The cellular mechanism responsible for decreased oxidative metabolism involves decreased mitochondrial enzyme activity, as demonstrated by citrate synthase and malate dehydrogenase studies. The activities of these enzymes decreased to relatively the same extent as oxidative metabolism decreased (fig. 4). There were no decline, however, in the activities of several cytosolic enzymes including creatine kinase, myosin ATP'ase, and phosphoribosyl transferase. Thus, decreased oxidative metabolism is a function of decreased mitochondrial oxidation of substrates[30,31,56].

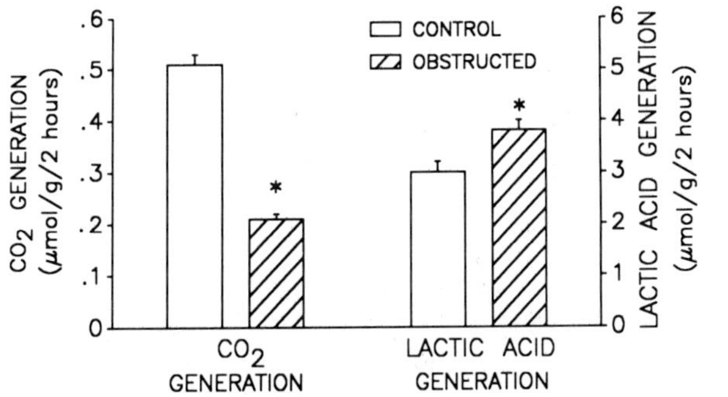

Figure 3. Effect of partial outlet obstruction on glucose metabolism

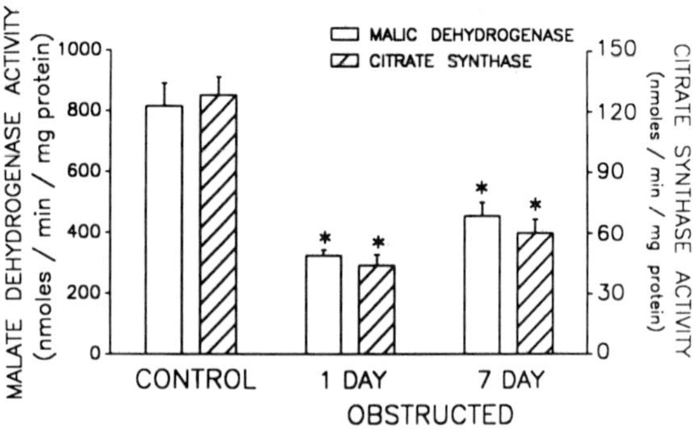

Figure 4. Effect of partial outlet obstruction on mitochondrial enzyme activity

c. Partial outlet obstruction induces a shift in the extracellular / intracellular calcium pools that mediate contractile response to neuronal stimulation. In normal bladder, ryanodine inhibition of low frequency stimulation is significantly greater than inhibition of high frequency stimulation (fig 5). Partial outlet obstruction increases significantly bladder sensitivity to ryanodine inhibition (fig 5). Increased sensitivity of the hypertrophied bladder to ryanodine correlates with a significant increase in ryanodine binding sites (fig. 6). Based on these results we propose that, associated with increased smooth muscle cell volume (hypertrophy), there is progressively increased intracellular calcium utilization (calcium-induced calcium release [CICR]) to support contractile response to stimulation. This theory is supported by studies on neonatal rabbit bladder development: Neonatal rabbit bladders exhibit an increased sensitivity to extracellular calcium in the absence of any changes in PN200 binding (to voltage operated calcium channels). Associated with this decreased sensitivity to extracellular calcium (with age) is a progressively increased ryanodine inhibition of contractile response to stimulation, which correlates with a significant increase in ryanodine binding sites[53,54].

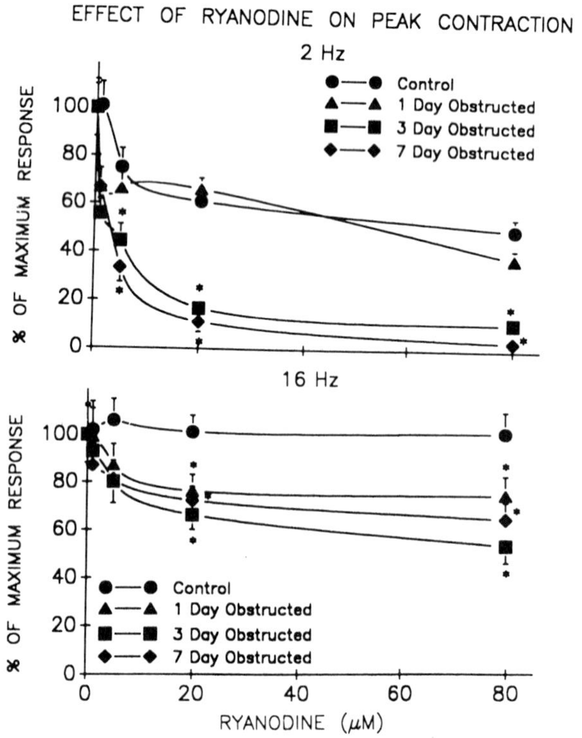

Figure 5. Effect of partial outlet obstruction on the effect of ryanodine on the response to field stimulation.

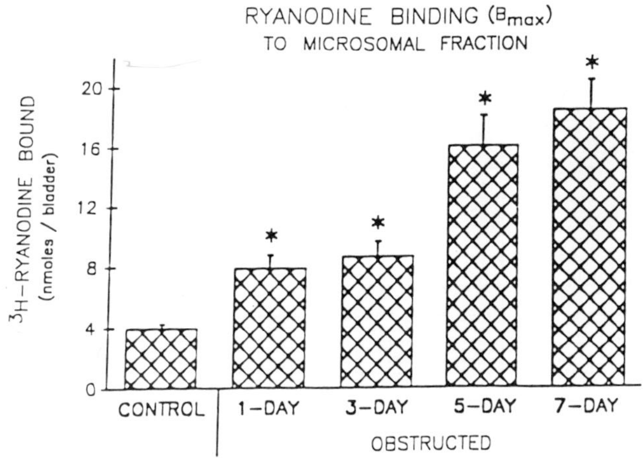

Figure 6. Effect of partial outlet obstruction on ryanodine binding

d. Partial outlet obstruction stimulates an immediate and rapid activation of DNA synthesis in the urothelium followed by activation of cellular elements in the intrinsic and extrinsic connective tissue compartments. ^{3}H-Thymidine was given (iv) to control rabbits and rabbits subjected to outlet obstruction at 1, 3, 5, 7, 14, and 21 days following partial outlet obstruction. After 3 hours, the bladders were excised and both quantitative ^{3}H-thymidine incorporation into DNA and its autoradiographic localization were determined. ^{3}H-thymidine incorporation into DNA increases by 7 fold at day 1, increases to a maximum at day 3, and decreases to stable levels between 14 and 21 days (at 5 fold above control levels) (table 1). Urothelium shows intense DNA synthesis at day 1, moderate incorporation at day 3, and no further incorporation thereafter. The lamina propria and connective tissue elements show increased activity at 3 days, reach a maximum at 5 days, and return to stable levels (above control levels) thereafter (table 1). No significant labeling of smooth muscle was observed. Acute overdistension (1 hour) initiates an immediate, marked activation of the urothelium and lamina propria; no labeling of smooth muscle or connective tissue elements was seen. In a related study, acute onset diuresis (furosamide implants) stimulated a progressive increase in bladder mass which correlated with increased ^{3}H-thymidine incorporation into urothelium at 1 day, into connective tissue between days 3 and 7, and into smooth muscle between days 3 and 14[57-59].

The net result of the dynamic DNA synthesis described above is a re-organization of the structural relationships among the smooth muscle cells, the surrounding connective tissue matrix, and the extrinsic connective tissue lamina. In response to partial obstruction, this structural re-arrangement increases the bladder's tensile strength, or "stiffness", and is responsible for its decreased compliance; which may relate to the obstructed bladder's markedly decreased ability to empty through structural impairment of its ability to change shape during micturition. The morphological response to diuresis results in increased compliance (no increase in extrinsic connective tissue), and increased contractile responses to field, bethanechol, and KCl stimulation.

In chronic partial outlet obstruction in both cats and rabbits, degree of functional impairment is directly proportional to increased bladder mass. As described previously, compensated function is characterized by intermediate capacities and mild to moderately increased organ mass and wall thickness. Decompensation results in one of two characteristic bladders; either a relatively small capacity, fibrous, thick-walled organ, or a dilated, large capacity, fibrous, thin-walled one. These bladders can be distiguished easily utilizing combined ultrasound and cystometry.

Table 1. Effect of partial outlet obstruction and acute overdistension on ^{3}H-thymidine incorporation into DNA.

	MUCOSA URO	LP	SUB-SM	MUCOSA ICT	ECT	fm ^{3}HTdr/ mg DNA
Control	±	-	±	±	NP	0.64±0.1
OB-1D	4+	-	-	2+	NP	4.74±0.1
OB-3D	2+	4+	±	3+	3+	18.0±2.7
OB-5D	1+	3+	-	2+	2+	5.95±0.5
OB-7D	±	2+	-	2+	1+	7.2±0.2
OB-14D	±	-	-	1+	1+	4.91±2.0
OB-21D	±	1+	-	1+	1+	3.44±.67
OD	3+	2+	-	1+	NP	1.8±0.17

OB = obstructed **URO = urothelium** **ICT = intrinsic connective tissue**
D = day **LP = lamina propria** **ECT = extrinsic connective tissue**
OD = overdistended **SM = smooth muscle** **Tdr = thymidine**
fm = femtomoles

± = 1-3 labelled nuclei / field, some fields
1+ = 1-3/field
2+ = 3-5/field
3+ = 5-10/field
4+ = >10/field
- = none observed
NP = tissue not present

The Genetic Response to Partial Outlet Obstruction and Acute Overdistension

Bladders were removed from normal rabbits and rabbits sacrificed progressively (2, 6, 10, 12, 24 and 48 hrs) after the production of partial outlet obstruction. Poly (A)+ RNA was extracted from individual specimens (representing the full complement of time points); comparatively electrophoresed (5 ug per lane) on formaldehyde-agarose gels, then transferred to nylon filters for Northern blot analysis. Several blots were generated; each blot was hybridized successively to three or four ^{32}P-labeled gene probes. All blots were hybridized to an 18S ribosomal RNA probe for normalization. The autoradiograms derived from each hybridization were analyzed by laser densitometry to quantitate band densities of bladder mRNA transcripts.

The results were as follows (Table 2): Transcription of mammalian 70K heat shock protein was increased 20-fold at 10-hrs following obstruction. (Subsequent studies demonstrated that HSP-70 translation increased temporally with increased transcription.) Within 24 hours post-obstruction, there was substantially increased bFGF expression and simultaneous, decreased transforming growth factor b (TGF-b) expression. TGF-b is most well known as an inhibitor of cellular proliferation (especially epithelial cells). Thus, the combined increased bFGF and decreased TGF-b expression may account for the increased ^{3}H-thymidine incorporation observed following partial outlet obstruction.

Over the same time period, there was increased expression of proto-oncogenes c-myc, N-, and Ha-ras; which probably are involved in the stimulated bladder growth and hyperplasia observed in urothelium, lamina propria, and connective tissue components.

In general, the initial period following the creation of a partial outlet obstruction includes a brief period of acute overdistension. In order to determine if this period of acute overdistension initiates the genetic signals observed during partial outlet obstruction, we have completed a second study on the genetic response to acute overdistension In this study, the bladder reacted with rapid, short-term increases in HSP-70 and bFGF transcripts and an associated phasic decrease in TGF-b transcripts; transcription returned to control levels significantly more rapidly than it did in response to partial outlet obstruction. The oncogenes expressed in response to partial outlet obstruction were not detected (i.e.: did not participate) in the genetic response to acute overdistension [60,61].

Table 2. Genetic response to partial outlet obstruction and acute overdistension: Northern Blot Assay of Rabbit Bladder RNA

PROBE	CONTROL	OBSTRUCTION	OVERDISTENSION
hsp-70	1.00*	23 (10 hours)	10 (1 hour)
bFGF	ND	400** (10 hours)	1.00* (1 hour)
TGF $_{-1}$	1.0*	0.4 (24 hours)	0.5 (1 hour)
c-fos	ND	ND	1.00* (1 hour)
c-jun	ND	ND	1.00* (1 hour)
N-ras	1.0*	900 (48 hours)	ND
cHa-ras	ND	600** (24 hours)	ND
c-myc	ND	450** (24 hours)	ND
c-sis	ND	ND	ND

*** Assigned an arbitrary value of 1.00. Used for relative comparison with obstructed and overdistension values for the same probe. This value is not necessarily the minimal detectible level. ** = Greater than minimum detectible level; ND = Not detected.**
Values in parentheses represent the time after obstruction or overdistension where signal change is maximal.

SUMMARY AND CONCLUSIONS

The two functions of the urinary bladder is to store urine at low intravesical pressures, and to periodically expell the urine through a coordinated contraction of the bladder and relaxation of the urethra. To a large extent, urinary bladder function depends upon the underlying structure of the organ as a whole, particularly on the inter-relationships among the smooth muscle, connective tissue, and neuronal elements. An alteration in the ratio of connective tissue to smooth muscle, for example, can significantly alter compliance and functional capacity, structurally impairing the bladder's ability to empty efficiently and fully. Thus, a change in structural compartmentation can affect bladder function independent of autonomic receptor density, response to receptor stimulation, and the contractile capabilities of the smooth muscle elements. Similarly, a selective alteration in either the afferent or efferent innervation of the bladder or urethra can induce significant alterations in the structural interrelationships between smooth muscle and connective elements. In addition, the bladder responds rapidly to alterations in urine volume and urethral resistance with marked changes in bladder and urethral structure and function, and these changes are under the controls of specific genes that are known to control cellular growth, hypertrophy, and hyperplasia. A knowledge of the mechanisms that control the response to specific forms of stress may lead to novel therapies for specific disease states.

LITERATURE CITED

1. Steers, W.D. Physiology of the urinary bladder. In Cambells Urology, (eds. Walsh, P.C., Retik, A.B., Stamey, T.A., and Vaughan, E.D. Jr.) Saunders, Phila., 1992, pp142-176.

2. Wein AJ, Levin RM, Barrett DM: Voiding function: Relevant anatomy, physiology, and pharmacology. In Gillenwater JY, Grayhack JT, Howards SS, Duckett JD (eds): Adult and Pediatric Urology, 2nd ed. Chicago, Year Book Medical Publishers, pp. 933-999, 1991.

3. Anderson, K.E. Pharmacology of lower urinary tract smooth muscles and penile erectile tissue. Pharmacol. Rev. 45: 253-307, 1993.

4. Levin RM, Longhurst PA, Monson FC, Kato K, Wein AJ: Effect of bladder outlet obstruction on the morphology, physiology, and pharmacology of the bladder. Prostate, Supp. 3: 9, 1990.

5. Levin, R.M., Longhurst, P.A., Monson, F.C., Haugaard, N., and Wein, A.J., Experimental studies on bladder outlet obstruction. In Prostate Diseases (H. Lepor MD, R.K. Lawson MD). W.B. Saunders Co., Philadelphia, pp 119-130, 1993.

6. Hinman Jr., F. (ed.) Benign Prostatic Hypertrophy, Springer-Verlag, N.Y. 1983.

7. Lepor, H. and Lawson, R.K. (eds). Prostate diseases, Saunders Co. Phila., 1993.

8. Grayhack, J.T. and Kozlowski, J.M. Benigh Prostatic Hyperplasia. In Adult and Pediatric Urology (Gillenwater, J.Y., Grayhack, J.T., Howards, S.S., and Duckett, J.W. eds.) Year Book Med. Pub. Inc. (Chicago), pp 1062-1126, 1987.

9. Levin, R.M., Haugaard, N., and Wein, A.J., Metabolic alterations induced by obstructive hypertrophy of the rabbit urinary bladder. in Proceedings of the International Symposium "Smooth Muscle", Jap. J. Pharmacol. 58 Supp. II, 1992, pp 341.

10. Mattiasson, A. and Uvelius, B.: Changes in contractile properties in hypertrophic rat urinary bladder. J. Urol. 128: 1340-1342, 1982.

11. Uvelius, B., Persson, L., and Mattiasson, A.: Smooth muscle cell hypertrophy and hyperplasia in the rat detrusor after short-time infravesical outflow obstruction. J. Urol. 131: 173-176, 1984.

12. Steers, W.D. and De Groat, W.C.: Effect of bladder outlet obstruction on micturition reflex pathways in the rat. J. Urol. 140: 864-871, 1988.

13. Kato K, Wein AJ, Kitada S, Haugaard N, Levin RM: The functional effect of mild outlet obstruction on the rabbit urinary bladder. J Urol 140:880, 1988.

14. Kato K, Wein AJ, Longhurst PA, Haugaard N, Levin RM: The functional effects of long-term outlet obstruction on the rabbit urinary bladder. J Urol 143:600, 1990.

15. Mayo, M.E. and Hinman, F.: Structure and function of the rabbit bladder altered by chronic obstruction or cystitis. Invest. Urol. 14: 6-9, 1976.

16. Brent, L. and Stephens, F.D.: The response of smooth muscle cells in the rabbit urinary bladder to outflow obstruction. Invest. Urol. 12: 494-502, 1975.

17. Ghoniem, G.M., Regnier, C.H., Biancani, P., Johnson, L., and Susset, J.G.: Effect of vesical outlet obstruction on detrusor contractility and passive properties in rabbits. J. Urol. 135: 1284-1289, 1986.

18. Mostwin, J.L., Karim, O.M.A., VanKoeveringe, G. and Brooks, E.L.: The guinea pig as a model of gradual urethral obstruction. J Urol 145: 854-858, 1991.

19. Seki, N., Karim , O.M., and Mostwin, J.L., Changes in electrical properties of guinea pig smooth muscle membrane by experimental bladder outflow obstruction, Am J Physiol, 262: F885-891, 1992.

20. Radzinski, C., McGuire, E.J., Smith, D., Wein, A.J., Levin, R.M., Miller, L.F., Meyer, S., and Elbadawi, A., Creation of a feline model of obstructive uropathy. J. Urol. 145, 859-863, 1991.

21. Levin, R.M., Longhurst, P.A., Barasha, B., McQuire, E.J., Elbadawi, A., and Wein, A.J. Studies on experimental bladder outlet obstruction in the cat: Long-term functional effects. J. Urol. 148:939-943, 1992.

22. Rohner, R.J.Jr., Hannigan, J.D., and Sanford, E.J.: Altered in vitro adrenergic responses of dog detrusor muscle after chronic bladder outlet obstruction. Urology 11: 357-361, 1978.

23. Speakman, M.J., Brading, A.F., Gilpin, C.J., Dixon, J.S., Gilpin, S.A., and Gosling, J.A.: Bladder outflow obstruction. A cause of denervation supersensitivity. J. Urol. 138: 1461-1466, 1987.

24. Speakman, M.J., Brading, A.F., Dixon, J.S., Gilpin, S.A., Gilpin, C.J. and Gosling, J.A., Cystometric, physiological and morphological studies after relief of bladder outflow obstruction in the pig. Br J Urol 68: 243-247, 1991.

25. Malkowicz, S.B., Wein, A.J., Elbadawi, A., Van Arsdalen, K., Ruggieri, M.R., and Levin, R.M.: Acute biochemical and functional alterations in the partially obstructed rabbit urinary bladder. J. Urol. 136: 1324-1329, 1986.

26. Uvelius, B. and Mattiasson, A.: Collagen content in the rat urinary bladder subjected to infravesical outflow obstruction. J. Urol. 132: 587-590, 1984.

27. Saito M., Wein, A.J., and Levin R.M. Effect of partial outlet obstruction on contractility: Comparison between severe and mild obstruction. Neurourol. Urodyn. 12: 573-584, 1993.

28. Mattiasson, A., Ekstrom, J., Larsson, B., and Uvelius, B.: Changes in the nervous control of the rat urinary bladder induced by outflow obstruction. Neurourol. Urodynam. 6: 37-45, 1987.

29. Levin, R.M., Ruggieri, M.R., Gill, H.S., Haugaard, N., and Wein, A.J.: Studies on the biphasic nature of urinary bladder contraction and function. Neurourol. Urodynam. 6: 339-350, 1987.

30. Haugaard, N., Potter, L., Wein, A.J., and Levin, R.M. Effect of partial obstruction of the rabbit urinary bladder on malate dehydrogenase and citrate synthase activity. J. Urol. 147:1391-1393, 1992.

31. Levin, R.M., Haugaard, N., Levin, S.S., and Wein, A.J. Creatine kinase activity in normal and hypertrophied bladder tissue. Mol. and Cell. Biochem. 106: 143-149, 1991.

32. Haugaard, N., McKenna, B.A.W., Wein, A.J., and Levin, R.M. Effect of partial urinary outlet obstruction in the rabbit on the incorporation of adenine into adenine nucleotides in bladder smooth muscle. Neurourology and Urodynamics 12: 473-480, 1993.

33. Levin RM, Ruggieri MR, Wein AJ: Functional effects of the purinergic innervation of the rabbit urinary bladder. J Pharmacol Exp Ther 236:452, 1986.

34. Elbadawi, A., Meyer, S. and Regnier, C.H.(1989) Role of ischemia in structural changes in the rabbit detrusor following partial bladder outlet obstruction: a working hypothesis and a biomechanical/structural model proposal, Neurourol. and Urodynam., 8(2), 151-162.

35. Tammela, T., Wein, A.J., and Levin, R.M., Effect of tetrodotoxin blockade as an in vitro model of bladder neuropathy in the rabbit. J. Urol. 148:1937-1940, 1992.

36. Levin, R.M., Saito, M., Wein, A.J., Packard, D., Cohen, A., and Haugaard, M. Effect of partial outlet obstruction on choline acetyltransferase activity in the rat and rabbit. Neurourol. Urodyn. 12:255-262, 1993.

37. Steers, W.D., Ciambotti, J., Erdman, S., and de Groat, W.C. Morphological plasticity in efferent pathways to the urinary bladder of the rat following urethral obstruction. J. Neurosci. 10: 1943-1951, 1990.

38. Steers, W.D., Ciambotti, J., Etzel, B., Erdman, S., and de Groat, W.C.: Alterations in afferent pathways from the urinary bladder of the rat in response to partial urethral obstruction. J. Comp. Neurol. 310: 401-410, 1991.

39. Steers, W.D., Kolbeck, S., Creedon, D., Tuttle, J.B.: Nerve growth factor in the urinary bladder of the adult regulates neuronal form and function. J. Clin. Invest. 88: 1709-1715, 1991.

40. Isenberg. G/. Wendt-Gallitelli, and Ganitkevich, V.: Contribution of Ca^{2+}-induced release to depolarization-induced Ca^{2+} transients of myocites from guinea-pig urinary bladder myocytes. Jpn. J. Pharmacol. 58: (Suppl. 2) 81P-86P, 1992.

41. Ganitkevich, V.Y. and Isenberg, G.: Contribution of Ca^{2+}-induced Ca^{2+} release to the Ca^{2+} transients in myocites from guinea-pig urinary bladder, J. Physiol. 458: 119-137, 1992.

42. Brading, A.F.: Ion channels and control of contractile activity in urinary bladder smooth muscle. Jpn. J. Pharmacol. 58: (Suppl. 2) 120P-127P, 1992.

43. Mostwin, J.L.: The action potential of guinea pig bladder smooth muscle. J. Urol. 135: 1299-1303,1986.

44. Foveus, M., Andersson, K-E., Batra, S., Morgan, E., and Sjogren, C.: Effects of calcium, calcium channel blockers, and Bay K8644 on contractions induced by muscarinic receptor stimulation of isolated bladder muscle from rabbit and man. J. Urol. 137:798-803, 1987.

45. Maggi, C.A., Giuliani, S., Patacchini, R., Turini, D., Barganti, G., Giachetti, A., and Meli, A.: Multiple sources of calcium for contraction of the human urinary tract. Br. J. Pharmacol. 98: 1021-1031, 1989.

46. Mostwin, J.L.: Receptor-operated intracellular calcium stores in the smooth muscle of guinea-pig bladder. J. Urol. 900-905, 1985.

47. Huddart, H. and Butler, D.J.: Field stimulation responses of rat urinary bladder detrusor-muscle. Dependence upon slow calcium channel activity determined by K^+ depolarization and calcium antagonists. Gen. Pharmacol. 17: 695-703, 1986.

48. Batra, S., Sjogren, C., Andersson, K.-E., and Fovaeus, M.: Source of calcium for contraction induced by depolarization and muscarinic receptor stimulation in rabbit urinary bladder. Acta Physiol. Scand. 130:545- 551, 1987.

49. Maggi, C.A., Manzini, S., Parlani, M., Conte, B., Giuliani, S., and Meli, A.: The effect of nifedipine on spontaneous, drug induced and reflexly-activated contractions of the rat urinary bladder: evidence for the participation of an intracellular calcium store to micturition contraction. Gen. Pharmacol. 19: 73-81, 1988.

50. Kishii, K.-I., Hisayama, T. and Takayanagi, I.: Comparison of contractile mechanisms by carbachol and ATP. Jpn, J. Pharmacol. 58: 219-229, 1992.

51. Longhurst, P.A., Eika, B., Leggett, R.E., and Levin, R.M. Comparison of urinary bladder function in 6 and 24 month male and female rats. J. Urol. 148: 1615-1620, 1992.

52. Levin, R.M., Hypolite, J., Longhurst, P.A., and Wein, A.J. Comparison of the contractile and metabolic effects of muscarinic stimulation with those of KCl. Pharmacol. 42: 142-150, 1991.

53. Levin, R.M., Zderic, S.A., Yoon, J-Y, Sillen, U., and Wein, A.J. Effect of ryanodine on the contractile response of the normal and hypertrophied rabbit urinary bladder to field stimualation. Pharmacol. 47: 244-251, 1993.

54. Levin, R.M., Levin, S.S., Zderic, S.A., Yoon, J-Y, and Wein, A.J.: Effect of partial outlet obstruction of the rabbit urinary bladder on ryanodine-inhibition of the contractile response to field stimulation and ryanodine binding to microsomal membranes. Gen. Pharmacol. (in press).

55. Zderic, S.A., Sillen, U., Liu, G-H., Snyder, H.McC., Duckett, J.W., Gong, C., and Levin, R.M. Developmental aspects of excitation contraction coupling of the rabbit urinary bladder smooth muscle (in press)

56. Kato, K., Lin, A.T-L., Wein, A.J., and Levin, R.M. Effect of outlet obstruction on glucose metabolism of the rabbit urinary bladder. J. Urol. 143: 844-847, 1990.

57. Monson, F.C., McKenna, B.A., Wein, A.J., and Levin, R.M. Effect of outlet obstruction on ^{3}H-Thymidine uptake and metabolism: A radiographic and biochemical study. J. Urol. 148: 158-162, 1992.

58. Saito, M. Longhurst, P.A., Murphy, M., Monson, F.C., Wein, A.J., and Levin, R.M. 3H-Thymidine uptake by the rat urinary bladder after partial outflow obstruction, Neurourol. Urodyn. 13: 63-70, 1994.

59. Monson, F.C., McKenna, B.A., Wein, A.J., and Levin, R.M. Effect of outlet obstruction on ^{3}H-Thymidine uptake and metabolism: A radiographic and biochemical study. J. Urol. 148: 158-162, 1992.

60. Buttyan, R., Jacobs, B., Blaivis, J.G., and Levin, R.M. The early molecular response to rabbit bladder outlet obstruction. Neurourol. Urodyn. 11: 253-260, 1992.

61. Chen, M-W, Krasnapolsky, L., Levin, R.M., and Buttyan, R. An early molecular response induced by acute overdistention of the rabbit urinary bladder. Mol. Cellular Biochem.(In press)

ELECTRICAL PROPERTIES OF OBSTRUCTED GUINEA PIG BLADDER

Jacek L Mostwin, Narihito Seki Omer Karim, and Gommert van Koeveringe

Brady Urological Institute
Johns Hopkins Medical Institutions
600 North Wolfe Street
Baltimore, MD 21287-2411

It is now well recognized that prolonged obstruction of the urinary bladder results in structural and urodynamic changes in bladder function. The structural changes include muscular hypertrophy and hyperplasia, increased collagen deposition and a proliferation of sensory afferent neurons. In this paper we would like to advance the thesis that the unstable contractions and poor contractility seen after bladder outlet obstruction are due, at least in part, to abnormalities in electrical properties of the muscle cells and basic changes in cell-to-cell propagation of electrical activity necessary for normal contraction.

THE GUINEA PIG MODEL OF OBSTRUCTION

We began by developing a small animal model of partial outlet obstruction to provide a continuous and reproducible source of obstructed bladder tissue for comparison to normal unobstructed tissue. We were successful in developing such a model in the guinea pig, using standard size jeweler's jump rings placed around the urethra of immature guinea pigs (250-290 gms) and allowing the animals to grow into the obstruction (Mostwin, Karim, van, & Brooks, 1991) . Analysis of proto-oncogenes associated with muscular growth showed that the bladder began to increase in size two weeks after the obstructing rings had been placed around the urethra (Karim, Seki, & Mostwin, 1992) , confirming our assumption that obstruction was gradual and that changes of sudden obstruction, over-distention, ischemia and infarction were minimal.

We found that the guinea pig bladder, like the bladders of almost all mammals tested, responded to partial outlet obstruction by an increase in muscle mass. We had expected that this muscle mass would be due to a mixture of collagen and muscular hypertrophy, but we found, to our surprise, that early increases in muscle mass in these immature animals were due to muscular hyperplasia, with hypertrophy occurring as a secondary phase [(Karim, Cendron, Mostwin, & Gearhart, 1993)].[1]

---

1. In another publication, we have advamced the thesis that this early surge in cell growth may be the result of purely mechanical stretchon the bladder cells, transmitted to the nuclear DNA by the cellular actin matrix. We have also suggested that the hyperplastic reposnse to stretch, and hence, to obstruction, is more pronounced in young animals, and less so in older animals. These finding have implications for age-specificity of animal models of obstruction although they are not specifically relevant to the thesis being advanced here.

Muscle, Matrix, and Bladder Function,
Edited by S. Zderic, Plenum Press, New York, 1995

We also found that the increase in collagen deposition, measured by hydroxyproline content and concentration was proportional to the increase in muscle mass (Karim & Mostwin, 1991).

We also found no significant decrease in the net number of nerves containing acetylcholine (by quantitative measurement of choline acetyl transferase), although the *concentration* of these nerves decreased as a result of muscle mass growth (Karim & Mostwin, 1990) . These rough measurements of tissue content suggested that muscle growth was the dominant response to obstruction in our model.

We identified significant urodynamic abnormalities in bladders that grew in weight. The urodynamic abnormalities were what we would consider clinically worse as bladder weight increased. With small weight increases, the bladders showed significantly higher pressure voiding. As weight increased further, a pattern of unstable contractions, similar to those seen in humans appeared. With further weight gain there was loss of compliance, and finally, with the maximum weight gain (about three times that of controls), there was loss of contractility with overflow at the end stage of filling (Figure 1) (Mostwin, et al., 1991) . Similar patterns of abnormality have been found in boys with persistent voiding difficulties after otherwise successful ablation of posterior urethral valves, a serious form of neonatal bladder outlet obstruction (Peters, Retick, Hendren, Mandel, Colodny, & Bauer, 1990) . The biochemical studies of tissue content mentioned above now became significant because they suggested that most of these voiding abnormalities could be due primarily to an increase in muscle mass, and changes in other cellular or interstitial elements of the bladder were not criticalin producing disordered voiding in this model. Put simply, disordered voiding following obstruction may be due to abnormalities of muscle size, shape and function alone.

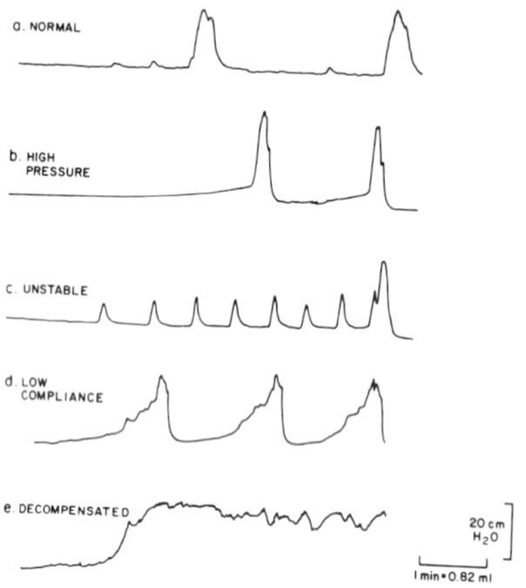

Figure 1. Representative *in vivo* cystometrograms (pressure v. volume) from urethane anesthetized guinea pigs following partial urethral obstruction. All tracings drawn to scale. Reference (Mostwin, et al., 1991)

We studied the velocity of contraction of muscle strips *in vitro* which had been obstructed and compared the results to the weight of the bladder. Using phase plot analysis, we found that the velocity of contraction, an indicator of contractile efficiency, decreased as the weight of the bladder increased. In other words, the heavier the bladder, the worse the contractile performance, and the worse the urodynamic behavior (van Koeveringe, Mostwin, R., & J., 1993) .

Once it became clear that disordered cell size shape and number was associated with decreased contractility and urodynamic performance, we began to look for underlying mechanisms.

Electrical Properties of Bladder Muscle Cells

We were already familiar with the basic electrical properties of the guinea pig bladder smooth muscle membrane. We had studied them by double sucrose gap technique and also intracellular microelectrode. We had shown that the action potential of the guinea pig bladder is calcium based, that the resting membrane potential is relatively low, i.e., -38 mV (Mostwin,

1986), and that neuromuscular transmission may involve a biphasic response to neurotransmitter release: there may be an early purinergic component resulting in rapid transient depolarization, and a longer, more latent muscarinic depolarization (Brading & Mostwin, 1989) . Using double sucrose gap technique we had shown discrete relations between individual action potentials, whether spontaneous or agonist-produced, and small phasic increases in muscle tone (Mostwin, 1988) . With prolonged or sustained firing of action potentials, these small contractions could summate, giving the appearance of a tetanic contraction with an overall increase in muscle tone (Figure 2). It appeared, therefore, that summation of individual muscle groups firing action potentials, could be a means by which overall muscular force was generated by the bladder.

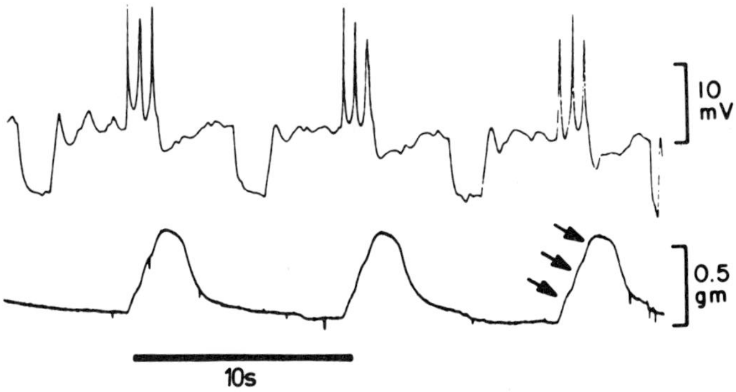

Figure 2. Response of guinea pig bladder muscle strip in response to current pulses of alternating polarity in the double sucrose gap.. The upper tracing represents membrane potential. Depolarization is shown as an upward deflection. The lower tracing represents mechanical contraction of the muscle strips. The increments in tension indicated by the arrows are associated with each of the three action potentials, producing the total contractile response. Contraction stops when stimulation ceases. Reference (Mostwin, 1988)

Since our initial experiments had been performed in normal guinea pigs, we were quite pleased to find that we were able to introduce intracellular microelectrodes into the cells of obstructed guinea pig bladders and study the effects of obstruction on the passive cable properties of the bladder muscle cells. For many years, underwater submarine cables have been used as a mathematical model of tubular excitable tissue such as nerve and muscle, and well established mathematical and electrical properties have been described using these theories. Electrical analysis is merely a way of describing these events and getting closer to the molecular mechanisms which underlie them, namely, changes in physical substrates, ionic gradients and channels in the membrane.

The electrical properties that we studied were the classic ones: passive cable constants which describe the spread of sub-threshold current from a point source in accordance with exponential decay in time and space, and the shape and properties of the action potential. which tell us something about the ionic currents which flow across the membrane when activity is at its peak, and the physical substrates which condition this activity.

We studied three fundamental passive properties of the obstructed muscle cell: the resting membrane potential, the time constant and the space constant.

Resting Membrane Potential

The resting membrane potential is the potential difference between the inside of the cell with respect to the outside of the cell, a small value usually measured in millivolts. It is recorded as soon as the electrode penetrates the cell membrane and is typically the signal used by the experimenter that a cell has been entered. The value is usually negative with respect to the reference electrode outside the cell, in the bathing medium. The value of the r.m.p. reflects a balance of forces by which charged particles are distributed across the cell membrane. The membrane is a multilaminated protein and lipid surface which separates charged particles in accordance with their size and concentration on either side of the membrane. Charged particles will move from regions of higher to lower concentrations until the concentration gradient is balanced by the separation of electrical charges in accordance with the Gibbs-Donnan equilibrium. Normally there is a high intracellular concentration of potassium and non-diffusable protein, and a large extracellular concentration of sodium. A metabolically active sodium/potassium pump exchanges intracellular sodium for extracellualr potassium and thus helps to maintain the gradient. The result is a charged biological membrane, much like a charged photographic flash unit, ready to fire an action potential, a signal, which can be propagated from cell to cell, then quickly be recharged for the next event (Ferreira & Marshall, 1985).

The resting membrane potential is also a reflection of the physical properties of the membrane: its resistance to the flow of intracellular and extracellualr current, and the capacitance, or ability to separate charge. This in turn, is due to the size and shape of the cells, the permeability of the membrane channels, and the active exchange mechanisms which maintain the gradients. Changes in the r.m.p. may by due to changes in these physical properties, again, due to a change in channel properties, cell size, shape, or molecular make-up, which affects the ionic distribution.

Electrical Properties of the Obstructed Bladder

We found that the resting membrane potential of obstructed bladder muscle was unchanged from normal (Seki, Karim, & Mostwin, 1992b) . This was an unexpected and significant finding because previous observations on vas deferens had suggested that the r.m.p. was somewhat depolarized, thus introducing a possible explanation for why denervated tissue might be supersensitive to neurotransmitters (Fleming & Westfall, 1975) . This is what we might have expected to find in accordance with the theory of denervation supersensitivity as a cause of unstable bladder activity (Speakman & Brading, 1987).

We also found a significant reduction in spontaneous activity of the muscle cell. Normally, action potentials are spontaneously produced by a pacemaker mechanism (the details of which are only partially understood in all smooth muscle cells). Sixty-eight per cent of normal cells showed spontaneous activity. In the obstructed bladder, only seventeen per cent of cells were spontaneously active (Seki, et al., 1992b) . Usually, spontaneous activity is suppressed is membrane hyperpolarization (the mechanism, for example, where by pinacidil or other potassium channel openers work). What we did find, as a possible explanation for the suppression of spontaneous activity was an excessively active sodium/potassium pump which suggested that increase intracellular concentrations of sodium were developing in our obstructed bladder muscle cells. This pointed to an abnormality of channel behavior and a decrease in membrane resistance, but in order to study this more fully, one would have to use the more modern method of patch clamp analysis, potentially very interesting experiments still waiting to be done. Therefore, we left these findings to pursue studies of the other passive properties, space and time constants, which require the use of multisyncytial preparations.

When current is introduced into a cell it is depolarized towards the threshold at which an action potential (AP) arises from self-propagating channel activation. Once an AP is fired, the activity propagates itself, and no further current injection is required. A great deal of excitable cell regulation, however, takes place at *sub-threshold* levels. At these levels, current injected into a cell propagates in three dimensions through the surrounding tissue, depolarizing (and/or hyperpolarizing, and thus modulating) other cells. The amplitude of the current decays exponentially with distance from the point of origin. The rate of decay is expressed as a *space constant*, the constant of the solution of the first order differential equation which describes the change in current amplitude as a function of distance from the source of the current. This is experimentally studied by recording the membrane response to current injection of constant amplitude as the recording electrode is moved further and further away from the source of stimulation. The space constant is a reflection of the intracellular and membrane resistance and the membrane capacitance of the cell. The rate at which the membrane is charged by the injected current, or decays from its starting conditions, is expressed as the *time constant*, the constant of the solution of the first order differential equation which describes the change in current amplitude as a function of time. The time constant is experimentally measured by recording the membrane response to a series of graded stimuli, plotting the log of change in membrane voltage as a function of time. The time constant is also a reflection of intracellular and extracellular resistance and capacitance, but in different mathematical relationships to each other (Ferreira & Marshall, 1985).

We found that the time and space constants were both reduced from normal following a period of obstruction (Seki, Karim, & Mostwin, 1992c) . After a period of reversal of obstruction, the space constant recovered almost completely, but the time constant recovered only partially (Figure 3 & Table I) . Since the time constant gives some idea of how rapidly the muscle membrane charges, one might suppose that a lower time constant might indicate a more easily charged membrane, and it is very tempting to speculate that such changes could contribute to the instability of contraction that is seen following obstruction. Similarly, a decrease in the space constant means that current does not spread as far in obstructed muscle tissue. Again, it is tempting to speculate that this might contribute to the reduced contractile efficiency seen in bladder muscle following obstruction.

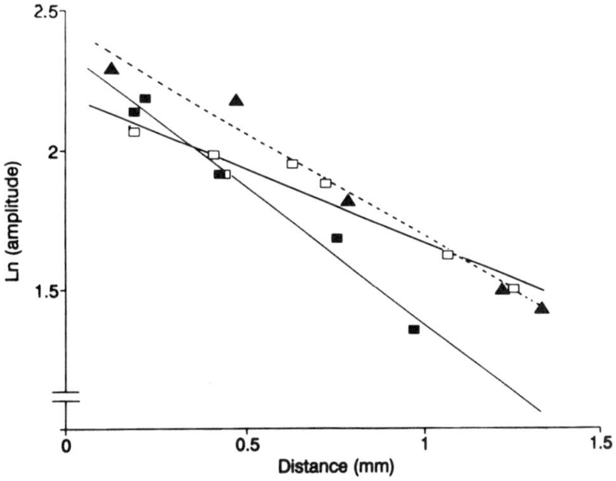

Figure 3. Determination of the *space constant.* The steady state of the electrotonic potential recorded in response to a fixed current square wave is plotted as a function of distance from the stimulating electrode. Open squares are control, solid squares represent obstructed animals, solid triangles represent animals released after obstruction. (Dashed line represent regression line calculated from released group). *Space constant* is the slope of the line. Reference (Seki, et al., 1992c).

TABLE 1

	Control	Obstructed	Control	Released
Membrane potential (mV)	-40.2+/-3.2 (70)	-40.8+/-3.1(96)	-39.9+/-3(71)	-39.5+/-2.5(89)
Time constant (sec)	203.1+/-25.4(49)	118.8+/-18.5(79)	204.5+/-22.8(24)	189.5+/-30(43)*
Space constant (mm)	1.73+/-0.28 (5)	0.84+/-0.27 (5)*	1.70+/-0.32 (5)	1.21+/-0.28(5)*

Values represent the mean plus or minus the standard error. Number in parentheses are the numbers of cells studied. Asterisks represent p<0.05 compared to the control values. Reference: (Seki, Karim, & Mostwin, 1992b).

Active Properties

Finally, we looked at the action potential of the obstructed bladder muscle and found that it is reversibly altered (Seki, Karim, & Mostwin, 1992a) The rate of rise of the action potential is decreased, and the duration of the action potential is prolonged (Figure 4 & Table II). The findings are similar to those seen in hypertrophied myocardial cells, in which conduction abnormalities are known to co-exist with impaired mechanical properties. In bladder muscle, the decrease in depolarization rate may indicate changes in the behavior of the

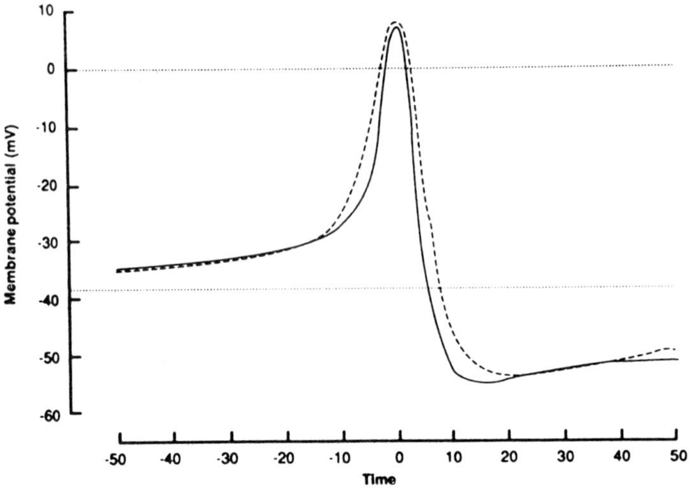

Figure 4. Representative action potentials from normal (solid line) and obstructed (dotted line) guinea pig bladders. Traces represent mean of 15 APs recorded from three animals. The *abscissa* shows time (ms) from the point of maximum depolarization. Horizontal lines show zero potential and mean r.m.p. Reference (Seki, et al., 1992a)

TABLE 2

	APA (mV)	AH (mV)	OS (mV)	Depo5 0% (ms)	Repo 50% (ms)	Repo 90% (ms)	dv/dt (V/S)	dr/dt (V/S)
control	46.2	14.5	7.8	5.0	4.0	5.9	7.1	11.5
	6.3	3.1	6.0	1.0	1.0	1.5	1.9	3.4
obstructed	45.4	13.6	8.2	6.5 *	4.7 *	7.6 *	5.8 *	8.7
	5.5	4.3	5.9	2.0	1.1	1.8	1.7	1.9

APA - actiona potential amplitude; AH - after hyperpolarization; OS - overshoot potential; Depo 50%, Repo 50%, and Repo 90% - action potential paramters measuring time in milli- seconds to 50% of maximal depolarization or repolarization; dv/dt - velocity of depolarization; dr/dt - velocity of repolarization. * indicates $p<0.05$ for control versus obstructed. 189 observations form 14 control animlas and 137 observations from 8 obstructed animals. Reference (Seki, Karim, & Mostwin, 1992b).

calcium channels, a redution of the membrane resistance, or alterations in the ionic gradients driving the action potential.

Summary and Clinical Significance

What then are the known electrical abnormalities produced by obstruction? First there is a reduction in the time and space constant of the bladder muscle membrane with a reduction in spontaneous activity, no change in the resting membrane potential, and an increase in sodium/potassium pump activity. These findings suggest the possibilities of changes in channel behavior or ionic gradients, and offer, perhaps too simply, some suggestions for how bladder instability and impaired contractility might be a result of changes in electrical properties of bladder muscle alone. Second, decreased velocity of depolarization and prolonged action potential duration also suggest disordered channel behavior or altered ionic gradients, and would be expected to result in impaired contractility, as in our phase plot experiments, in the form of decreased velocity of contraction, and in impaired urodynamic behavior.

In summary, we have studied the normal and obstructed guinea pig bladder model for several years and have found evidence of partially reversible changes in active and passive electrical properties which are consistent with many tissue and organ behavioral changes familiar to most investigators. The singular significance of our findings is that the guinea pig bladders we have studied are not obviously scarred nor do they have excessive amounts of collagen. The overall hydroxyproline content is not elevated out of proportion to the amount of growth, and the total number of cholinergic nerves is not reduced. While work continues on all cellular components of the bladder, including collagen, intracellular and extracellualr matrix, sensory and motor nerves, our work serves as a reminder that the function of the muscle itself cannot be assumed to be "normal." Bladder smooth muscle need not be considered a "black box" in the overall evaluation of disordered voiding, but its specific abnormalities can be studied, measured, and the findings can be included in any natural history or pathophysiology of the disease process.

REFERENCES

Brading, A. F., & Mostwin, J. L. (1989). Electrical and mechanical responses of guinea-pig bladder muscle to nerve stimulation. *British Journal of Pharmacology,* 98(4), 1083-90.

Ferreira, H. G., & Marshall, M. W. (1985). *The biophysical basis of excitability.* Cambridge: Cambridge University Press.

Fleming, W. W., & Westfall, D. P. (1975). Altered resting membrane p[otential in the supersensitive vas deferens of the guinea-pig. *J. Pharmacol Exp Therap,* 192, 381-389.

Karim, O. M., Cendron, M., Mostwin, J. L., & Gearhart, J. P. (1993). Developmental alterations in the fetal lamb bladder subjected to partial urethral obstruction in utero. *Journal of Urology,* 150(3), 1060-3.

Karim, O. M., Seki, N., & Mostwin, J. L. (1992). Detrusor hyperplasia and expression of "immediate early" genes with onset of abnormal urodynamic parameters. *American Journal of Physiology,* 263(6 Pt 2).

Karim, O. M. A., & Mostwin, J. L. (1990). Changes in bladder choline acetyltransferase activity following short term obstruction and release of obstruction. *Neurourology and Urodynamics,* 9, 427.

Karim, O. M. A., & Mostwin, J. L. (1991). Bladder outflow obstruction and release in the guinea pig: effect on urodynamic parameters and biochemical markers of detrusor structure. *Journal or Urology,* 145, 387A.

Mostwin, J. L. (1986). The action potential of guinea pig bladder smooth muscle. *Journal of Urology,* 135(6), 1299-303.

Mostwin, J. L. (1988). Electrical membrane events underlying bladder muscle contractility. *Neurourology and Urodynamics,* 6, 429-437.

Mostwin, J. L., Karim, O. M., van Koeveringe G., & Brooks, E. L. (1991). The guinea pig as a model of gradual urethral obstruction. *Journal of Urology,* 145(4), 854-8.

Peters, C. A., Retick, A. B., Hendren, W. H., Mandel, J., Colodny, A. H., & Bauer, S. D. (1990). The urodynamic consequences of posterior urethral vlaves. *Journal of Urology,* 144, 122-126.

Seki, N., Karim, O. M., & Mostwin, J. L. (1992a). Changes in action potential kinetics following experimental bladder outflow obstruction in the guinea pig. *Urological Research,* 20(6), 387-92.

Seki, N., Karim, O. M., & Mostwin, J. L. (1992b). Changes in electrical properties of guinea pig smooth musclc membrane by experimental bladder outflow obstruction. *American Journal of Physiology,* 262(5 Pt 2).

Seki, N., Karim, O. M., & Mostwin, J. L. (1992c). The effect of experimental urethral obstruction and its reversal on changes in passive electrical properties of detrusor muscle. *Journal of Urology,* 148(6), 1957-61.

Speakman, M. J., & Brading, A. F. (1987). Bladder outflow obstruction - a cause of denervation supersensitivity. *Journal of Urology,* 138, 1461-1466.

Van Koeveringe, G. A., Mostwin, J. L., van Mastrigt, R. & van Koeveringe, B. (1993). Effect of partial urethral obstruction on force development of the guinea pig bladder. *Neurology & Urodynamics,* 12(6), 555-66.

METABOLISM OF DETRUSOR SMOOTH MUSCLE IN NORMAL AND OBSTRUCTED URINARY BLADDER

Bengt Uvelius[1] and Anders Arner[2]

[1]Department of Urology,
Lund University Hospital, S-221 85 Lund
[2]Department of Physiology,
Sölvegatan 19, S-223 62 Lund, Sweden

INTRODUCTION

The detrusor smooth muscle cells are arranged in bundles of various directions, diameters and lengths (Gabella and Uvelius, 1990). In the rat roughly 85 per cent of a bundle is occupied by smooth muscle cells (Arner, Malmqvist and Uvelius, 1990), the remainder of the bundle constitutes the intercellular space, containing e.g. collagen fibrils. Non-muscle cells are rare but occasional fibroblasts can be found. The blood supply to the urinary bladder is mainly through the superior vesical arteries originating from the internal iliac arteries. The corresponding veins run in parallel with the arteries. In the rat the major intramural vessels are situated in the connective tissue between the epithelium and the muscle. Capillaries are found between but not in the muscle bundles (Gabella and Uvelius, 1990). The distance for diffusion of oxygen and energy substrates from the capillaries to the smooth muscle cells and for waste products from the muscle cells to the capillaries is thus considerable. The extracellular space would be expected to restrict the rate of diffusion within the bundle for hydrophilic substances.

The urinary bladder is most of the time in the relaxed state, and is slowly filled with urine. When the bladder is filled above a certain degree a micturition is initiated, and the bladder empties. We have found that a rat voids about 25 times during a 24 hour period. Each micturition lasts less than 10 seconds, and the total time the bladder contracts a day is thus less than five minutes. As the micturition is so fast it seems unlikely that any major increase in perfusion will take place during the micturition. The oxygen available for the smooth muscle cells during contraction is, due to the considerable diffusion distance from the capillaries, such that has been transported to the the muscle cells and their intercellular space when they were relaxed. In an unpublished study we mounted rat bladder strips in vitro and stimulated them by five minute periods in K^+-high solution separated by 10 minute relaxation periods. If glucose was omitted from the bathing medium the strips could contract for hours, but if the oxygen supply was also cut the force was abolished within 30 seconds. This indicates that the smooth muscle cells have stores of substrates which, when metabolized in the presence of oxygen, are sufficient for maintaining an appropriate ATP supply to the contractile machinery for a long time, and also indicates that the contractile process is dependent on the instantaneous production of ATP.

METABOLISM IN DETRUSOR MUSCLE FROM NORMAL URINARY BLADDERS

Metabolic Properties of Bladder Muscle in Relation to Other Types of Smooth Muscle

Bladder muscle can metabolize both glucose and fatty acids (Hypolite et al., 1989) for the production of ATP. Ketone bodies can also be used by smooth muscle (Lövgren and Hellstrand, 1987), but in the non-diabetic animal the available amount of this substrate is limited. The rate of fatty acid degradation is much slower than that of glucose (Hypolite et al., 1989).

The metabolism of glucose can be considered to take place in two steps. First via glycolysis to pyruvate, a chain of reactions not dependent on oxygen and with a low net production of ATP. Then via the citric acid cycle with CO_2 and H_2O as end products, and with a high oxygen dependent production of ATP. The glycolysis is dependent on access to NAD. This is normally generated in the citric acid cycle, and also by the conversion of pyruvate to lactate, a reaction catalyzed by lactate dehydrogenase. A peculiarity in smooth muscle metabolism is that lactate production is high even during optimal oxygen supply. This "aerobic glycolysis" has been suggested to generate ATP for various cellular pumps (Paul, Bauer and Pease, 1979; Paul, 1983) and is dependent on extracellular glucose (Hellstrand, 1977, Lynch and Paul, 1983). In relaxed rat detrusor muscle aerobic glycolysis accounts for 42 per cent of glucose metabolized (Arner, Malmqvist and Uvelius, 1990). Corresponding figures for rabbit detrusor are 37 (Wendt and Gibbs, 1987) and 81 per cent (Haugaard, Wein and Levin, 1987). Due to the low yield of ATP, glycolysis will only account for 4-6 per cent of the total energy flux (Wendt and Gibbs, 1987; Arner, Malmqvist and Uvelius, 1990). The higher figure (59 per cent) in the study by Haugaard, Wein and Levin (1987) might be due to a larger cross-sectional area of their preparations, resulting in a higher rate of anaerobic glycolysis due to diffusion limitation of oxygen.

In general smooth muscle has low phosphagen reserves when compared with skeletal muscle. This is mainly due to the low levels of creatine phosphate (Butler and Davies, 1980). The rabbit bladder contains 1.2 µmol/g ATP and 1.5 µmol/g creatine phosphate (Kato et al., 1990). A basal ATP turnover rate of 0.9 µmol/min at 27°C (which would correspond to 2 µmol/g x min at 37°C) for the rabbit bladder (Wendt and Gibbs, 1987) and 3 µmol/g x min for the rat detrusor (Arner, Malmqvist and Uvelius, 1990) have been reported. The turnover rate is considerably higher in the contracting muscle.

The small phosphagen reserves indicate that the activity of the detrusor muscle is dependent on the instantaneous production of ATP. There are some glycogen reserves in smooth muscle. Rat portal vein contains, when expressed in glucose units, 6 µmol glycogen /g tissue (Hellstrand, Johansson and Norberg, 1977) and rabbit urinary bladder contains 4.4 µmol/g (Levin et al., 1988). The low rate of fat oxidation (Hypolite et al., 1989) indicates that under normal conditions the major source for energy production in bladder muscle is glucose. In conditions with a deficiency of glucose the situation might be different. The 50 per cent decrease in glycogen concentration observed in the rat portal vein after 2 hours in glucose-free medium (Hellstrand, Johansson and Norberg, 1977) is clearly lower than that necessary for the ATP supply to the basal metabolism of the vessel , even if completely oxidized. Also, no lactate is produced in normoxia if glucose is omitted from the bathing medium (Hellstrand, 1977). This suggests that under conditions of glucose deficiency ATP can be produced by utilization of other endogenous stores. In the rabbit aorta (Odessey and Chace, 1982) utilization of tissue lipid (but not protein) could account for the ATP production under such circumstances. We have found (unpublished observations) that contractile force of rat detrusor muscle strips activated by high-K^+ solution (5 min. contraction, 10 min. relaxation), for several hours in glucose free medium, stays constant after an initial decrease of about 50 per cent. This might indicate that fat oxidation, even if slow (c.f. Hypolite et al., 1989) could be important in situations with glucose deficiency.

In situations with lack of oxygen the smooth muscle cells will be totally dependent on glycolysis for their instantaneous ATP production, as no other ATP-producing metabolic pathways are available. Consequently, ATP level and contractile force rapidly decrease in rabbit detrusor muscle if both oxygen and glucose is removed from the medium (Levin, High and Wein, 1981). In the rat detrusor the fall in force was even faster (unpublished observations), and the preparations were damaged irreversibly.

Basal and Tension-dependent Energy Metabolism in Normoxia

The basal metabolic rate is the rate of energy consumed in the relaxed muscle. It represents the activity of several energy-consuming processes such as cell pumps maintaining cell homeostasis. Fig. 1 shows that the metabolism of the relaxed rat bladder in normoxia involves both oxygen consumption and lactate production. Due to the low ATP yield the lactate production corresponds to only 6 per cent of the ATP production. Our figure of basal metabolic rate corresponds well with those in rabbit bladder (Wendt and Gibbs, 1987) and rat portal vein (Arner and Uvelius, 1981) when the relative amount of smooth muscle in these preparations has been taken into account.

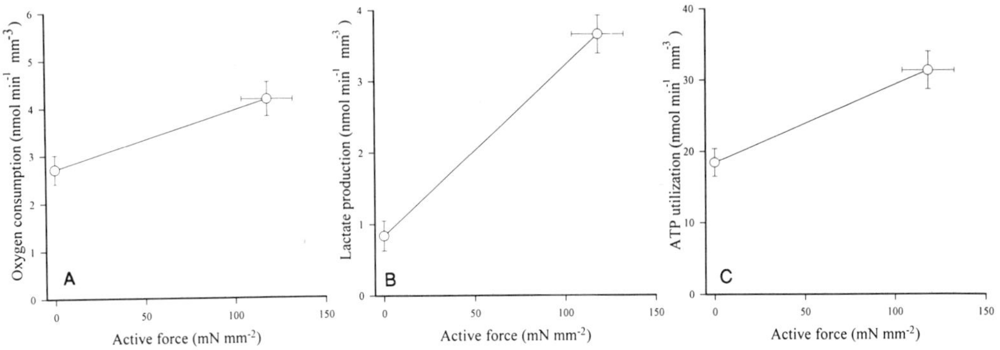

Figure 1. Oxygen consumption (panel A), lactate production (panel B), and calculated ATP turnover rate (panel C) in relation to active force for control rat detrusor muscle. Determinations were made in Ca^{2+}-free Krebs solution and during the plateau phase of contractions induced by K^{+}-high solution. Force and metabolic rates are given relative to smooth muscle area and mass, respectively (n=6). Adapted from Arner et al., 1990.

Both oxygen consumption and aerobic glycolysis increase (fig. 1) when the bladder muscle is activated and contracting (Arner, Malmqvist and Uvelius, 1990). In rabbit bladder the suprabasal oxygen consumption and lactate production increase linearly with mean force (Wendt and Gibbs, 1987). Whether the increased aerobic glycolytic activity contributes to the energy supply of the contractile proteins, or to other energy requiring functions (activation of pumps?) is not known.

The slope of the relation between active force and ATP hydrolysis rate is called the metabolic tension cost. Although several energy-consuming processes involved in activation and contraction might influence the contraction-induced increase in metabolism, the major part is considered to originate in the cross-bridge turnover of the actomyosin system (Paul, 1980) and to reflect the intrinsic rate of cross-bridge cycling (Rüegg, 1971). A high cross-bridge turnover rate per unit force produced gives a steeper slope and a high metabolic tension cost, whereas a slow turnover rate has the reverse effects. The tension cost of the rat detrusor muscle is comparable to that of the rat portal vein (Arner and Hellstrand, 1981, Arner and Uvelius, 1981) and considerably higher than that of rat aorta (Arner and Hellstrand, 1981). There is a good correlation between the tension costs and the unloaded shortening velocities (Uvelius and Hellstrand, 1980; Arner and Uvelius, 1982; Munro and Wendt, 1993) of these preparations. The tension cost (Arner, Malmqvist and Uvelius, 1990) and the heat production (Wendt and Gibbs, 1987) during contraction does not change with time, suggesting that the urinary bladder muscle has no time dependent decrease in cross-bridge turnover rate, with the modes of activation used.

Oxygen Dependence of Bladder Contraction

Figure 2 shows typical recordings of K^+ induced contractions of a detrusor strip from a rat bladder at high PO_2 and in anoxia. At PO_2 680 mm Hg, the output force increases rapidly

Figure 2. Original recordings of high-K^+-induced force in a detrusor muscle strip from a control rat bladder at PO_2 of 680 and 0 mm Hg. Note the pronounced decrease of force during contraction in anoxia. Adapted from Arner et al., 1990.

to level out at a plateau. The amplitude of the steady force is clearly lower than at high PO_2. If glucose is present the detrusor strips can produce reproducible contractions for long time in anoxia (Arner, Malmqvist and Uvelius, 1990). Thus the response to anoxia in fig. 2 is not an illustration of a transient phenomenon but represents the force output in a preparation in steady state. It is clearly seen that the contraction is biphasic in anoxia. The peak force of the first phase is higher than the steady state force which thus seems to be more sensitive to oxygen deficiency. Similar results have been reported for rabbit bladder (Levin, High and Wein, 1981). One might speculate that the first phase utilizes ATP produced during the relaxation period, while the second phase is dependent on the instantaneous ATP production. The anoxia does not have any deleterious effects on the rat detrusor muscle; if oxygen is again given to the strips, contractile force normalizes.

The reduction of force associated with inhibition of cell respiration in smooth muscle from the rat portal vein is considered to be due to metabolic effects on at least two steps in the activation of the contractile proteins, namely calcium influx and myosin light chain phosphorylation (Swärd et al., 1993). The metabolic inhibition of the K^+-contractures in the detrusor muscle in anoxia may involve those mechanisms.

Down to PO_2 of about 100 mm Hg variations in oxygen tension have little effect on contractile force of the muscle strips (fig. 3). If PO_2 is further decreased contractile force will fall. At 0 mm Hg the first phase amounts to about 60 per cent and the sustained contraction to about 35 per cent, of the corresponding at PO_2 680 mm Hg.

Lactate Production and ATP Production in Anoxia

If the oxygen tension in the bath is decreased from 680 mm Hg to zero, the oxidative metabolism stops. In the relaxed muscle the lactate production increases seven-fold (cf. figs 1 and 4). This clearly shows that the detrusor muscle cells can turn on the glycolysis when conditions are such that the oxidative metabolism is inhibited. This is not a property unique for the bladder, and is e.g. also seen in the rat portal vein (Lövgren and Hellstrand, 1985). The basal metabolic rate is (fig. 4) considerably lower in the anoxic situation. It seems likely that in anoxia the detrusor cells turn down (or are forced to turn down) activities responsible for part of the basal metabolism.

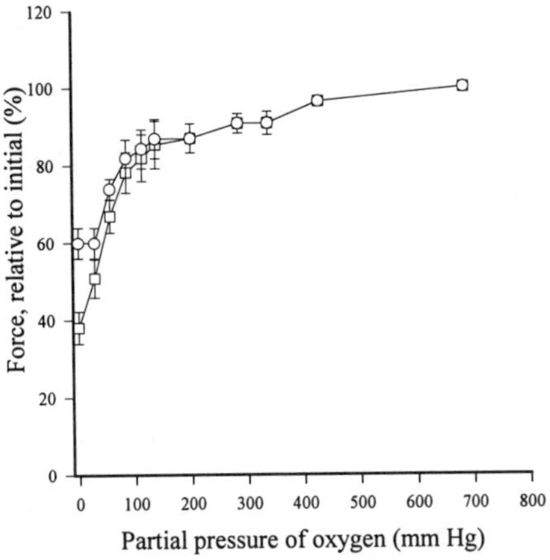

Figure 3. Active force at peak (circles) and plateau(squares) of high-K^+-induced contractions at varied partial pressure of oxygen, in percent of the value at 680 mm Hg. Control rat bladder, n=6. Adapted from Arner et al., 1990.

At normoxia it is difficult to assess whether ATP produced by the increased glycolysis during contraction is utilized by the contractile proteins. In anoxia it is definitely so as no other source of ATP is available. In this situation glycolysis is thus not exclusively used for pump activity. The decreased tension cost in anoxia shown in fig. 4 can not at present be explained with certainty, but suggests a lowered cross-bridge turnover rate.

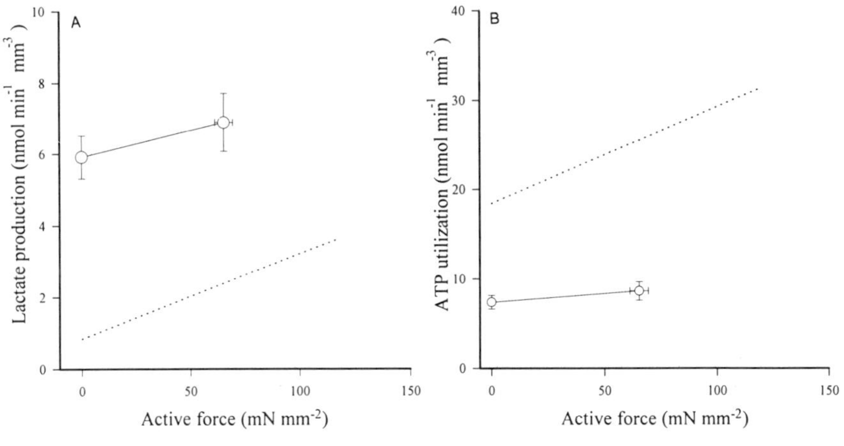

Figure 4. Lactate production (panel A), and calculated ATP turnover rate (panel B) in relation to active force for control rat detrusor muscle at PO_2 0 mm Hg. The measurements were made at the plateau of the contractions. Broken lines show the corresponding relations obtained from fig. 1. Adapted from Arner et al., 1990.

Lactate Dehydrogenase Activity and Isoform Distribution

The transformation of pyruvate to lactate is catalyzed by lactate dehydrogenase (LDH). The enzyme consists of four subunits composed of two different polypeptide chains (denoted H and M, see Everse and Kaplan, 1973). Five different forms exist (M4, M3H, M2H2, MH3, H4) which differ in electrophoretic mobility. Compared to M, the H subunit has a higher degree of substrate inhibition by pyruvate, and a higher product inhibition by lactate (Stambaugh and Post, 1966). With increasing amount of H subunits the enzyme is thus less capable of forming lactate from pyruvate. The distribution of LDH isoforms (Malmqvist, Arner and Uvelius, 1991; Polyanska et al., 1993) in the rat detrusor muscle with M amounting to 69 per cent of (M+H) is favourable for glycolysis to occur during anoxia, with its possibility of accumulation of pyruvate and lactate.

EFFECTS OF INFRAVESICAL OUTLET OBSTRUCTION ON DETRUSOR MUSCLE METABOLISM

Effect of Infravesical Obstruction on Bladder Structure and Function

Experimental infravesical outlet obstruction (for technical details see Mattiasson and Uvelius, 1982) in the rat rapidly leads to a compensatory growth of the detrusor smooth muscle cells (Gabella and Uvelius, 1990). The bladder weight can increase by a factor of 10 after 6 weeks. In such animals the detrusor pressure during micturition is much higher than in the controls (in fact not far from the systolic pressure of the animal), and the voiding is slow and prolonged (Malmgren et al., 1987). Due to a considerable residual urine in such bladders the volume at onset of micturition is higher than normal. This is an unfavourable situation according to the law of Laplace, as the active tension the smooth muscle cells must develop to produce a certain increase in pressure is higher in the distended bladder. It is not clear how residual urine is developed. Mattiasson and Uvelius (1981) suggested that the rightward shift of the length-active tension relation in the obstructed bladder with a decreased ability to produce force and pressure at low volumes would lead to the development of residual urine. Kato et al. (1990), on the other hand, suggested that the residual urine was a consequence of a decreased aerobic metabolism.

The metabolic energy consumed during a micturition is much higher in obstructed than in control bladders. Despite a considerable remodelling of the resistance vessels in the obstructed bladder (Boels et al., 1994) to a larger diameter, the resting blood flow per unit bladder weight is significantly decreased (Lin et al., 1993). No studies exist on the effect of chronic obstruction on bladder perfusion during micturition, but acute obstruction leads to a more than 60 per cent decrease in blood flow during micturition and oxygen tension fell more than during a control micturition (Siroky et al., 1993). It seems likely that the blood (and oxygen) supply is insufficient during micturition in the obstructed bladder. The detrusor smooth muscle cells in the obstructed bladder would then be more dependent on glycolysis for ATP production than those in the control bladder.

Decreased Oxygen Dependence of Active Force in Obstructed Bladder

Detrusor muscle from obstructed bladders is considerably less dependent on oxygen for force production (fig. 5) than the control detrusor strips. Compared to the value at PO_2 680 mm Hg, peak force at zero oxygen was 80 and 60 per cent, for obstructed and control bladder muscle, respectively. After the peak the force of the control muscles fell to a plateau level that at zero PO_2 was only 35 per cent of the value obtained at PO_2 680 mm Hg. In the strips from the obstructed bladders the force decrease after the peak was much less pronounced than for control muscle. The plateau amplitude was actually, even at zero PO_2, not significantly lower than the peak. Possible explanations are decreased tension cost, and better metabolic supply to the contractile proteins. It is to be pointed out that the strips from control and obstructed bladder used in all metabolic experiments were dissected to have similar cross-sectional areas (about 0.5 mm^2) when mounted in the organ bath.

Basal and Tension-dependent Energy Metabolism in Obstructed Bladder Muscle

In the obstructed detrusor the basal ATP turnover rate is slightly lower at PO_2 680 mm Hg than the corresponding for control muscle (fig. 6). In anoxia basal metabolism, although lower than at PO_2 680 mm Hg, is similar between the two groups. The slope of the ATP turnover rate vs. force is at PO_2 680 mm Hg similar for control and hypertrophic detrusor muscle (fig. 6, left panel). The metabolic tension costs are thus similar. This indicates that the changes in isoform pattern of actin and of myosin heavy chains in the obstructed bladder (Malmqvist, Arner and Uvelius, 1991) have no effect on the intrinsic cross-bridge ATP turnover under isometric conditions. At PO_2 680 mm Hg maximum active force per unit muscle cell area is, however, lower in the hypertrophic detrusor (fig. 6, left panel). A lower active force per unit cell area leads, if the tension economy is similar, to a decreased ATP turnover rate of the contracting muscle (fig. 6, left panel).

The decreased active force in obstructed bladder muscle correlates well with our reported decrease in myosin concentration and the increased actin/myosin ratio (Malmqvist, Arner and Uvelius, 1991a). A similar change in actin/myosin ratio is also found in bladder muscle from patients with outlet obstruction due to prostatic enlargement (Malmqvist, Arner and Uvelius, 1991b).

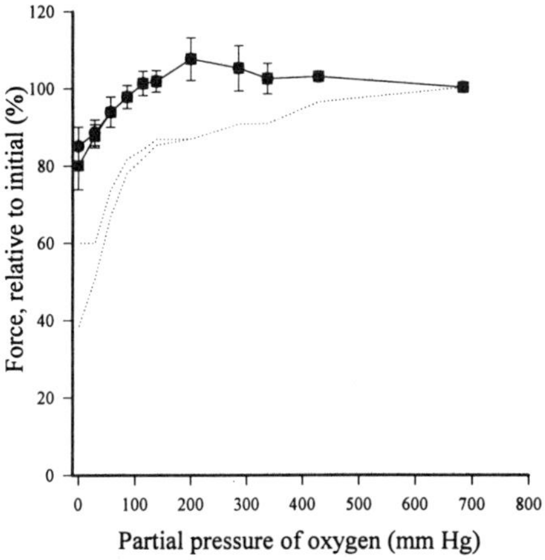

Figure 5. Active force at peak (circles) and plateau (squares) of high-K^+-induced contractions at varied partial pressure of oxygen, in percent of the value at 680 mm Hg. Obstructed bladder, n=6. Broken lines show the corresponding relations for control detrusor, obtained from fig. 3. Adapted from Arner et al., 1990.

An alternative explanation for the decreased active force could be that the metabolic supply of ATP to the contractile proteins at PO_2 680 mm Hg is inhibited in the hypertrophic muscle. In obstructed rabbit bladder a decreased degree of emptying in vitro (Kato et al., 1988), a decreased glucose oxidation, an increased lactate production (Kato et al., 1990), and an inhibition (Haugaard et al., 1992) of key enzymes in the citric acid cycle have been reported. We found (Arner, Malmqvist and Uvelius, 1990) that the relation between ATP produced by glycolysis and glucose oxidation in relaxed muscle was 6 per cent in control and 10 per cent in hypertrophic rat detrusor. During contraction the corresponding figures were 13 and 15 per cent, respectively. The shift of metabolism towards glycolysis is thus during contraction, in obstructed rat detrusor muscle, not greater than for control muscle. That would otherwise be expected if mitochondrial function was impaired. The relative volume of the smooth muscle cells accounted for by mitochondria is unchanged in the obstructed rat detrusor (Gabella and Uvelius, 1990).

The difference in relative force between control detrusor and detrusor from obstructed bladder becomes more pronounced the lower the oxygen tension is in the bathing medium (fig. 5). It is therefore reasonable to assume that the ATP production in the control strips is affected more by hypoxia than that in strips from obstructed bladder. At anoxia the active force per unit cell area of the control strips has fallen so much that it is now the same as for the strips from the obstructed bladders (fig. 6). The maximally activated control muscle strips thus develop in anoxia no more force per cell area than the strips from obstructed bladders despite the higher concentration of myosin and consequently, of crossbridges (Malmqvist, Arner and Uvelius, 1991b).

Changes in LDH Activity and Isoform Distribution in Obstructed Bladder Muscle

The functional results described above are obtained from in vitro experiments with well controlled changes of the bathing medium. The preparations from control and obstructed bladders have also been of a similar size (cross-sectional area 0.5 mm^2). The blood flow studies found in the literature (Lin et al. 1993; Siroky et al., 1993) suggest that the obstructed bladders have a compromized blood flow. Gabella (1984) found a decreased density of capillaries in hypertrophic intestinal muscle. Corresponding measurements have not been made in the obstructed bladder. However, the muscle bundles in the control rat bladders do not contain any capillaries, and the cross-sectional areas of the bundles have increased in the obstructed bladders (Gabella and Uvelius, 1990). One might then expect that the diffusion distance in the in vivo situation has increased in the muscle of the obstructed bladder.

It seems clear that the cells of the obstructed detrusor function in conditions of a more limited support of e.g. oxygen than the control muscle. Also, the increased micturition time, the increased voiding pressure, and the bladder distension will increase the work performed during a micturition. Fig. 5 shows that the obstructed bladder muscle has adapted to work under conditions with insufficient oxygen supply.

The glycolysis constitutes an important source of ATP production in the bladder muscle. Even at PO_2 680 mm Hg the glycolysis accounts for 20 (control muscle) and 24 (hypertrophic muscle) per cent of the increased ATP production during contraction in the rat bladder (Arner, Malmqvist and Uvelius, 1990). As mentioned earlier, lactate formation is catalyzed by lactate dehydrogenase (LDH).

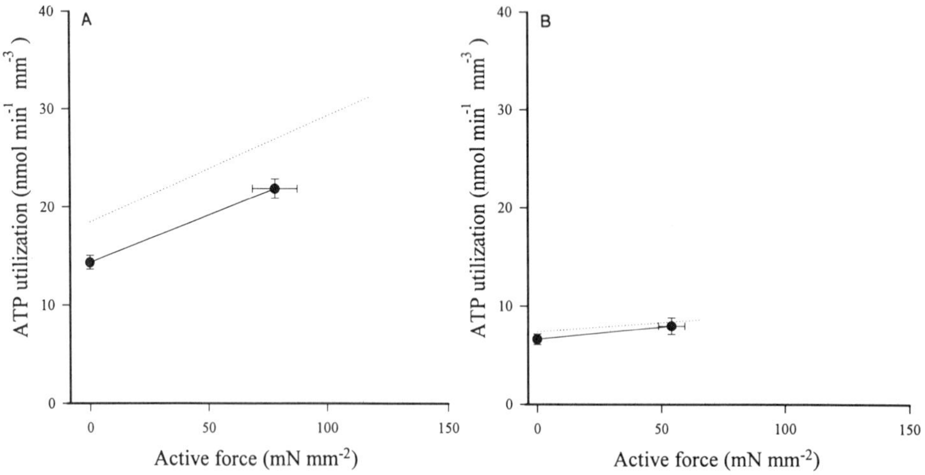

Figure 6. Calculated ATP turnover rate in relation to active force for detrusor muscle from obstructed rat bladder. Determinations were made in Ca^{2+}-free Krebs solution and during the plateau phase of contractions induced by K^+-high solution. Force and metabolic rate are given relative to smooth muscle area and mass, respectively (n=7). Panel A: PO_2 680 mmHg. Panel B PO_2 0 mm Hg. Broken lines show the corresponding relations for control detrusor, obtained from figs. 1 and 4. Adapted from Arner et al., 1990.

The LDH isoform distribution is, already in the rat control detrusor, favourable for lactate formation in hypoxia (high proportion of M subunit). The Table shows that the M3H isoform is the most abundant, followed by M4. Obstruction induces a rapid change in isoform distribution with a relative increase in M subunits. The increase is apparent already after 3 days of obstruction. In the muscle of the obstructed bladders the M4 isoform is the dominant. After 90 days more than 50 per cent is M4. LDH activity per unit detrusor weight increases slowly with time, and is after 90 days of obstruction more than two times higher than in controls (Polyanska et al., 1993). If the obstruction is removed isoform pattern and LDH activity rapidly normalizes (Polyanska et al., 1993).

The changes in LDH activity and isoform pattern would increase the capacity for lactate production under hypoxic conditions. Are the changes unspecific reactions to hypertrophy or do they represent specific adaptations of the smooth muscle of the obstructed bladder to cope with its changed working conditions? Diabetes induces in the rat a tenfold increase in 24 hours micturition. The micturition pressure (Malmgren, Andersson and Uvelius, 1989) is, however, normal, and there is no sign of infravesical obstruction. Following the increased functional demands detrusor muscle from the diabetic bladders have increased about twofold in weight, a figure comparable to those of the obstructed detrusors used in the study by Arner, Malmqvist and Uvelius (1990). The diabetic detrusors have a similar LDH isoform pattern as the controls, and the relation between oxygen tension and active force is little affected (Arner et al., 1993). The change in LDH isoform pattern and the increased ability to contract under hypoxic conditions are thus specific phenomena for the obstructed detrusors. It is interesting to note that vascular muscle from rats living in a hypoxic environment has similar changes in LDH isoform pattern (Lindy et al., 1974).

Table 1. LDH isoenzyme pattern

	M_4	M_3H	M_2H_2	MH_3	H_4	M/(M + H)
Controls (7)	29.8 ± 2.18	34.3 ± 2.3	22.8 ± 2.2	8.7 ± 1.1	4.3 ± 1.4	69.1 ± 1.4
3 days obstruction (5)	44.3 ± 1.0*	35.0 ± 0.8	15.4 ± 1.2	5.2 ± 0.6	0	79.1 ± 0.5*
10 days obstruction (5)	39.1 ± 3.0	35.4 ± 1.3	14.7 ± 1.1	5.5 ± 1.4	5.2 ± 0.9	74.5 ± 2.0
42 days obstruction (5)	49.6 ± 4.0*	32.7 ± 1.5	11.6 ± 2.4	3.1 ± 1.4	3.5 ± 1.6	80.4 ± 3.4*
90 days obstruction (6)	54.8 ± 2.8*	30.9 ± 1.4	12.0 ± 1.9	2.3 ± 0.9	0.6 ± 0.4	84.6 ± 1.3*
10 days obstruction + 32 days deobstruction (6)	32.6 ± 1.8	30.9 ± 0.7	19.2 ± 1.6	8.2 ± 1.9	9.1 ± 1.1	67.4 ± 1.4

LDH isoenzyme activities are expressed as percentages of the respective total activities. Numbers of bladders are given in paranthesis. Asterisk denotes significant difference to the respective control value.

CONCLUSION

Outlet obstruction results in a prolonged micturition and a pronounced increase in micturition pressure. This imposes a considerably increased functional demand on the detrusor smooth muscle cells which hypertrophy. Apart from the morphological changes found in the obstructed detrusor, functional changes are also found. Active force of control detrusor muscle is fairly sensitive to decreased oxygen tension in vitro. Force of detrusor muscle from obstructed bladders is markedly less sensitive to anoxia. Glucose metabolism is the major source of ATP production in detrusor muscle. In conditions with oxygen deficiency ATP production is mainly by glycolysis and lactate formation, catalyzed by lactate dehydrogenase (LDH). The increased resistance of active force (and ATP production) to oxygen deficiency is associated with a changed distribution of LDH isoforms towards M forms, which better maintain the enzyme activity during lactate accumulation.

ACKNOWLEDGEMENT

The Swedish Medical Research Council (grants 14x-9479 and 04x-8268) and the Medical Faculty, Lund University have supported the studies by us that are presented here.

REFERENCES

Arner, A. and Hellstrand, P. , 1981, Energy turnover and mechanical properties of resting and contracting aortas and portal veins from normotensive and spontaneously hypertensive rats. *Circulat. Res.* 48:539-548.

Arner, A. and Uvelius, B. , 1981, Oxygen dependence and energy turnover in normal and hypertrophic rat portal vein. *Acta Physiol. Scand.* 113:341-348.

Arner, A. and Uvelius, B. , 1982, Force-velocity characteristics and active tension in relation to content and orientation of smooth muscle cells in aortas from normotensive and spontaneously hypertensive rats. *Circulat. Res.* 50:812-821.

Arner, A., Malmqvist, U. and Uvelius, B., 1990, Metabolism and force in hypertrophic smooth muscle from rat urinary bladder. *Am. J. Physiol.* 258:C923-C932.

Arner, A., Malmqvist, U., Österman, Å. and Uvelius, B. , 1993, Energy turnover and lactate dehydrogenase activity in detrusor smooth muscle from rats with streptozotocin-induced diabetes. *Acta Physiol. Scand.* 147:375-383.

Boels, P., Arner, A., Malmqvist, U. and Uvelius, B. , 1994, Structure and mechanics of growing arterial microvessels from hypertrophied urinary bladder in the rat. Pfluegers Arch. (*Eur. J. Physiol.*) in press.

Butler, T.M. and Davies, R.E. , 1980, High-energy phosphates in smooth muscle. In: Handbook of Physiology, The Cardiavascular System. Vascular Smooth Muscle. Bethesda, MD: Am. Physiol. Soc., sect. 2, vol. 2, chapt. 10, pp. 237-252.

Everse, J. and and Kaplan, N.O. , 1973, Lactate dehydrogenases: structure and function. In: Nord, F.F. (ed.) Advances in Enzymology. Wiley, New York, pp. 37:61-133.

Gabella, G. , 1984, Hypertrophic smooth muscle. V. Collagen and other extracellular materials. Vascularization. *Cell Tissue Res.*235:275-283.

Gabella, G. and Uvelius, B. , 1990, Urinary bladder of rat: fine structure of normal and hypertrophic musculature. *Cell Tissue Res.*262:67-79.

Haugaard, N., Wein, A.J. and Levin, R.M. , 1987, In vitro studies of glucose metabolism of the rabbit urinary bladder. *J. Urol.* 137:782-784.

Haugaard, N., Potter, L., Wein, A.J. and Levin, R.M. , 1992, Effect of partial obstruction of the rat urinary bladder on malate dehydrogenase and citrate synthase activity. *J. Urol.* 147:1391-1393.

Hellstrand, P. , 1977., Oxygen consumption and lactate production of the rat portal vein in relation to its contractile activity. *Acta Physiol. Scand.* 100:91-106

Hellstrand, P., Johansson, B. and Norberg, K. 1977., Mechanical, electrical and biochemical effects of hypoxia and substrate removal on spontaneously active vascular smooth muscle. *Acta Physiol. Scand.* 100:69-83.

Hypolite, J.A., Haugaard, N., Wein, A.W., Ruggieri, M. and Levin, R.M. , 1989, Comparison of palmitic acid and glucose metabolism in the rat urinary bladder. *Neurourol. Urodyn.* 8:599-606.

Kato, K., Wein, A.J., Kitada,S., Haugaard, N. and Levin, R.M. , 1988, The functional effect of mild outlet obstruction on the rabbit urinary bladder. *J. Urol.* 140:880-884.

Kato, K., Lin, A.T.L., Haugaard, N., Longhurst, P., Wein, A.J. and Levin, R.M. , 1990, Effects of outlet obstruction on glucose metabolism of the rabbit urinary bladder. *J. Urol.* 143:844-847.

Levin, R.M., High, J. and Wein, A.J. , 1981, Metabolic and contractile effects of anoxia on the rabbit urinary bladder. *J. Urol.* 128:194-196.

Levin, R.M., Ruggieri, M.R., Gill, H.S., Haugaard, N. and Wein, A. , 1988, Effect of bethanechol on glycolysis and high energy phosphate metabolism of the rabbit urinary bladder. *J. Urol.* 139:646-649.

Lin, A.T.L., Chen, M.T., Yang, C.H.and Chang, L.S. , 1993, Blood flow of the urinary bladder - effects of outlet obstruction and correlation with energetic metabolism. Abstract 685, American Urological Association annual meeting, San Antonio, May 15-20, 1993. *J. Urol.* 149:384A.

Lindy, Turto, H., Uitto, J., Garbarsch, C., Helin, P. and Lorenzen, I. , 1974, The effect of chronic hypoxia on lactate dehydrogenase in rabbit arterial wall. *Atherosclerosis* 20:295-301.
Lövgren B, and Hellstrand, P. , 1985, Graded effects of oxygen and respiratory inhibitors of cell metabolism and spontaneous contractions in smooth muscle of the rat portal vein. *Acta Physiol. Scand.* 123:485-495.
Lövgren, B. and Hellstrand, P. , 1987, Functional role of aerobic glycolysis in rat portal vein. *Acta Physiol. Scand.* 129:211-219.
Lynch, R.M. and Paul, R.J. , 1983, Compartmentation of glycolytic and glycogenolytic metabolism in vascular smooth muscle. *Science* 222:1344-1346.
Malmgren, A., Sjögren, C., Uvelius, B., Mattiasson, A., Andersson, K.-E. and Andersson, P.O. , 1987, Cystometrical evaluation of bladder instability in rats with infravesical outflow obstruction. *J Urol* 137:1291-1294.
Malmgren, A., Andersson, P.O. and Uvelius, B. 1989. Bladder function in rats with short- and long-term diabetes; effects of age and muscarinic blockade. *J. Urol.* 142:1608-1614.
Malmqvist, U., Arner, A. and Uvelius, B. Lactate dehydrogenase activity and isoform distribution in normal and hypertrophic smooth muscle tissue from the rat. Pfluegers Arch. (Eur. J. Physiol.) 419:230-234 1991a.
Malmqvist, U., Arner, A. and Uvelius, B. Contractile and cytoskeletal proteins in smooth muscle during hypertrophy and its reversal. Am. J. Physiol. 260:C1085-C1091, 1991b.
Malmqvist, U., Arner, A., and Uvelius, B. Cytoskeletal and contractile proteins in detrusor smooth muscle from patients with bladder outlet obstruction. Scand. J. Urol. Nephrol. 25:261-267, 1991c.
Mattiasson, A. and Uvelius, B. , 1982, Changes in contractile properties in hypertrophic rat urinary bladder. J Urol 128:1340-1342.
Munro, D.D. and Wendt, I. , 1993, Contractile and metabolic properties of longitudinal smooth muscle from rat urinary bladder and the effects of ageing. J. Urol. 150:529-536.
Odessey, R. and Chace, K.W. , 1982., Utilization of endogenous lipid, glycogen, and protein by rabbit aorta. *Am. J. Physiol.* 243:H128-H132.
Paul, R.J. , 1980, Chemical energetics of vascular smooth muscle. In: Handbook of Physiology. The Cardiovascular System. Vascular Smooth Muscle. Bethesda MD: *Am. Physiol. Soc.*, 1980, sect. 2, vol. 2, pp. 201-235.
Paul, R.J. , 1983, Functional compartmentalization of oxidative and glycolytic metabolism in vascular smooth muscle. *Am. J. Physiol.* 244:C399-C409.
Paul, R.J., Bauer, M and Pease, W. , 1979, Vascular smooth muscle: Aerobic glycolysis linked to sodium and potassium transport processes. *Science* 206:1414-1416.
Polyanska, M., Arner, A., Malmqvist, U. and Uvelius, B. , 1993., Lactate dehydrogenase activity and isoform distribution in the rat urinary bladder: *Effects of outlet obstruction.* J. Urol. 150:543-545.
Rüegg, J.C. , 1971, Smooth muscle tone. *Physiol. Rev.* 51:201-247.
Siroky, M.B., Krane, R.J., Pontari, M. and Azadzoi, K. , 1993, Effect of bladder filling and contraction on bladder microcirculation. *Neurourol. Urodyn.* 12:400-401 (abstract).
Stambaugh, R. and Post, D. , 1966, Substrate and product inhibition of rabbit muscle lactic dehydrogenase heart (H4) and muscle (M4) isozymes. *J. Biol. Chem.* 241:1462-1467
Swärd, K., Josefsson, M., Lydrup, M.-L. and Hellstrand, P. , 1993, Effects of metabolic inhibition on cytoplasmic calcium and contraction in smooth muscle of rat portal vein. *Acta Physiol. Scand.* 148:265-272.
Uvelius, B. and Hellstrand, P. , 1980, Effects of phasic and tonic activation on contraction dynamics in smooth muscle. *Acta Physiol Scand* 109:399-406.
Wendt, I.R. and Gibbs, C.L. , 1987, Energy expenditure of longitudinal smooth muscle of rabbit urinary bladder. *Am. J. Physiol.* 252:C88-C96.

THE NEURONAL RESPONSE TO BLADDER OUTLET OBSTRUCTION, A ROLE FOR NGF

Mary C. Dupont, Katarina Persson, John Spitsbergen, Jeremy B. Tuttle, and William D. Steers

Department of Urology, University of Virginia, Charlottesville, Virginia

INTRODUCTION

Theories of Irritative Voiding Symptoms in Response to Obstruction

Bladder outflow obstruction can give rise to irritative voiding symptoms characterized by urinary frequency, urgency and nocturia. Some of these symptoms have been linked to involuntary detrusor contractions during the filling phase of cystometry. Several hypotheses have been advanced to account for the irritative voiding symptoms and involuntary detrusor contractions present with bladder outlet obstruction.

Altered electromechanical coupling in smooth muscle cells has been demonstrated in animal models. These alterations are only partially reversible after relief of the outlet obstruction (Seki et al, 1992). The authors postulate that persistent defects in cell-to-cell transfer of electrical activity hamper the detrusor muscle's ability to function as a single synctium and thus give rise to multiple foci of electrical activity.

In other studies (Speakman et al, 1987), muscle strips of obstructed bladders had an increased response over controls to exogenously applied acetylcholine. This muscarinic supersensitivity was attributed to partial denervation of smooth muscle. In muscle strips obtained from patients with documented detrusor instability, the muscarinic antagonist atropine is less effective at inhibiting contractions than in control muscle, in which complete blockade can be achieved (Sjogren et al, 1982). Although more recent data suggests that atropine resistance in the human bladder is a function of the pattern or frequency of nerve firing, increased nonadrenergic, noncholinergic transmission may contribute to functional disturbances. Such transmission has been attributed to a variety of purines, peptides, and nitric oxide.

In addition to hypotheses based on altered electromechanical coupling and denervation supersensitivity, the functional abnormalities observed following obstruction may be due to alterations in the neural pathways to the bladder. Autonomic reflexes control urine storage and release. These reflex pathways consist of afferent (sensory), central, and efferent (motor) components. Parasympathetic efferents are responsible for voiding.

Muscle, Matrix, and Bladder Function,
Edited by S. Zderic, Plenum Press, New York, 1995

Conversely, sympathetic efferents facilitate bladder storage. Voiding dysfunction may develop secondary to alterations in afferents and efferents in one or both of these pathways. For example, adrenergic innervation is normally sparse and limited to the bladder base and proximal urethra (Benson et al., 1979; Sundin et al., 1977, Ek et al., 1977). However, in pathological states, such as denervation, decentralization, and outlet obstruction, changes are manifested in sympathetic adrenergics. This article will focus on the experimental and clinical evidence that supports the notion that alterations in neural pathways occur after obstruction of the bladder. Furthermore, experimental data suggests that a portion of this neural plasticity may be orchestrated by nerve growth factor (NGF).

Clinical Evidence for a Neurogenic Etiology of Irritative Voiding

Irritative voiding symptoms are more bothersome than obstructive symptoms and thus more likely to result in surgical treatment of benign prostatic hypertrophy Irritative voiding symptoms have been attributed to an abnormality in the sensory nerves supplying the bladder. This hypothesis is supported by several clinical observations. Lidocaine, injected into the lateral lobes of the prostate in patients with detrusor instability, eliminates irritative symptoms (Chalfin and Bradley, 1982). Capsaicin, a potent neurotoxin for some unmyelinated (C -fiber) afferents, also abolishes symptoms of detrusor hyperreflexia when instilled intravesically(Fowler et al., 1992; Maggi et al, 1989). Similarly, patients with sensory neuropathies, such as advanced diabetes mellitus, tabes dorsalis (Hattori et al., 1990) and congenital sensory neuropathy, often lack irritative voiding symptoms. Further circumstantial evidence that irritative voiding symptoms involve altered neural pathways is provided by prospective studies showing a lower incidence of prostatectomy among diabetics and alcoholics (Sidney, 1991). The mechanisms for irritative voiding symptoms remain unclear and may not be identical to those for the detrusor instability. Irritative symptoms do not always correlate with the degree of bladder outlet obstruction or the presence of involuntary contractions on cystometrogram (Christensen and Bruskewitz, 1990). Surgical relief of obstruction often fails to alleviate symptoms (Frimodt-Møller et al., 1984; McGuire, 1987). In addition, symptoms may not be specific for prostatism. An age-matched population of women are likely to complain of the same symptoms as men thought to have prostatism(Lepor and Machi, 1993).

HYPERTROPHIC SMOOTH MUSCLE STATES AND NEURAL RESPONSE

Insight into the mechanisms responsible for alterations in myoneuronal function following bladder outlet obstruction may be gained by comparison to other hypertrophic smooth muscle conditions.

Smooth Muscle Hypertrophy in Hypertension

The spontaneously hypertensive rat (SHR) and the genetically similar normotensive Wistar-Kyoto (WKY) strain have been used to study structural and functional changes that appear during the development of hypertension. Hypertensive animals develop hyperplasia and hypertrophy of the medial layer of the vessel wall, believed to be a critical event in the etiology of hypertension (Owens *et al.*, 1981; Lee *et al.*, 1987; Owens *et al.*, 1988). These vessels contain an increased number of noradrenergic nerves.

Sympathetic neurons that innervate the vasculature become dependent on NGF during development (Thoenen and Barde, 1980, Levi-Montalcini, 1987). Korshing and Thoenen (1983) demonstrated that the level of NGF expressed by a given tissue correlates to the density of sympathetic innervation. Indeed, an increase of NGF protein (Donohue *et al.*,

1989; Zettler and Rush, 1993) and NGF gene expression (Falckh *et al.*, 1992) has been reported in the vasculature of SHR compared to WKY tissues. Lee *et al.* (1987) found that sympathectomy of neonatal SHR's, using a monoclonal antibody against NGF in combination with guanethidine which destroys sympathetic neurons, prevented the development of hypertension. Conversely, exposure to higher than normal levels of NGF during development leads to increases in neuronal size and number in sympathetic ganglia and in catecholamine content of sympathetically innervated tissues (Zettler *et al.*, 1991). Thus, the SHR model implies that an association exists between vascular smooth muscle hypertrophy, enhanced levels of NGF, and hypernoradrenergic innervation in the development of hypertension.

Congenital Megacolon/Bowel Obstruction

Other viscera also undergo smooth muscle hypertrophy in response to chronic, partial obstruction. Intestinal obstruction elicits primarily hypertrophic growth, although hyperplasia is noted in some fully differentiated muscles (Gabella, 1990). Similarly, smooth muscle hypertrophy and hyperplasia of the intestine occurs following inflammation (Blennerhassett et al., 1992) and in response to humoral factors (Johannson, 1984).

Obstruction-induced hypertrophy of the rat ileum causes changes in the nerves of the intestinal wall (Jew et al, 1989; Williams and Jew, 1991) as well as in the dorsal root ganglia (Williams et al., 1993). Following mechanical obstruction, an increase in the cross sectional area of neurons is observed in the myenteric plexus as well as decreased transmitter expression (Gabella, 1984; Jew et al., 1989). Dorsal root ganglion cells (Williams et al., 1993) supplying the obstructed ileum also hypertrophy. The authors postulated that enlargement of neurons was due to elaboration of trophic factors or other signals by a variety of cells within the intestine. A functional form of bowel obstruction occurs in Hirschsprung's disease due to an aganglionic segment of bowel. The disease is characterized by an absence of ganglion cells and an increase in adrenergic and nonadrenergic fibers in the involved segment. Peptidergic innervation is also affected (Larsson et al., 1991; Johanson et al., 1991). In support of a trophic factor involvement, nerve growth factor (NGF) has been identified in the rat ileum (Weskamp and Otten, 1987) and increases with gut hypertrophy. Elevated intestinal levels of NGF and the mRNA for the NGF receptor have been identified in Hirschsprung's disease (Kuroda, 1994).

In summary, hypertrophy of intestinal and vascular smooth muscle, is accompanied by changes in the density, size and function of innervating neurons. Does the urinary bladder respond to obstruction in a similar fashion?

RESPONSE OF BLADDER NERVES TO OBSTRUCTION

Anatomical Changes

Examination of obstructed bladder tissue reveals smooth muscle cell hypertrophy, increases in connective tissue (Gosling and Dixon, 1980; Levin et al.,1984; Uvelius et al., 1984), and a decreased density of autonomic innervation (Gosling et al., 1986). Retrograde axonal tracing experiments in rats document that afferent cell bodies lie in the L6 and S1 dorsal root ganglia (Steers et al., 1991), while efferent soma reside in the major pelvic ganglia (Steers et al., 1990). Following obstruction, the cross sectional areas of these sensory and motoneurons nearly double in size (Steers et al., 1991; Gabella et al., 1992). Within the sacral spinal cord, there is a 40-60% increase in labeling of bladder afferents with WGA-HRP, especially in the vicinity of the sacral parasympathetic nucleus (Steers et al., 1991).

GAP-43 is a G-protein substrate associated with axonal growth after injury (Jacobson et al., 1986; Benowitz and Perrone-Bizzozero, 1991). Immunohistochemical staining reveals an increased expression of GAP-43 along the sacral parasympathetic nucleus of the L6 and S1 spinal cord of the obstructed rat. This finding supports the notion that a reorganization and modification of neuronal connections in afferent pathways may take place (Steers et al, 1994).

Biochemical Analysis

In addition to the morphological plasticity noted with obstruction, there is evidence of changes in the amount of nerves in the bladder and prostate. Tissue levels of norepinephrine in the prostate are inversely correlated to the severity of symptoms of prostatism and prostate size (Lepor et al., 1990). Chapple and colleagues (1991) describe a reduction in the density of all neuronal populations, including acetylcholinesterase-positive, adrenergic and putative neuropeptidergic nerves, in the central portion the prostate. However, increased density of varicosities containing these substances is detected in the capsule in the prostates of patients with bladder outlet obstruction. Andersson et al. (1988) find a decrease in substance P and an increase in vasoactive intestinal peptide in the bladder following outlet obstruction. Although the total amount of noradrenaline content and choline acetyltransferase (ChAT) content have been shown to increase after obstruction, it represents only 30% of control values after correction for hypertrophy (Mattiasson et al., 1987). ChAT activity, a marker of acetylcholine nerve density, decreased after obstruction relative to total protein. Yet, total ChAT activity within the bladder is unchanged (Karim and Mostwin,1992). The authors postulate that there may be a spatial redistribution of nerves without an actual decrease in their numbers after short term obstruction. Thus, true denervation may not occur. It should also be kept in mind that different responses to bladder outlet obstruction may be seen among experimental animal models. ChAT activity decreases in the rabbit while it remains constant in the rat (Levin et al., 1993). Whether total neurotransmitter content or neurotransmitter density is more relevant in interpreting the significance of these changes remains to be determined. From a clinical perspective, the effect these alterations have on bladder function or detrusor instability seen with obstruction is yet unclear.

Physiological Evidence

Bladder outlet obstruction alters neural pathways regulating micturition. Voiding frequency increases in awake animals with partial urethral obstruction (Steers et al., 1991). Cystometrograms and electrophysiologic studies provide a useful model for investigating the neural control of the bladder and changes associated with outlet obstruction (Maggi et al., 1986; Steers et al., 1987). Urethral ligation of unanesthetized rats produces unstable bladder contractions similar to those seen in humans following obstruction (Malmgren et al., 1987). These low amplitude contractions have been shown to be myogenic in origin (Igawa et al., 1992). However, irritative voiding symptoms do not necessarily directly correspond to bladder contractions (Christensen and Bruskewitz, 1990).

Electrophysiologic evaluation of bladder pathways following obstruction have been used to provide insight into changes in afferent and efferent nerves. These studies in combination with ultrastructural studies have identified both myelinated (A-delta) and unmyelinated (C fibers) afferents from the bladder. Two parasympathetic reflex pathways to the bladder have been described in the rat (Mallory et al., 1989) and cat (de Groat, 1975). A spinobulbospinal reflex is responsible for micturition in normal animals. In contrast, a spinal reflex regulates voiding after spinal injury. While both reflexes can be demonstrated in normal animals, obstructed rats appear to have an enhanced spinal reflex (Steers and de Groat, 1988). Increased neural activity, decreased inhibitory mechanisms, or altered synaptic transmission between afferents and preganglionic neurons may explain the enhancement of a spinal micturition reflex. It is tempting to postulate a link detrusor instability with an alteration in sacral pathways since a spinal reflex appears in spinalized and obstructed animals. Both of which display hyperactive voiding.

ETIOLOGY OF NEURAL CHANGES WITH OBSTRUCTION

Nerve Growth Factor (NGF)

Experiments have been performed to determine the cellular and molecular events associated with the morphological and functional changes seen in neural pathways during obstruction. The major pelvic ganglion (MPG) of the rat is unusual in that it contains neuronal cell bodies of the sympathetic and parasympathetic pathways. Neurons in both pathways probably undergo hypertrophy after obstruction (Steers et al., 1990,1991; Tuttle and Steers,1992). The neuronal hypertrophy measured in the MPG following obstruction is not prevented by transection of preganglionic input to the MPG (Steers et al., 1990). Therefore, the bladder or nearby supportive cells must provide the signal for these changes. One of the best characterized molecular signal that effects neuronal size is nerve growth factor (NGF). This neurotrophic protein promotes the selective growth and survival of neural crest-derived afferents and adrenergic efferents.

NGF is essential for the survival of dorsal root ganglion (DRG) and sympathetic cells during embryonic and postnatal life (Johnson et al., 1980; Thoenen et al., 1985;Levi-Montalcini, 1987). Exposure of rats in-utero to antisera against NGF results in the death of most DRG neurons (Johnson et al., 1980). However,NGF deprivation fails to induce cell death of mature sensory neurons (Gorin and Johnson, 1980). In mature cells, NGF regulates normal function (Johnson et al., 1986). For example in the skin, the actions of NGF and anti-NGF are specific for high threshold mechanical nociceptors, a subclass of sensory neurons. Chronic treatment with NGF, in the early postnatal period, results in decreases in the thresholds of cutaneous mechanoreceptors (Aδ afferents) and behavioral hyperalgesia to noxious stimuli (Ritter et al., 1993). NGF treatment in adult animals results in a behavioral hyperalgesia and reduces the threshold for afferent firing. Therefore, NGF's regulation of afferents, may effect the sensitivity of bladder afferents to normally innocuous stimuli.

In the lower urinary tract, NGF messenger RNA content of the prostate and seminal vesicles is high (MacGrogan et al, 1991, 1992), thus raising the possibility that NGF plays a role in the regulation of normal cell growth and function in these structures. Experimental studies in rats and humans show that hypertrophied bladders, due to outlet obstruction, contain significantly more NGF than normal bladders (Steers et al., 1991). NGF content in the bladder peaks shortly after obstruction and precedes the morphologic changes seen in the MPG. In order to establish that NGF is at least one of the neurotrophic factors involved in altering micturition pathways, additional studies were performed that disrupted neuronal access to NGF. Endogenous antibody blocks the NGF receptor (Steers et al., 1991). The NGF receptor consists of two subunits: a low affinity subunit termed p75 and a high affinity tyrosine kinase subunit termed trkA, which is responsible for the growth and survival effects of NGF(Barker et al., 1993). In rats immune to NGF, obstruction does not produce hypertrophy of NGF-responsive neurons, increased WGA-HRP labeling in the sacral spinal cord, enhancement of a spinal reflex, GAP-43 staining and urinary frequency(Steers et al.,1993). Relief of obstruction causes a partial reversal of NGF elevations as well as of neuronal hypertrophy (Steers et al., 1991). These changes were not completely reversed after relief of obstruction. This incomplete reversibility of neural plasticity may be responsible for continued symptoms and bladder dysfunction (Abrams, 1985).

Response of Cultured MPG Cells to Exogenous NGF

Survival of MPG neurons in culture exposed to NGF is increased by 8-35% compared to without exogenous NGF. In contrast to MPG neurons in vivo that possess a single axon and one or no dendrites, increased sprouting of neuronal processes occurs in vitro with the addition of NGF.. Mean neuronal size is unchanged. However, increased populations of both large and small cells are observed in cultures containing NGF. NGF responsiveness of the obstructed cells is enhanced compared to controls . This observation implies that an alteration in the NGF receptor or its access to neurotrophin develops in neurons from obstructed animals. Blocking NGF with monoclonal and polyclonal anti-NGF antibodies decreases survival of pelvic neurons. As expected, the action of NGF is selective for adrenergic neurons. Without the addition of NGF to the culture medium, there is a selective loss of tyrosine hydroxylase (TH) immunoreactive cells.

Other Neurotrophins

Not all neurons are supported by NGF. Parasympathetic, non-neural crest-derived sensory neurons and spinal motor neurons are responsive to other neurotrophic factors and/or other members of the neurotrophin family, including brain derived growth factor (BDNF), neurotrophin-3 (NT-3), and neurotrophin-4/5 (Davies et al., 1993; Escandon et al., 1993). Spinal cholinergic motoneurons respond to these neurotrophins by increasing their choline acetyltransferase (ChAT) activity. Ciliary neurotrophic factor which bears no sequence similarity to NGF, acts synergistically with BDNF, NT-3, and NT-4/5 in increasing ChAT activity .

Characterization of the early molecular responses of the bladder to outlet obstruction in rabbits revealed an early and substantial increase in the mRNA encoding basic fibroblast growth factor (bFGF). Transforming growth factor-β (TGF-β), a growth inhibitory substance, decreased to 50% of control values within 24 hours of the onset of obstruction (Buttyan et al., 1992). The alterations seen in these growth factors, after partial urethral obstruction provide additional evidence that growth factors in the bladder change with obstruction. A substantial number of MPG cells do not respond to exogenous NGF in culture and blocking NGF action does not completely prevent the neuronal hypertrophy in vivo (Tuttle and Steers, 1992; Steers and Tuttle, 1991). The response of MPG cells to exogenous BDNF and CNTF has been measured in culture using a neuronal survival bioassay. Increased survival of MPG neurons exposed to either BDNF or CNTF was observed (Tuttle et al., 1992) suggesting that, in addition to NGF, other trophic factors may be important in the study of urologic disease.

STIMULI OF NGF SYNTHESIS

Laboratory studies of the lower urinary tract and its' innervation have identified several potential stimuli that increase NGF.

Denervation

Benign prostatic hypertrophy is associated with detrusor instability in 49 to 80% of obstructed patients (Abrams, 1985). Denervation, and the resultant hypersensitivity to cholinergic agonists, has been proposed to explain the development of detrusor instability. Denervation, induced by removal of the MPG in the rat, produces a temporary loss of adrenergic and ChAT activity in the ipsilateral bladder wall (Elkstrom and Elmer, 1977). Detrusor hypertrophy is observed without any change in the cystometric or voiding frequency parameters (Tuttle et al., 1994). Ultrastructural evaluation of bladder smooth muscle, three days after unilateral ganglionectomy, reveals a significant decrease in the number and function of motor nerve endings (Berggren et al., 1993). Weeks later, the bladder becomes reinnervated (Elkstrom and Elmer, 1977). After unilateral removal of the MPG, a rise in bladder NGF is seen prior to the enlargement of neurons in the contralateral MPG and dorsal root ganglion (Tuttle et al., 1994). Thus, reinnervation may be orchestrated by neurotrophic factors produced by the bladder (Berggren et al., 1992).

Inflammation

Benign prostatic hypertrophy has also been associated with urinary tract infections and acute and chronic inflammation (Hasner, 1962). Chronic inflammation may, in turn, lead to bladder hypertrophy. Chemical inflammation, produced by intravesical instillation of formalin, results in increases in bladder NGF content and enlargement of retrogradely labeled bladder neurons in the MPG (Dupont et al., 1994). Within four to eight hours after chemical inflammation, mRNA for NGF is detected in bladder urothelium (Andreev et al., 1993). Likewise, mechanical inflammation of the bladder, produced by placement of a foreign body within the bladder, elicits a more pronounced increase in bladder NGF content as well as an increase in cell areas for bladder neurons in the MPG (Dupont et al., 1994). In contrast to the time course for NGF elevation in denervated and obstructed, tissue levels rise slowly the first several weeks after the inflammatory stimulus. Thus inflammation can stimulate production of NGF and the structure of bladder neurons.

Mechanical Stretch

Indeed, if the increase in NGF in bladders of obstructed rats is the direct result of mechanical stress, stretch of smooth muscle should evoke a rise in this neurotrophin. Stretching bladder smooth muscle cells in vitro on a deformable membrane stimulates NGF secretion (Herrell et al., 1993; Persson et al., 1994). Actinomycin-D, a transcription-inhibitor, and cycloheximide, a protein synthesis inhibitor, inhibit this stretch-induced NGF secretion. These observations verify that NGF is newly synthesized and not released from stored sources. RNA-based PCR reveals that mRNA expression for NGF in stretched cells is higher than for unstretched controls as early as two hours, then decreases to near control levels. Consistent with this observation, secretion of NGF, above levels for unstretched cells, is detected after four to six hours. Thus, mechanical stretch also activates the cellular machinery for the production and release of NGF.

CELLULAR AND MOLECULAR MECHANISMS

Calcium Dependence

The cellular mechanisms coupling external stretch and NGF production have recently been investigated in urinary tract smooth muscle cells (Persson et al., 1994). Depletion of intracellular IP-3 (inositol trisphosphate) -sensitive calcium stores by thapsigargin totally blocks basal and stretch-induced NGF production by cultured bladder smooth muscle cells. In contrast, blockade of influx of extracellular calcium- through voltage-operated -channels using nifedipine and verapamil fails to prevent basal or increased NGF secretion with stretch. Likewise, blockade of a subset of stretch-activated cation channels with gadolinium fails to curtail NGF secretion. In vivo, calcium-channel antagonists partially prevent NGF production, neuronal changes and bladder hypertrophy in obstructed rats treated with verapamil and diltiazem (Steers et al., 1994). This discrepancy may be explained if the decrease in NGF production following voltage-gated calcium channel antagonists in vivo is secondary to a decrease in overall cellular growth. Indeed, bladder weights in obstructed animals maintained on calcium channel blockers are significantly lower than controls (Steers et al.,1994)

Role of Protein Kinase C

Isoforms of protein kinase C are important in cellular growth. The phorbol ester, phorbol myristate acetate (PMA) is an activator of protein kinase C (PKC). PMA stimulates NGF production by urinary tract smooth muscle cells (Persson et al., 1994). Chronic treatment with PMA for 24 hours down-regulates PKC. This methodology was utilized to test the hypothesis that stretch-induced NGF production is PKC dependent. Downregulation of PKC with PMA reduced NGF production in response to stretch by approximately 75%, suggesting that activation of the PKC signal transduction pathway is important for stretch-induced NGF production.

The PKC enzyme family consists of calcium-dependent (α, βI, βII, γ) and -independent (δ, ε, ζ, η, θ) isozymes that respond differently to different input signals (Nishizuka, 1992). Stretched cells from "downregulated" and control cultures were lysed and separated on a SDS polyacrylamide gel for Western blot analysis. The amount of the isozymes α, δ and ζ were found to be markedly lower in downregulated cells, raising the possibility that one or several of these PKC isoenzymes regulate stretch-induced NGF expression.

Influence on Other Growth Factors

Platelet-derived growth factor-β (PDGF-β), which acts through the tyrosine kinase signal pathway, is a potent stimulus for NGF secretion in urinary tract smooth muscle cells (Persson et al., 1994). The tyrosine kinase inhibitor, genistein, reduces stretch-induced NGF output. This suggests that stretch-mediated NGF production partially depends on tyrosine kinase activity. Interestingly, downregulation of PKC also reduces NGF secretion in response to PDGF. Thus, NGF secretion, following activation of tyrosine kinase coupled receptors, seems to involve PKC as well. Studies are currently in progress to investigate if stretched cells secrete growth factors that via an autocrine mechanism initiate NGF expression.

OBSTRUCTION RECAPITULATES ONTOGENY

During development, the bladder becomes innervated by sympathetic and parasympathetic nerves. A dense cholinergic innervation, with a correspondingly high density of muscarinic receptors, is present in bladders soon after birth (Levin et al., 1981). Few adrenergic fibers and alpha and beta adrenergic receptors are detected in the early neonatal period. A progressive increase in the density of adrenergic innervation occurs during the first six weeks after birth while little or no change occurring in cholinergics. Studies in newborn rats found that catecholamine fluorescent fibers were absent in fetal and newborn bladders (unpublished observations). The density of adrenergic fibers gradually increases and peaks at three and four weeks. Total NE bladder content followed a pattern similar to that of fiber density with low levels at one week rising to peak by three weeks. Additionally, there is maximal TH staining, a marker for adrenergic cells, in the MPG at three weeks. The area of TH positive neurons in the MPG also peaks by three weeks. Examination of bladder NGF content showed relatively high levels soon after birth, a decrease at two weeks, and a second rise at three and four weeks (unpublished observations). One stimulus for increased NGF production may be distention of the bladder after closure of the urachus. Thus, NGF levels in the bladder can be correlated to NE content and adrenergic innervation in the newborn rat. These findings support results obtained from other sympathetically innervated tissues, indicating that sympathetic neurons are dependent on NGF for survival (Thoenen and Barde, 1980; Levi-Montalcini, 1987). Furthermore, the density of adrenergic innervation is dependent on tissue NGF (Korshing and Thoenen, 1983). The progression of changes that occur during development including bladder hypertrophy, the pattern of elevations in NGF and NE content, and the subsequent neuronal hypertrophy within the MPG, are remarkably similar to the sequence of events following obstruction.

FUTURE DIRECTIONS AND SUMMARY

Obstruction results in neuroanatomical and electrophysiologic alterations in reflex pathways supplying the bladder. As a result of sustained mechanical stretch, the bladder produces increased amounts of NGF. Stretch-induced NGF production probably relies on activation of tyrosine kinase receptors and PKC signal transduction pathways. Growth of bladder afferents and efferents as well as an enhanced spinal micturition reflex may contribute to the detrusor hyperactivity and irritative symptoms following obstruction. Myogenic and neurogenic mechanisms have been proposed as the cause of detrusor dysfunction. However, an interdependent relationship exists between the bladder and its innervation. The changes seen in the micturition pathways as a result of obstruction are not unique. Remodeling of the micturition pathways occurs in inflammation, diabetes, and denervation of the bladder. The identification of other potential target organ-derived neurotrophic factors will allow us to better understand the pathophysiology of bladder dysfunction. Future diagnostic methods and treatments of bladder pain, irritative symptoms and hyperactivity may be based on these mechanisms.

REFERENCES

1. Abrams, P.H., Farrar, D.J., Turner-Warwick, R., Whiteside, C.G., and Feneley, R.C.L., 1979, The results of prostatectomy: A symptomatic and urodynamic analysis of 152 patients. *J. Urol.* 121:640.

2. Abrams, P.H., 1985, Detrusor instability in bladder outlet obstruction. *Neurourol. Urodynam.* 4:317.

3. Andersson, P.0., Andersson, K-E., Fahrenkrug, J., Mattiasson, A., Sjogren, C., and Uvelius, B., 1988, Contents and effects of substance P and vasoactive intestinal polypeptide in the bladder of rats with and without infravesical outflow obstruction. *J. Urol.* 140(1):168.

4. Andreev, N.Y., Bennett, D., Priestley, J., Rattray, M., and McMahon, S. B., 1993, NGF m-RNA is increased by experimental inflammation of adult rat urinary bladder. *Society for Neuroscience Abstracts* 19(1):248.

5. Barker, P.A., Lomen-Hoerth, C., Gensch, E.M., Meakin, S.O., Glass, D.J., and Shooter, E.,1993, Tissue-specific alternative splicing generates two isoforms of the trkA receptor. *J. Biol. Chem.* 268(20):15150.

6. Benowitz, L.I. and Perrone-Bizzozero, N.I., 1991, The expression of GAP-43 in relation to neuronal growth and plasticity. When, where, how and why? *Prog. Brain Res.* 89:69.

7. Benson, G.S., McConnell, J.A. and Wood, J.G.,1979, Adrenergic innervation of the human bladder body. *J. Urol.* 122:189.

8. Berggren, T., Gabella, G., Malmgren, A., and Uvelius, B., 1993, Effects of unilateral pelvic ganglionectomy on urinary bladder function in the male rat. Scand. J. Urol. Nephrol. 27(2):181.

9. Blennerhassett, M.G., Vignjevic, P., Vermillion, D.L., and Collins, S.M.1992, Inflammation causes hyperplasia and hypertrophy in smooth muscle of rat small intestine. *Am. J. Physiol.* 262:G1041.

10. Buttyan, R., Jacobs, B.Z., Blaivis, J.G., and Levin, R.M.,1992, Early molecular response to rabbit bladder outlet obstruction. *Neurourol. and Urodynamics.* 11:225.

11. Chalfin, S.A., and Bradley, W.E., 1982, The etiology of detrusor hyperreflexia in patients with infravesical obstruction. *J. Urol.* 127:938.

12. Chapple, C.R., Crowe, R., Gilpin, S.A., Gosling, J., and Burnstock, G., 1991, The innervation of the human prostate gland:The changes associated with benign enlargement. *J. Urol.* 146:1637.

13. Christensen, M.M., and Bruskewitz, R.C., 1990, Clinical manifestations of benign prostatic hyperplasia and indications for therapeutic intervention, *in*: "Urologic Clinics of North America," W.B. Saunders, Philadelphia.

14. Davies, A.M., Horton, A., Burton, L.E., Schmelzer, C., Vandlen, R., and Rosenthal, A.,1993, Neurotrophin-4/5 is a mammalian-specific survival factor for distinct populations of sensory neurons. *J. Neurosci.* 13(11):4961.

15. de Groat, W.C., 1975, Nervous control of the urinary bladder of the cat, *Brain Res.* 87:201.

16. Donohue, S.J., Head, R.J. and Stitzel, R.E., 1989, Elevated nerve growth factor levels in young spontaneously hypertensive rats. *Hypertension.* 14(4):421.

17. Dupont, M.C., Steers, W.D., McCarty, R.C., and Tuttle, J.B., 1994, Neural plasticity and alterations in nerve growth factor and norepinephrine in response to bladder inflammation. *J. Urol.* 151: (Abstract #2004).

18. Ek, A., Alan, P., Hendersson, K.E. and Persson, C.G., 1977, Adrenergic and cholinergic nerves of the human urethra and urinary bladder: A histochemical study. *Acta Physiol. Scand.* 99:345.

19. Elkstrom, J., and Elmer, M.,1977, Choline acetyltransferase activity in the denervated urinary bladder of the rat. *Acta. Physiol. Scand.* 101:58.

20. Escandon, E., Burton, L.E., Szonyi, E., and Nikolics, K., 1993, Characterization of neurotrophin receptors by affinity crosslinking. *J. Neurosci. Res.* 34 (6):601.

21. Falckh, P.H., Harkin, L.A. and Head, R.J., 1992, Resistance vessel gene expression of nerve growth factor is elevated in young spontaneously hypertensive rats. *J. Hypertension.* 10:913.

22. Fowler, C.J., Jewkes, D., McDonald, W.I., Lynn, B., and de Groat, W.C., 1992, Intravesical capsaicin for neurogenic bladder dysfunction (letter). *Lancet.* 339(8803):1239.

23. Frimodt-Møller, P.C., Jensen, K.M−E., Iversen, P., et al., 1984, Analysis of presenting symptoms in prostatism. *J. Urol.* 132:272.

24. Gabella, G., 1984, Size of neurons and glial cells in the intramural ganglia of the hypertrophic intestine of the guinea pig. *J. of Neurocytol.* 13:73.

25. Gabella, G., 1990, Hypertrophy of visceral smooth muscle. (Review). *Anat. Embryol (Berl).* 182(5):409.

26. Gabella, G., Berggren, T., and Uvelius, B., 1992, Hypertrophy and reversal of hypertrophy in rat pelvic ganglion neurons. *J. Neurocytol.* 21:649.

27. Gorin, P.D. and Johnson, E.M., Jr., 1980, Effects of long-term nerve growth factor deprivation on the nervous system of the adult rat: An experimental autoimmune approach. *Brain Res.* 198:27.

28. Gosling, J.A. and Dixon, J.S., 1980, Structure of trabeculated detrusor smooth muscle in cases of prostatic hypertrophy. *Urol. Int.* 35:351.

29. Gosling, J.A., Gilpin, S.A., Dixon, J.S. and Gilpin, C.J., 1986, Decrease in the autonomic innervation of human detrusor muscle in outflow obstruction. *J. Urol.* 136 (2):501.

30. Hasner, E., 1962, Prostatic urinary infection. *Acta Chir Scand Suppl.,* 285:1.

31. Hattori, T., Yasuda, K., Kita, K., and Hirayama, K., 1990, Disorders of micturition in tabes dorsalis. *Br. J. Urol.*.65(5):497.

32. Herrell, D. and Steers, W.D., 1993, Effect of rate versus magnitude of cyclic mechanical deformation on NGF production and growth properties of bladder smooth muscle in vitro. *J. Urol.* 340A.

33. Igawa, Y., Mattiasson, A., and Andersson, K-E., 1992, Is bladder hyperactivity due to outlet obstruction in the rat related to changes in reflexes or to myogenic changes in the detrusor? *Acta. Physiol. Scand.* 146:409.

34. Jacobson, R.D., Virag, I., and Skene, J.H.P., 1986, A protein associated with axon growth, GAP-43, is widely distributed and developmentally regulated in the rat CNS. *J. Neurosci.* 6:1843.

35. Jew, J.Y., Williams, T.H., Gabella, G., and Zhang, M-Q., 1989, The intestine as a model for neuronal plasticity. *Arch Histol Cytol.* 52 (suppl):167.

36. Johannson, B., 1984, Different types of smooth muscle hypertrophy. *Hypertension Dallas* 6 (Suppl. III):64.

37. Johanson, J.F., Carney, J.A., Go, V.L., and Koch, T.R., 1991, Segmental distribution of colonic neuropeptides in Hirschsprung's disease. *Regulatory Peptides.* 36(1):59.

38. Johnson, E.M., Jr., Gorin, P.D., Brandeis, L.D., Pearson, J., 1980, Dorsal root ganglion neurons are destroyed by exposure in utero to maternal antibody to nerve growth factor. *Science.* 219:916.

39. Johnson, E.M., Rich, K.M., and Yip, H.K., 1986, Role of nerve growth factor in sensory neurons in vivo. *Trends Neurosci.* 9:33.

40. Karim, O.M.A., and Mostwin, J.L., 1992, Changes in bladder choline acetyltransferase activity following short term obstruction and release of obstruction. *Neurourol. and Urodynamics.* 8: 80A.

41. Korshing, J. and Thoenen, H., 1983, Nerve growth factor in sympathetic ganglia and corresponding target organs of the rat. *Proc. Nat. Acad. Sci.* (U.S.A.). 80:3513.

42. Kuroda, T., Ueda, M., and Nakano, M., 1994, Altered production of nerve growth factor in aganglionic intestines. *J. Ped. Surg.* In Press.

43. Larsson, L.T., Malmfors, G., Ekblad, E. Ekman, R., and Sundler, F., 1991, NPY hyperinnervation in Hirschsprung's disease: both adrenergic and nonadrenergic fibers contribute. *J. Pediatr. Surg.*. 26(10):1207.

44. Lee, R.M.K.W., Triggle, C.R., Cheung, D.W.T. and Coughlin, M.D., 1987, Structural and functional consequence of neonatal sympathectomy on the blood vessels of spontaneously hypertensive rats. *Hypertension.* 10(3):328.

45. Lepor, H., Shapiro, E., Bowsher, R.R., and Henry, D.P., 1990, Determination of norepinephrine levels in the adult human prostate. *J. Urol.* 144(5):1263.

46. Lepor, H. and Machi, G., 1993, Comparison of AUA symptom index in unselected males and females between fifty-five and seventy-nine years of age. *Urology..* 42(1):36.

47. Levi-Montalcini, R., 1987, The nerve growth factor 35 years later. *Science.* 237:1154.

48. Levin, R.M., High, J., and Wein, A.J., 1984, The effect of short-term obstruction on urinary bladder function in the rabbit. *J. Urol.* 132:789.

49. Levin, R.M., Saito, M., Wein, A.J., Packard, D., Cohen, A., and Haugaard, N., 1993, Effect of partial outlet obstruction on choline acetyltransferase activity in the rat and rabbit. *Neurourol. and Urodynamics.* 12:255.

50. Levin,R.M., Malkowitz,S.B., Jacobowitz,D., and Wein, A.J.,1981, The ontogeny of the autonomic innervation and contractile response of the rabbit urinary bladder. *J. Pharmacol. Exper. Ther. 219:250.*

51. MacGrogan, D., Despres, G., Romand, R., and Dicou, E., 1991, Expression of the beta-nerve growth factor gene in male sex organs of the mouse, rat, and guinea pig. *J. Neurosci. Res.* 28(4):567.

52. MacGrogan, D., Saint-Andre, J.P. and Dicou, E., 1992, Expression of nerve growth factor and nerve growth factor receptor genes in human tissues and in prostatic adenocarcinoma cell line. *J. Neurochem.* 59(4):1381.

53. Maggi, C.A., Santicioli, P. and Meli, A., 1986, The nonstop transvesical cystometrogram in urethane-anesthetized rats: a simple procedure for quantitative studies on the various phases of urinary bladder voiding cycle. *J. Pharm. Meth.* 15:157.

54. Maggi, C.A., Barbanti, G., Santicioli, P., et al., 1989, Cystometric evidence that capsaicin-sensitive nerves modulate the afferent branch of micturition reflex in humans. *J. Urol.* 142:150.

55. Mallory, B., Steers, W.D., and de Groat, W.C., 1989, Electrophysiological study of micturition reflexes in rats. *Am. J. Physiol.* 257(2 Pt 2):R410.

56. Malmgren, A., Sjogren, C., Uvelius, B., Mattiasson, A., Andersson, K-E. and Andersson, P.O., 1987, Cystometrical evaluation of bladder instability in rats with infravesical outflow obstruction. *J. Urol.* 137:1291.

57. Mattiasson, A., Ekström, J., Larsson, B., and Uvelius, B., 1987, Changes in the nervous control of the rat urinary bladder induced by outflow obstruction. *Neurourol and Urodynamics,* 5:37.

58. McGuire, E., 1987, Detrusor response to obstruction. *Department of Health and Human Services, NIH Publication* No. 87-2881:227.

59. Nishizuka, Y., 1992, Intracellular signaling by hydrolysis of phospholipids and activation of protein kinase C. (Review). *Science.* 258(5082):607.

60. Owens, G.K., Rabinovich, P.S. and Schwartz, S.M., 1981, Smooth muscle cell hypertrophy versus hyperplasia in hypertension. *Proc. Nat. Acad. Sci.* (USA) 78:7759.

61. Owens, G.K., Schwartz, S.M. and McCanna, M. , 1988, Evaluation of medial hypertrophy in resistance vessels of spontaneously hypertensive rats. *Hypertension.* 11:198.

62. Persson, K., Herrell, D., Tuttle, J.B., and Steers, W.D. 1994, Regulation of NGF secretion by urinary tract smooth muscle cells. *J. Urol.* 151 (Abstract # 1162).

63. Ritter, A.M., Lewin, G.R., and Mendell, L.M., 1993, Regulation of myelinated nociceptor function by nerve growth factor in neonatal and adult rats. *Brain Res.* 30:245.

64. Seki, N., Karim, O.M. and Mostwin, J.L., 1992, The effect of experimental urethral obstruction and its reversal on changes in passive electrical properties of detrusor muscle. *J. Urol.* 148:1957.

65. Sidney, S., Quesenberry, C., Jr., Sadler, M.C., et al., 1991, Risk factors for surgically treated benign prostatic hyperplasia in a prepaid health care plan. *Urology.* 38 (1 suppl):13.

66. Sjogren, C., Andersson, K.E., Husted, S., Mattiasson, A., Moller-Madsen, B., 1982, Atropine resistance of transmurally stimulated human bladder muscle. *J. Urol.* 128:1368.

67. Speakman, M.J., Brading, A.F., Gilpin, D.J., Dixon, J.S., Gilpin, S.A. and Gosling, J.A., 1987, Bladder outflow obstruction--a cause of denervation supersensitivity. *J. Urol.* 138:1461.

68. Steers, W.D., Mallory, B.S. and de Groat, W.C., 1987, Combined CMG and electrophysiological analysis of the micturition reflex in the rat. *J. Urol.* 137:108A.

69. Steers, W.D. and de Groat, W.C., 1988, Effect of bladder outlet obstruction on micturition reflex pathways in the rat. *J. Urol.* 140:864.

70. Steers, W.D., Ciambotti, J., Erdman, S., and de Groat, W.C., 1990, Morphological plasticity in efferent pathways to the urinary bladder following urethral obstruction in the rat. *J. Neurosci.* 10:1943.

71. Steers, W.D., Ciambotti, J., Etzel, B., Erdman, S., and de Groat, W.C., 1991, Alterations in afferent pathways from the urinary bladder of the rat in response to partial urethral obstruction. *J. Comp. Neurol.* 310:401..

72. Steers, W.D., Kolbeck, S., Creedon, D., and Tuttle, J.B., 1991, Nerve growth factor in the urinary bladder of the adult regulates neuronal form and function. *J.Clin. Invest.* 88:1709.

73. Steers, W.D., Albo, M., and Tuttle, J.B., 1994, Calcium channel antagonists prevent urinary bladder growth and neuroplasticity following mechanical stress. *Am. J. Physiol.* 266: In Press.

74. Sundin, T., Dahlstrom, A., Norlen, L., and Svedmyr, N., 1977, The sympathetic innervation and adrenoreceptor function of the human lower urinary tract in the normal state and after parasympathetic denervation. *Invest. Urol.* 14:322.

75. Thoenen, H., Korsching, S., Heumann, R., and Acheson, A. ,1985, In: "Growth factors in Biology and Medicine," Ciba. ed. by Pitman., London: 133-138.

76. Thoenen, H. and Barde, Y-A., 1980, Physiology of nerve growth factor. *Physiol. Rev.* 60: 1284.

77. Tuttle, J.B. and Steers, W.D., 1992, Nerve growth factor responsiveness of cultured major pelvic ganglion neurons from the adult rat. *Brain Res.* 588:29.

78. Tuttle, J.B., Steers, W.D., Albo, M., and Nataluk, E., 1994, Neural input regulates tissue NGF and growth of the adult urinary bladder. *J. Autonom. Nerv. Sys., In Press.*

79. Tuttle, J.B., Mackey, T., and Steers, W.D., 1994, Non-NGF growth factors regulate survival of major pelvic ganglion cells in the rat. *Neurosci. Lett., In Press.*

80. Uvelius, B., Persson, L. and Mattiasson, A., 1984, Smooth muscle hypertrophy and hyperplasia in the rat detrusor after short-time infravesical outflow obstruction. *J. Urol.* 131: 183.

81. Weskamp, G., and Otten, U., 1987, An enzyme-linked immunoassay for nerve growth factor (NGF): a tool for studying regulatory mechanisms involved in NGF production in brain and in peripheral tissues. *J. Neurochem.* 48(6):1779.

82. Williams, T.H., Ming-Qun, Z., and Jean, J., 1993, Hypertrophy of rat sensory ganglion neurons following intestinal obstruction. *Gastroenterology*. 105:8.

83. Williams, T.H. and Jew, J.Y., 1991, Neuronal responses to gut hypertrophy. *J. Auton. Nerv. Sys.*:33.

84. Zettler, C. and Rush, R.A., 1993, Elevated concentrations of nerve growth factor in heart and mesenteric arteries of spontaneously hypertensive rats. *Brain Res.* 614:15.

85. Zettler C., Head R.J., Rush, R.A., 1991, Chronic nerve growth factor treatment of normotensive rats. *Brain Res.* 538:251.

CONTRACTILE PROTEINS AND THEIR RESPONSE TO BLADDER OUTLET OBSTRUCTION

Samuel Chacko[1] and Penelope A. Longhurst[2]

[1]Department of Pathobiology
[2]Division of Urology
University of Pennsylvania
Philadelphia, PA 10104

CHANGES IN BLADDER MUSCLE CONTRACTION ASSOCIATED WITH OUTLET OBSTRUCTION

Urinary outlet obstruction causes rapid increases in bladder mass; the extent of the increase in mass is directly proportional to the severity of the obstruction[1-3]. Although there may be some species differences, in general, the changes in bladder contractility associated with outlet obstruction are also dependent upon the degree of increase in bladder mass. Bladder strips from animals with "mild" obstruction (less than 3-fold increase in bladder mass) respond to electrical field stimulation or contractile agents with no change or increases in contractile response compared to controls[1-5]. In contrast, bladder strips from animals with "severe" obstruction, where the bladder mass increases more than 6-fold, respond to contractile stimuli with decreases in contractile responses and an impaired ability of the bladder to empty[2,3,6-10]. Depending on the duration and severity of obstruction, removal of the obstruction generally causes reversal of the contractile dysfunction[11,12]. The mechanisms responsible for the changes in bladder contractility associated with bladder outlet obstruction are still unclear, but as described in other chapters in this book, may include alterations in glucose metabolism[13-19], in bladder innervation and/or action potential properties[20-25], or alterations in the contractile proteins responsible for smooth muscle contraction[26-31]. The latter mechanism will be discussed in this chapter.

CONTRACTILE PROTEINS IN NORMAL AND OBSTRUCTED BLADDER

Changes in Myosin Isoforms

Myosin is the major component of the thick-filaments in all types of muscle. Conventional myosin is a long asymmetrical molecule containing a globular amino-terminal (N-terminal) head and a rod-like coiled coil carboxyl-terminal (C-terminal) tail[32]. Smooth muscle myosin is composed of two heavy chains of around 200 kDa each and two pairs of light chains, located on the head region of the myosin molecule. One of the light chains

Muscle, Matrix, and Bladder Function,
Edited by S. Zderic, Plenum Press, New York, 1995

(20 kDa) is called the regulatory (phosphorylatable) light chain, since it regulates the actin-activated ATPase activity of myosin *via* phosphorylation by a Ca^{2+}-calmodulin dependent light chain[32]. The other light chain (17 kDa) has been suggested to play a role in the enzymatic activity of myosin[33]. The ATP binding sites, enzymatic activity, and the actin binding site reside on the head region of the molecule. Crystallized myosin head (subfragment-1) has provided a structural basis for understanding the molecular interaction between myosin, ATP, and actin during muscle contraction[34]. The characteristic of the myosin molecule to assemble into thick-filaments is attributed to the rod portion of the molecule[32].

Polymorphism of myosin is known to exist in smooth muscle cells. Two myosin heavy chain (MHC) isoforms, SM1 (204 kDa) and SM2 (200 kDa), are present in smooth muscle cells[35-38]; however, their ratios vary in smooth muscles from different sources[35,39,40]. These two isoforms differ in the amino acids at the C-terminal tail region[41]. The mRNAs for these heavy chains originate from a single smooth muscle MHC gene through an alternative splicing mechanism at the 3'-end of the coding region[36]. The protein products encoded by these mRNA can be separated into two distinct bands of 204 kDa and 200 kDa by electrophoresis on a highly porous SDS-polyacrylamide gel, and the MHC isoforms can be distinguished also by antibodies against peptides specific to the C-terminal regions of the MHC. Myosin in adult bovine aorta is shown to be homodimers of SM1 or SM2[42].

A myosin heavy chain which contains an insert of seven amino acids in the ATP binding site in the myosin head is present in intestinal smooth muscle myosin, but is absent in vascular smooth muscle myosin[43]. It has been suggested that the presence or absence of this insert is responsible for the functional diversity of smooth muscles in different parts of the body[43].

Two isoforms of 17 kDa light chain (LC17a and LC17b), which separate on isoelectric focusing have been reported[44]. The presence of these isoforms is correlated with the actin-activated ATPase activity of smooth muscle myosin isolated from various sources[33]. It is proposed that the reduced level of actin-activated ATPase activity of arterial myosin, compared with that of visceral smooth muscle, is due to a difference in the composition of the 17 kDa light chain isoforms[33]. In a recent study, Trybus used expressed truncated smooth muscle myosin molecules that are homogeneous with respect to 17 kDa light chain content in an *in vitro* motility assay and showed that the 17 kDa light chain is not required to obtain phosphorylation-dependent actin movement[45].

There are several studies on the expression of myosin isoforms[46,47] during growth and differentiation of vascular smooth muscle cells *in vivo* and *in vitro*[38,48]. The smooth muscle myosin heavy chain isoforms SM1 and SM2 are regulated during development[46]. A third type of myosin heavy chain isoform exists in embryonic and prenatal vascular smooth muscle[46]. There is very little information on the regulation of expression of myosin isoforms in the urinary bladder muscle during smooth muscle hypertrophy induced by increased functional demand. Studies on uterine smooth muscle indicate that the composition of the myosin isoforms is altered during hypertrophy [49]. We demonstrated that arterial myosin from normotensive and hypertensive dogs shows no difference in the composition of the SM1 and SM2 isoforms, which differ only in the amino acid sequence of the C-terminus[50]. However, finger printing analysis of the peptides from myosin heavy chains reveals quantitative and qualitative differences in arterial myosin from normotensive and hypertensive dogs, indicating a difference in the myosin heavy chain at regions other than the C-terminus[50].

The hypertrophy of the urinary bladder induced by partial outlet obstruction is associated with changes in the expression of smooth muscle myosin heavy chain; the SM1 to SM2 ratio changes from 1:3 to 1:1[29]. Down-regulation of the SM2-isoform has been found by other groups after obstruction of the rat[28] and rabbit[27] bladder. The myosin heavy chain expression returns to normal during regression of the smooth muscle hypertrophy upon removal of the obstruction[28,51]. Obstruction-induced hypertrophy activates a up-regulating mechanism for SM1 heavy chain[51]. A difference in the amounts of myosin isoforms, SM1 and SM2, may have an effect on the assembly of myosin molecules into thick-filaments, a major component of the contractile apparatus. The initial factor(s) responsible for the modulation of the expression of myosin heavy chain is not known.

Changes in Actin Isoforms

Actin is the major component of the thin-filament, which is an integral part of the contractile apparatus. Actin is also important for the maintenance of the cytoskeleton of smooth muscle and non-muscle cells. Functionally, actin is involved in muscle contraction and actin-myosin mediated motility in the cell[52]. In mammals, actin consists of at least six isoforms: α-skeletal muscle, α-cardiac muscle, α-smooth muscle, γ-smooth muscle, β-cytoplasmic, and γ-cytoplasmic actin[53]. These actin isoforms originate from a multigene family[54]. Although highly homologous in primary structure, they differ in their amino acid sequence at the NH_2 terminal end. The actin isoforms are highly conserved across species lines; however, they are tissue specific[53]. This pattern of conservation implies distinct functions for various actin isoforms in different tissues or cell types.

Monomeric actin has a single polypeptide chain of 46 kDa, and is globular[55]. Each molecule of monomeric actin (G-actin) binds to an ATP molecule with high affinity, and in the presence of divalent cations (Ca^{2+}or Mg^{2+}) the binding of ATP is followed by polymerization to form filamentous actin (F-actin). For each molecule of G-actin added to the F-actin, one molecule of ATP is hydrolyzed[56]. The resulting ADP is bound to the actin monomer in the F-actin. The F-actin binds to myosin and activates the Mg-ATPase of myosin under physiological conditions. In the muscle cells, the two strands of F-actin are coiled about each other in thin-filaments. It has not been unequivocally shown whether the thin-filaments in smooth muscle are polymers of one actin isoform or copolymers consisting of two or more isoactins.

In general, smooth muscle cells contain α- and γ-actins in addition to small amounts of the cytoplasmic type of β- and γ-actins[57]. The amount of actin isoform present in smooth muscle cells varies according to the source of the smooth muscle tissues[58]. The dominant type of actin in the smooth muscle of chicken gizzard, urinary bladder, uterus, intestine, and several other visceral organs in mammals is the γ-actin, whereas the major actin type in vascular smooth muscle is the α-actin[53,54,59]. Although the relevance for the multiplicity of isoactins in vertebrates is not clear, there is some evidence to indicate that each isoform has specific roles either in cellular motility and contraction or in the maintenance of cytoskeletal structure[60,61]. The relative amount of each isoactin in a cell is also influenced by changes in cellular activities, such as proliferation[62,63], differentiation[64], hormonal stimulation[65,66], or pathological stress[57,67,68].

Using a rabbit model of partial outlet obstruction, we demonstrated that obstruction-induced bladder hypertrophy is associated with increases in both γ-actin and γ-actin specific mRNA, and with decreases in β-actin and β-actin mRNA[30,31]. Similar results have been found in the obstructed rat bladder[28]. Based on immunohistochemical studies, using antibodies against muscle proteins, Buoro et. al. suggested that the mesenchymal cells in the bladder serosa transform into myofibroblasts which differentiate into fetal type smooth muscle cells during remodelling of the bladder wall following outflow obstruction[26].

Changes in the Regulatory Protein, Caldesmon

Caldesmon, an actin/calmodulin binding protein, is believed to play a role in the regulation of the actin-myosin interaction and the actin-activated ATP hydrolysis in smooth muscle[69-73]. Two isoforms of caldesmon, *h*-caldesmon (90 kDa) and *l*-caldesmon (75 kDa), have been reported[69,74]. The *h*-caldesmon migrates as a 125 kDa protein on SDS-PAGE, and is unique to smooth muscle cells. We previously demonstrated that the kinetics of actin-activated ATP hydrolysis in the presence of *h*-caldesmon is modulated by tropomyosin[75], a protein that is important in the regulation of striated muscle contraction through the troponin complex [32]. Inhibition of actomyosin ATPase with or without tropomyosin does not require the entire caldesmon molecule, since a 28 kDa fragment from the C-terminus region of caldesmon is inhibitory[72,75,76]. The two isoforms show the following differences: 1) a central region consisting of 247 amino acids present in the *h*-caldesmon is absent in the *l*-caldesmon. 2) there is an insertion of Ala-508 in *l*-caldesmon, and 3) the *l*-form has 17 amino acids in the N-terminal region instead of the 25 amino acids that are present in the *h*-form, and the sequence in this region is completely different[77,78]. Two species of caldesmon mRNAs, 4.8 kb for h-caldesmon and 4.1 kb for *l*-caldesmon, may be derived from a single gene[77] by alternative splicing. The functional relationship between the *l*- and *h*-forms of caldesmon is not known, although they both bind to actin and calmodulin[69,79]. In addition, phosphorylation of the caldesmon decreases the binding of caldesmon both to the S2 region of the myosin and to actin[80]. *l*-Caldesmon is present in non-muscle cells and in un-differentiated smooth muscle cells[69,74,81,82]. The *h*-caldesmon isoform has been detected in rabbit bladder smooth muscle[83], however, the *l*-caldesmon isoform could not be detected in pig bladder smooth muscle (Horiuchi & Chacko, unpublished observation). In chick gizzard, expression of *l*-caldesmon decreases with development, while the *h*-caldesmon increases[84]. The expression of the two forms of the caldesmon is regulated by phenotypic modulation of the smooth muscle cells[83,84]. It is not known if the expression of *h*-caldesmon in the smooth muscle cell is correlated with the expression of smooth muscle specific actin or myosin.

In addition to its ability to bind to both actin and calmodulin, smooth muscle caldesmon interacts with tropomyosin[85-87], and with the S-2 region of myosin[72,88], which represents sequences between the myosin head and light meromyosin (LMM). Caldesmon binds at its N-terminus to the myosin S-2 portion[89-92], and this binding produces a "non-productive" attachment between myosin and actin[72]. Hemric and Chalovich suggest that this phenomenon is a possible mechanism for the maintenance of "latch", a state in which force is maintained with very low levels of myosin light chain phosphorylation[93], and the cross-bridges turn over very slowly[94]. Using an *in vitro* motility assay, Horiuchi and Chacko showed that the movement of actin filament over myosin head is impeded by caldesmon if phosphorylated myosin heads are intermixed with either unphosphorylated myosin heads or myosin rod containing the S-2 region of the myosin, but devoid of myosin heads[95]. This finding provides direct evidence for the inhibition of movement due to an interaction between the N-terminal region of caldesmon and the S-2 region of myosin.

Role of Monomeric Myosin I in Bladder Muscle Cell Function

Single headed, monomeric myosin species have been considered as possible motors in actin-based movements associated with several cellular activities such as exocytosis, endocytosis and modulation of cell shapes[96]. The monomeric myosin I belongs to the myosin super-family which consists of several mechanochemical proteins, including the double headed conventional myosin, that are characterized by force-generating ATPase activity upon interaction with actin. Several species of monomeric myosin with minor differences in their structure have been isolated from eukaryotic cells[97].

Myosin I species purified from various sources have a head and a tail composed of a single heavy chain (110-120 kDa) and one or more light chains (14-30 kDa) located in the neck region between the head and tail[98-100]. Myosin I species isolated from vertebrates (mammalian adrenal medulla, bladder smooth miuscle, and avian intestinal brush border) have calmodulin, the ubiquitous Ca^{2+} binding protein, as their light chains[98-103].

Calmodulin plays an important role in the calcium regulation of intestinal brush border myosin I[100-102,104]. The role of calcium in the regulation of myosin I is controversial. Calcium inhibits the actin-activated ATPase activity of myosin I and the *in vitro* motility of avian brush border myosin I [104]. Collins et al. proposed that the calmodulin light chains of brush border myosin I are dissociated by Ca^{2+} and that the Ca^{2+} regulation is thereby inhibitory[104,105]. In contrast, Mooseker et al. reported that the motile activity of myosin I requires Ca^{2+} [106]. A recent report by Wolenski et al. shows a reversible slowing of the motility of brush border myosin I by Ca^{2+} in the presence of exogenous calmodulin[107]. On the other hand, the motility is stopped by Ca^{2+} without exogenous calmodulin. Factors other than Ca^{2+} and calmodulin are suggested to regulate the mechanochemical properties of brush border myosin I[107].

We[103,108] showed that a myosin I with characteristics similar to those reported in avian intestinal brush border[102,105,106] and bovine adrenal medulla[98] exists in pig urinary bladder smooth muscle. Exogenous calmodulin binds to myosin I heavy chain, and the binding of exogenous calmodulin is associated with enhancement of the actin activation of the myosin I Mg-ATPase activity. Furthermore, caldesmon, a calmodulin/actin binding protein that inhibits the actin-activated ATPase activity of smooth muscle myosin II[69,70,109], also inhibits the actin-activated ATPase activity of smooth muscle myosin I. The inhibition of the actin-activated ATPase of myosin I by caldesmon is reversed by calmodulin in the presence of Ca^{2+}. Therefore, caldesmon in smooth muscle appears to play a role in the regulation of myosin I. It is not known if *l*-caldesmon, which is increased in obstruction-induced bladder hypertrophy, affects the regulation of myosin I. Partial outlet obstruction increases the sensitivity of bladder strips to inhibition of contraction by ryanodine, and increases the number of ryanodine binding sites[110,111]. It will be interesting to know if the alteration in ryanodine binding sites, calcium uptake by smooth muscle cells, calcium release from the sarcoplasmic reticulum, and other signal transduction processes are associated with an alteration in the regulation of myosin I, the motor responsible for cytoplasmic motility.

Acknowledgements.

This work was supported in part by NIH Grants DK 39740 and DK47514.

REFERENCES

1. K. Kato, F.C. Monson, P.A. Longhurst, A.J. Wein, N. Haugaard, and R.M. Levin, The functional effects of long-term outlet obstruction on the rabbit urinary bladder, *J. Urol.* 143:600 (1990).
2. R.M. Levin, P.A. Longhurst, B. Barasha, E.J. McGuire, A. Elbadawi, and A.J. Wein, Studies on experimental bladder outlet obstruction in the cat: long-term functional effects, *J. Urol.* 148:939 (1992).
3. M. Saito, A.J. Wein, and R.M. Levin, Effect of partial outlet obstruction on contractility: comparison between severe and mild obstruction, *Neurourol. Urodynam.* 12:573 (1993).
4. M. Saito, P.A. Longhurst, T.L.J. Tammela, A.J. Wein, and R.M. Levin, Effects of partial outlet obstruction of the rat urinary bladder on micturition characteristics, DNA synthesis and the contractile response to field stimulation and pharmacological agents, *J. Urol.* 150:1045 (1993).
5. M. Saito, A.J. Wein, and R.M. Levin, Effect of partial outlet obstruction of the rat urinary bladder on the response to alterations in the concentrations of potassium and calcium, *Gen. Pharmacol.* 24:1495 (1993).
6. R.M. Levin, J. High, and A.J. Wein, The effect of short-term obstruction on urinary bladder function in the rabbit, *J. Urol.* 132:789 (1984).
7. S.B. Malkowicz, A.J. Wein, A. Elbadawi, K. Van Arsdalen, M.R. Ruggieri, and R.M. Levin, Acute biochemical and functional alterations in the partially obstructed rabbit urinary bladder, *J. Urol.* 136:1324 (1986).
8. S. Kitada, A.J. Wein, K. Kato, and R.M. Levin, Effect of acute complete obstruction on the rabbit urinary bladder, *J. Urol.* 141:166 (1989).

9. A. Malmgren, K.-E. Andersson, P.O. Andersson, M. Fovaeus, and C. Sjögren, Effects of nifedipine and Bay K8644 on normal and hypertrophied rat bladder, *Neurourol. Urodynam.* 7:27 (1988).
10. A. Malmgren, K.-E. Andersson, P.O. Andersson, M. Fovaeus, and C. Sjögren, Effects of cromakalim (BRL 34915) and pinacidil on normal and hypertrophied rat detrusor in vitro, *J. Urol.* 143:828 (1990).
11. A. Malmgren, B. Uvelius, K.-E. Andersson, and P.O. Andersson, On the reversibility of functional bladder changes induced by infravesical outflow obstruction in the rat, *J. Urol.* 143:1026 (1990).
12. R.M. Levin, S.B. Malkowicz, A.J. Wein, M.A. Atta, and A. Elbadawi, Recovery from short-term obstruction of the rabbit urinary bladder, *J. Urol.* 134:388 (1985).
13. A. Arner, U. Malmqvist, and B. Uvelius, Metabolism and force in hypertrophic smooth muscle from rat urinary bladder, *Am. J. Physiol.* 258:C923 (1990).
14. A. Bilgen, A.J. Wein, N. Haugaard, D. Packard, and R.M. Levin, Effect of outlet obstruction on pyruvate metabolism of the rabbit urinary bladder, *Molec. Cell. Biochem* 117:159 (1992).
15. N. Haugaard, L. Potter, A.J. Wein, and R.M. Levin, Effect of partial obstruction of the rabbit urinary bladder on malate dehydrogenase and citrate synthase activity, *J. Urol.* 147:1391 (1992).
16. N. Haugaard, B.A.W. McKenna, A.J. Wein, and R.M. Levin, Effect of partial urinary outlet obstruction in the rabbit on the incorporation of adenine into adenine nucleotides in bladder smooth muscle, *Neurourol. Urodynam.* 12:473 (1993).
17. K. Kato, A.T.-L. Lin, N. Haugaard, P.A. Longhurst, A.J. Wein, and R.M. Levin, Effects of outlet obstruction on glucose metabolism of the rabbit urinary bladder, *J. Urol.* 143:844 (1990).
18. U. Malmqvist, A. Arner, and B. Uvelius, Lactate dehydrogenase activity and isoform distribution in normal and hypertrophic smooth muscle tissue from the rat, *Pflüger's Arch.* 419:230 (1991).
19. M. Polyanska, A. Arner, U. Malmquist, and B. Uvelius, Lactate dehydrogenase activity and isoform distribution in the rat urinary bladder - effects of outlet obstruction and its removal, *J. Urol.* 150:543 (1993).
20. N. Seki, O.M.A. Karim, and J.L. Mostwin, The effect of experimental urethral obstruction and its reversal on changes in passive electrical properties of detrusor muscle, *J. Urol.* 148:1957 (1992).
21. N. Seki, O.M.A. Karim, and J.L. Mostwin, Changes in electrical properties of guinea pig smooth muscle membrane by experimental bladder outflow obstruction, *Am. J. Physiol.* 262:F885 (1992).
22. N. Seki, O.M.A. Karim, and J.L. Mostwin, Changes in action potential kinetics following experimental bladder outflow obstruction in the guinea pig, *Urol. Res.* 20:387 (1992).
23. M.J. Speakman, A.F. Brading, C.J. Gilpin, J.S. Dixon, S.A. Gilpin, and J.A. Gosling, Bladder outflow obstruction. A cause of denervation supersensitivity, *J. Urol.* 138:1461 (1987).
24. W.D. Steers, S. Kolbeck, D. Creedon, and J.B. Tuttle, Nerve growth factor in the urinary bladder of the adult regulates neuronal form and function, *J. Clin. Invest.* 88:1709 (1991).
25. W.D. Steers, J. Ciambotti, B. Etzel, S. Erdman, and W.C. De Groat, Alterations in afferent pathways from the urinary bladder of the rat in response to partial urethral obstruction, *J. Comp. Neurol.* 310:401 (1991).
26. S. Buoro, P. Ferrarese, A. Chiavegato, M. Roelofs, M. Scatena, P. Pauletto, G. Passerini-Glazel, F. Pagano, and S. Sartore, Myofibroblast-derived smooth muscle cells during remodelling of rabbit urinary bladder wall induced by partial outflow obstruction, *Lab. Invest.* 69:589 (1993).
27. A. Chiavegato, M. Scatena, M. Roelofs, P. Ferrarese, P. Pauletto, G. Passerini-Glazel, F. Pagano, and S. Sartore, Cytoskeletal and cytocontractile protein composition of smooth muscle cells in developing and obstructed rabbit bladder, *Exp. Cell Res.* 207:310 (1993).
28. U. Malmqvist, A. Arner, and B. Uvelius, Contractile and cytoskeletal proteins in smooth muscle during hypertrophy and its reversal, *Am. J. Physiol.* 260:C1085 (1991).
29. M. Samuel, Y. Kim, K.Y. Horiuchi, R.M. Levin, and S. Chacko, Smooth muscle myosin isoform distribution and myosin ATPase in hypertrophied urinary bladder, *Biochem. Int.* 26:645 (1992).
30. Y.S. Kim, R.M. Levin, and S. Chacko, Alteration of the composition of actin isoforms in urinary bladder hypertrophy, *FASEB J.* 5:A1738 (1991).
31. Y.S. Kim, Z. Wang, R.M. Levin, and S. Chacko, Alterations in the expression of the β-cytoplasmic and the γ-smooth muscle actins in hypertrophied urinary bladder smooth muscle, *Molec. Cell. Biochem* 131:115 (1994).
32. R.S. Adelstein and E. Eisenberg, Regulation and kinetics of the actin-myosin-ATP interaction, *Annu. Rev. Biochem.* 49:921 (1980).
33. Y. Hasegawa and F. Morita, Role of 17-kDa essential light chain isoforms of aorta smooth muscle myosin, *J. Biochem.* 111:804 (1992).
34. I. Rayment, H.M. Holden, M. Whittaker, C.B Yohn, et al., Structure of the actin-myosin complex and its implications for muscle contraction, *Science* 261:58 (1993).

35. A.S. Rovner, M.M. Thompson, and R.A. Murphy, Two different heavy chains are found in smooth muscle myosin, *Am. J. Physiol.* 250:C861 (1986).
36. P. Babij and M. Periasamy, Myosin heavy chain isoform diversity in smooth muscle is produced by differential RNA processing, *J. Mol. Biol.* 210:673 (1989).
37. S. Kawamoto and R.S. Adelstein, Characterization of myosin heavy chains in cultured aorta smooth muscle cells, *J. Biol. Chem.* 262:7282 (1987).
38. S. Sartore, N. DeMarzo, A. Borrione, A. Zanellato, L. Saggin, L. Fabbri, and S. Chiaffino, Myosin heavy chain isoforms in human smooth muscle, *Eur. J. Biochem.* 179:79 (1989).
39. M. Mohammod and M.P. Sparrow, The heavy chain stoichiometry of smooth muscle myosin is a characteristic of smooth muscle tissues, *Aust. J. Biol. Sci.* 41:409 (1988).
40. N.L. Stephens, A.J. Halayko, and B. Swynghedauw, Myosin heavy chain isoform distribution in normal and hypertrophied rat aortic smooth muscle, *Can. J. Physiol. Pharmacol.* 69:8 (1991).
41. T.J. Eddinger and R.A. Murphy, Two smooth muscle myosin heavy chains differ in their light meromyosin fragment, *Biochemistry* 27:3807 (1988).
42. C.A. Kelley, J.R. Sellers, P.K. Goldsmith, and R.S. Adelstein, Smooth muscle myosin is composed of homodimeric heavy chains, *J. Biol. Chem.* 267:2127 (1992).
43. C.A. Kelley, M. Takahashi, J.H. Yu, and R.S. Adelstein, An insert of seven amino acids confers functional differences between smooth muscle myosins from the intestines and vasculature, *J. Biol. Chem.* 268:12848 (1993).
44. D.J. Helper, J.A. Lash, and D.R. Hathaway, Distribution of isoelectric variants of the 17,000-Dalton myosin light chain in mammalian smooth muscle, *J. Biol. Chem.* 263:15748 (1988).
45. K.M. Trybus, Regulation of expressed truncated smooth muscle myosins, *J. Biol. Chem.* 269:20819 (1994).
46. M. Kuro-o, R. Nagai, H. Tsuchimochi, H. Katoh, Y. Yazaki, A. Ohkubo, and F. Takaku, Developmentally regulated expression of vascular smooth muscle myosin heavy chain isoforms, *J. Biol. Chem.* 264:18272 (1989).
47. M. Kuro-o, R. Nagai, K. Nakahara, H. Katoh, R. Tsai, H. Tsuchimochi, Y. Yazaki, A. Ohkubo, and F. Takaku, cDNA cloning of a myosin heavy chain isoform in embryonic smooth muscle and its expression during vascular development and in arteriosclerosis, *J. Biol. Chem.* 266:3768 (1991).
48. S. Kawamoto and R.S. Adelstein, Chicken nonmuscle myosin heavy chains: differential expression of two mRNAs and evidence for two different polypeptides, *J. Cell Biol.* 112:915 (1991).
49. M.J. Lema, E.D. Pagani, R. Shemin, and F.J. Juli, Myosin isozymes in rabbit and human smooth muscles, *Circ. Res.* 59:115 (1986).
50. A. Upadhya, M. Samuel, R.H. Cox, R.J. Bagshaw, and S. Chacko, Characteristics of arterial myosin in experimental renal hypertension in the dog, *Hypertension* 21:624 (1993).
51. Z. Wang, R. Levin, and S. Chacko, Expression of myosin isoforms during bladder obstruction and reversal, *J. Urol.* 151:482A (1994).
52. T.D. Pollard and J.A. Cooper, Actin and actin-binding proteins: critical evaluation of mechanisms and functions, *Annu. Rev. Biochem.* 55:987 (1986).
53. J. Vandekerckhove and K. Weber, At least six different actins are expressed in a higher mammal; an analysis based on the amino acid sequence of the amino-terminal tryptic peptide, *J. Mol. Biol.* 126:783 (1978).
54. J. Vandekerckhove and K. Weber, Mammalian cytoplasmic actins are the products of at least two genes and differ in primary structure in at least 25 identified positions from skeletal muscle actins, *Proc. Natl. Acad. Sci. USA*. 75:1106 (1978).
55. M. Elzinga, J.H. Collins, W.M. Kuehl, and R.S. Adelstein, Complete amino acid sequence of actin of rabbit skeletal muscle, *Proc. Natl. Acad. Sci. U.S.A.* 70:2687 (1973).
56. E.D. Korn, M.-F. Carlier, and D. Pantaloni, Actin polymerization and ATP hydrolysis, *Science* 238:638 (1982).
57. G. Gabbiani, O. Kocher, W.S. Bloom, J. Vandekerckhove, and K. Weber, Actin expression in smooth muscle cells of rat aortic intimal thickening, human atheromatous plaque, and cultured rat aortic media, *J. Clin. Invest.* 73:148 (1984).
58. V. Fatigati and R.A. Murphy, Actin and tropomyosin variants in smooth muscles. Dependence on tissue type, *J. Biol. Chem.* 259:14383 (1984).
59. J. Vandekerckhove and K. Weber, The complete amino acid sequence of actins from bovine aorta, bovine heart, bovine fast skeletal muscle and rabbit slow skeletal muscle, *Differentiation* 14:123 (1979).
60. D. De Nofrio, T.C. Hoock, and I.M. Herman, Functional sorting of actin isoforms in microvascular pericytes, *J. Cell Biol.* 109:191 (1989).

61. G. Schevzov, C. Lloyd, and P. Gunning, High level expression of transfected beta- and gamma-actin genes differentially impacts on myoblast cytoarchitecture., *J. Cell Biol.* 117:775 (1992).
62. G.K. Owens, A. Loeb, D. Gordon, and M.M. Thompson, Expression of smooth muscle α-isoactin in cultured vascular smooth muscle cells: Relationship between growth and cytodifferentiation, *J. Cell Biol.* 102:343 (1988).
63. F. Barja, C. Coughlin, D. Berlin, and G. Gabbiani, Actin isoform synthesis and mRNA levels in quiescent and proliferating rat aortic smooth muscle cells in vivo and in vitro, *Lab. Invest.* 55:226 (1986).
64. A.R. Strauch, J.D. Offord, R. Chalkley, and P.A. Rubenstein, Characterization of actin mRNA levels during BC3H1 cell differentiation, *J. Biol. Chem.* 261:849 (1986).
65. C.-Y.J. Hsu and F.D. Frankel, Effect of estrogen on the expression of mRNAs of different actin isoforms in immature rat uterus: cloning of α-smooth muscle actin message, *J. Biol. Chem.* 262:9594 (1987).
66. P.K. Elder, L.J. Schmidt, T. Ono, and M.J. Getz, Specific stimulation of actin gene transcription by epidermal growth factor and cycloheximide, *Proc. Natl. Acad. Sci. USA.* 81:7476 (1984).
67. P.R. Bakerman, K.R. Stenmark, and J.H. Fisher, α-skeletal actin messenger RNA increases in acute right ventricular hypertrophy, *Am. J. Physiol.* 258:L173 (1990).
68. J. Leavitt, P. Gunning, L. Kedes, and R. Jariwalla, Smooth muscle α-actin is a transformation-sensitive marker for mouse NIH 3T3 and rat-2 cells, *Nature* 716:840 (1985).
69. K. Sobue, K. Takahashi, and I. Wakabayashi, $Caldesmon_{150}$ regulates the tropomyosin-enhanced actin-myosin interaction in gizzard smooth muscle, *Biochem. Biophys. Res. Commun.* 132:645 (1985).
70. K.Y. Horiuchi, H. Miyata, and S. Chacko, Modulation of smooth muscle actomyosin ATPase by thin filament associated proteins, *Biochem. Biophys. Res. Commun.* 136:962 (1986).
71. J.A. Lash, J.R. Sellers, and D.R. Hathaway, The effects of caldesmon on smooth muscle actoheavy meromyosin ATPase activity and binding of heavy meromyosin to actin, *J. Biol. Chem.* 261:16155 (1986).
72. M.E. Hemric and J.M. Chalovich, Effect of caldesmon on the ATPase activity and the binding of smooth and skeletal myosin subfragments to actin, *J. Biol. Chem.* 263:1878 (1988).
73. S.B. Marston and C.S. Redwood, Inhibition of actin-tropomyosin activation of myosin MgATPase activity by the smooth muscle regulatory protein caldesmon, *J. Biol. Chem.* 267:16796 (1992).
74. A. Bretscher, Thin filament regulatory proteins of smooth- and non-muscle cell, *Nature* 321:726 (1986).
75. K.Y. Horiuchi, M. Samuel, and S. Chacko, Mechanism for the inhibition of acto-heavy meromyosin ATPase by the actin/caldesmon binding domain of caldesmon, *Biochemistry* 30:712 (1991).
76. R. Dabrowska, A. Goch, B. Galazkiewicz, and H. Osinska, The influence of caldesmon on ATPase activity of the skeletal muscle actomyosin and bundling of actin filaments, *Biochim. Biophys. Acta* 842:70 (1985).
77. J. Bryan, M. Imai, R. Lee, P. Moore, R.G. Cook, and W.G. Lin, Cloning and expression of a smooth muscle caldesmon, *J. Biol. Chem.* 264:13873 (1989).
78. R.E. Novy, J.L.-C. Lin, and J.J.-C. Nin, Characterization of cDNA clones encoding a human fibroblast caldesmon isoform and analysis of caldesmon expression in normal and transformed cells, *J. Biol. Chem.* 266:16917 (1991).
79. A.C.-L. Wang, Photocrosslinking of calmodulin and/or actin to chicken gizzard caldesmon, *Biochem. Biophys. Res. Commun.* 156:1033 (1988).
80. C. Southerland and M.P. Walsh, Phosphorylation of caldesmon prevents its interaction with smooth muscle myosin, *J. Biol. Chem.* 264:578 (1989).
81. K. Hayashi, Y. Fujio, I. Kato, and K. Sobue, Structural and functional relationships between *h*- and *l*-caldesmon, *J. Biol. Chem.* 266:355 (1991).
82. A. Bretscher and W. Lynch, Identification and localization of immunoreactive forms of caldesmoon in smooth and nonmuscle cells: a comparison with the distribution of tropomyosin and α-actinin, *J. Cell Biol.* 100:1656 (1985).
83. V.K. Lin, I.L. Lee, and J.D. McConnell, Expression of non-muscle caldesmon in obstruction-induced detrusor hypertrophy is regulated at mRNA level, *J. Urol.* 147:314A (1992).
84. N. Ueki, K. Sobue, K. Kanda, and T. Hada, Expression of high and low molecular weight caldesmons during phenotypic modulation of smooth muscle cells, *Proc. Natl. Acad. Sci. USA.* 84:9049 (1987).
85. K.Y. Horiuchi and S. Chacko, Interaction between caldesmon and tropomyosin in the presence and absence of smooth muscle actin, *Biochemistry* 27:8388 (1988).

86. P. Graceffa, Evidence for interaction between smooth muscle tropomyosin and caldesmon, *FEBS. Lett.* 218:139 (1987).
87. T. Fujii, J. Ozawa, Y. Ogoma, and Y. Kondo, Interaction between chicken gizzard caldesmon and tropomyosin, *J. Biol. Chem.* 262:2757 (1988).
88. M. Ikebe and S. Reardon, Binding of caldesmon to smooth muscle myosin, *J. Biol. Chem.* 263:3055 (1988).
89. T. Fujii, M. Imai, G.C. Rosenfeld, and J. Bryan, Domain mapping of chicken gizzard caldesmon, *J. Biol. Chem.* 262:2757 (1987).
90. E.H. Ball and T. Kovala, Mapping of caldesmon: relationship between the high and low molecular weight forms, *Biochemistry* 27:6093 (1988).
91. A. Bertegi, A. Fattoum, and R. Kassab, Cross-linking of smooth muscle caldesmon to the NH_2-terminal region of skeletal actin, *J. Biol. Chem.* 265:2231 (1990).
92. A.C.-L. Wang, L.-W.C. Wang, S. Xu, R.C. Lu, V. Saavedra-Alanis, and J. Bryan, Localization of the calmodulin- and the actin- binding sites of caldesmon, *J. Biol. Chem.* 266:9166 (1991).
93. P.F. Dillon, M.O. Aksoy, S.P. Driska, and R.A. Murphy, Myosin phosphorylation and cross-bridge cycle in arterial smooth muscle, *Science* 211:495 (1981).
94. T.M. Butler, M.J. Siegman, and S.U. Mooers, Chemical energy usage during shortening and work production in mammalian smooth muscle, *Am. J. Physiol.* 244:C234 (1983).
95. K. Horiuchi and S. Chacko, Effect of unphosphorylated smooth muscle myosin on caldesmon-mediated regulation of actin filament velocity, *J. Muscle Res. Cell. Motil.* in press: (1994).
96. T.D. Pollard, S.K. Doberstein, and H.G. Zot, Myosin-I, *Annu. Rev. Physiol.* 53:653 (1991).
97. R.E. Cheney and M.S. Mooseker, Unconventional myosins, *Curr. Opin. Cell Biol.* 4:27 (1992).
98. B. Barylko, M.C. Wagner, R. Offer, and J.P. Albanesi, Purification and characterization of a mammalian myosin I, *Proc. Natl. Acad. Sci. USA.* 89:490 (1992).
99. Y. Lin, H. Takano-Ohmuro, and K. Kohama, Myosin I-like protein in gizzard smooth muscle, *Proc. Jap. Acad.* 65:203 (1989).
100. J.H. Collins and C.W. Borysenko, The 110,000-dalton actin-binding and calmodulin-binding protein from intestinal brush border is a myosin-like ATPase, *J. Biol. Chem.* 259:14128 (1984).
101. K.A. Conzelman and M.S. Mooseker, The 110-KD protein calmodulin complex of intestinal microvillus is an actin-activated MgATPase, *J. Cell Biol.* 105:313 (1987).
102. L.M. Coluccio and A. Bretscher, Calcium-regulated cooperative binding of microvillar 110K-calmodulin complex to F-actin - formation of decorated filaments, *J. Cell Biol.* 105:313 (1987).
103. S. Chacko, S. Jacob, and K.Y. Horiuchi, Myosin I in mammalian smooth muscle is regulated by caldesmon-calmodulin, *Biophys. J.* 64:A145 (1992).
104. K. Collins, J.R. Sellers, and P. Matsudaira, Calmodulin dissociation regulates brush border myosin I (110-kD-calmodulin) mechanochemical activity in vitro, *J. Cell Biol.* 110:1137 (1990).
105. H. Swanljung-Collins and J.H. Collins, Ca^{2+} stimulates the Mg^{2+}-ATPase activity of brush border myosin I with three or four calmodulin light chains but inhibits with less than two bound, *J. Biol. Chem.* 266:1312 (1991).
106. M.S. Mooseker, K.A. Conzelman, T.R. Coleman, J.E. Heuser, and M.P. Sheetz, Characterization of intestinal microvillar membrane discs: detergent-resistant membrane sheets enriched in associated brush border myosin I (110K-caldesmon), *J. Biol. Chem.* 109:1153 (1989).
107. J.S. Wolenski, S.M. Hayden, P. Forscher, and M.S. Mooseker, Calcium-calmodulin and regulation of brush border myosin-I Mg ATPase and mechanochemistry, *J. Cell Biol.* 122:613 (1993).
108. S. Chacko, S.S. Jacob, and K.Y. Horiuchi, Myosin I from mammalian smooth muscle is regulated by caldesmon-calmodulin, *J. Biol. Chem.* 269:15803 (1994).
109. S.B. Marston, W. Lehman, C.J. Moody, and C.W.J. Smith, Ca^{2+}-dependent regulation of smooth muscle thin filaments by caldesmon, *in*: "Advances in Protein Phosphatases II," W. Merlevede and J. DiSalvo, eds., Leuven University Press, Leuven (1985).
110. R.M. Levin, S.A. Zderic, J.-Y. Yoon, U. Sillen, and A.J. Wein, Effect of ryanodine on the contractile response of the normal and hypertrophied rabbit urinary bladder to field stimulation, *Pharmacology* 47:244 (1993).
111. R.M. Levin, S.S. Levin, S.A. Zderic, M. Saito, J.Y. Yoon, and A.J. Wein, Effect of partial outlet obstruction of the rabbit urinary bladder on ryanodine binding to microsomal membranes, *Gen. Pharmacol.* 25:421 (1994).

MOLECULAR ASPECTS OF BLADDER OUTLET OBSTRUCTION

Victor K. Lin, Ph.D., John D. McConnell, M.D.

Division of Urology
The University of Texas Southwestern Medical Center
Dallas, Texas, 7235-9110

INTRODUCTION

Lower urinary tract obstruction is a complex disease process wherein interactions between the partially obstructed urethra, detrusor, and central nervous system leads to a constellation of bladder function abnormalities that may not be totally reversible after relief of the obstruction. Although the precise molecular pathophysiology of obstruction-induced bladder dysfunction is still somewhat uncertain, obstruction of the urethra induces gross anatomical, histological, cellular, and molecular alterations in the bladder wall which impairs both bladder storage and emptying. [73,38,9,19,42]

In the early phases of outflow obstruction, hypertrophy of the detrusor permits a compensatory increase in detrusor pressure to maintain flow in the face of increased outflow resistance. As obstruction persists, however, increasing amounts of extracellular matrix (e.g. collagen) are seen in the bladder wall which leads to decreased compliance and impaired emptying.[58,52] The phenomenon of acute urinary retention may occur anywhere along this spectrum and may not be related to "bladder failure" in all cases. In long-term obstruction, alteration in extracellular matrix (ECM) expression is likely to be the predominant pathophysiological feature. However, during the early phases of obstruction alteration in smooth muscle contractile function is likely. Moreover, smooth muscle cells are the likely source of ECM production, in both the normal and abnormal bladder.

A variety of mechanisms have been proposed to explain the decreased contractility seen with experimental hypertrophy in cardiac and non-bladder smooth muscle tissues. These include: (1) alterations in contractile protein expression and/or regulation[41,39,37] proteins, such as intermediate filaments, which provide an internal resistance to shortening[7,19], (3) impaired excitation-contraction coupling[4], (4) increases in extracellular collagen[54], (5) remodelling (reorientation) of muscle cells[3], or (6) alterations in myofilament length or orientation within the cell.[76,7] However, no single mechanism for altered force generation has been elucidated in these systems.

There is compelling evidence that Ca^{2+} mediated myosin phosphorylation plays the central role in the regulation of smooth muscle contraction. Most investigators are in agreement that myosin phosphorylation is a prerequisite for tension development and that there is a correlation between the level of myosin phosphorylation and the maximum shortening velocity (crossbridge cycling rate).[48,2,17] Acting through membrane-associated mechanisms, extracellular effectors increase intracellular Ca^{2+}, which in turn activates and regulates the thick- and thin-filament contractile elements of the smooth muscle (excitation-contraction coupling). In thick-filament regulation, Ca^{2+} binds to calmodulin which subsequently binds to and activates myosin light chain kinase (MLCK). MLCK-activation results in the phosphorylation of the 20 kDa light chain subunit of myosin (P-MLC), and the stimulation of actin-activated Mg^{2+}-ATPase activity of smooth muscle myosin, which generates force by interaction with the major thin-filament protein, actin.[29,80] Impaired MLC phosphorylation could be a potential explanation for decreased contractility in hypertrophied bladder.

Alteration of bladder SM contractile protein isoform distribution including myosin heavy chain (MHC)[42,61] and actin[42,32] has been reported following bladder outlet obstruction in rabbit[61,32], rat[42] and human[42]. Although, the functional consequence of an altered MHC isoform (SM1 and SM2, 204 kDa and 200 kDa, respectively) ratio is not clear, the overall myosin ATPase and the actin-activated Mg^{2+}-ATPase activities are not altered despite the relative decrease of SM2 in rabbit hypertrophied bladder.[61] Furthermore, in the hypertrophied rat detrusor no change in metabolic tension cost (ATP turnover rate/active force), which reflects the intrinsic crossbridge turnover, was observed compared to control SM, in spite of a decrease in the relative amount of SM2.[42,5]

In addition to MLC phosphorylation and MHC activation, secondary Ca^{2+}-dependent mechanisms that can modulate or fine-tune the contractility of the smooth muscle cell have also been postulated.[29,80] One of the Ca^{2+}-dependent actin binding proteins, caldesmon, has been thought to regulate actin-myosin interaction in smooth muscle.[66-68,70] Caldesmon is a thin-filament associated protein originally identified in chicken gizzard smooth muscle[66] and later in many non-muscle cells[51]. Present in a 1:16 molar ratio with actin[80], caldesmon can reversibly regulate the Mg^{2+}-ATPase activity of myosin in vitro in a Ca^{2+}/calmodulin-dependent manner[70,50]. Sobue et al. reported that caldesmon inhibits actin-myosin interaction and that this inhibitory effect of caldesmon is overcome by calmodulin.[67] Therefore, caldesmon is suggested to regulate the actin-myosin interaction in a Ca^{2+}- and calmodulin-dependent "flip-flop" manner.[68] It has been demonstrated that caldesmon can enhance crossbridge activity between actin and myosin which inhibits actomyosin ATPase activity.[33,27] Furthermore, Sutherland and Walsh have speculated that the ability of caldesmon to bind to both myosin and actin may support tension in smooth muscle fibers and may be partly responsible for the intriguing "latch" state (the maintenance of smooth muscle tension without myosin phosphorylation).[22] A force generation study[30] using a hyperpermeable single cell model of the ferret aorta[11] and synthetic peptides derived from chicken caldesmon sequence[83] suggests that caldesmon regulates contraction in vascular SM by providing a basal resting inhibition of muscular tone. The presence of an actin-linked mechanism has also been demonstrated in non-muscle cells.[69,18,53] These studies support the idea that caldesmon is a common actin-linked regulatory protein in smooth muscle and non-muscle actomyosin systems. Since a caldesmon-mediated Ca^{2+}-calmodulin dependent mechanism may play an important role in regulation of smooth muscle contractility, altered caldesmon expression could explain decreased contractility in the hypertrophied bladder.

HYPOTHESIS

The major hypothesis of our laboratory is that obstruction induces a phenotypic modulation in the detrusor smooth muscle cell which initially alters the contractile function of the cell, and then ultimately affects its expression of extracellular matrix. The discussion, herein, will center upon the early phases of smooth muscle cell phenotypic alteration in the bladder, where obstruction is likely to affect smooth muscle contractile protein expression. Since studies of human tissue do not provide good insight into the early phases of obstruction, our laboratory as well as others, have turned to animal models to study what effects early, short-term obstruction has on gene expression [43,42].

Physiologic Studies in the Levin-Wein Rabbit Model

Partial ligation of the urethra in the adult rabbit produces a significant increase in bladder weight, secondary to smooth muscle hypertrophy.[35] Despite this increase in smooth muscle mass, bladder emptying and force production per cross-sectional area is decreased.[35] Theoretically, the increase in muscle mass observed could be due to hyperplasia (an increase in cell number), hypertrophy (an increase in single cell size) or polyploidy (the formation of large cells with multiple nuclei). Polyploidy is known to occur in vascular smooth muscle in response to spontaneous hypertension.[56,55] However, true hyperplasia also occurs in the rat aorta in some hypertensive models.[3] In the rabbit model of obstruction, we have demonstrated that the four-to-five fold increase in bladder mass observed is due mainly to cellular hypertrophy.[57] Single cell measurements of DNA, as well as flow cytometric evaluations demonstrate no evidence of increase in cellular DNA content or an increased number of proliferating cells. These studies were performed in adult animals, and it does not rule out the possibility that hyperplasia may play a role in the increased muscle mass seen in flow obstruction in young animals.

Theoretically, this decrease in force generation per cross-sectional area of muscle could be due to either a decrease in the number of force-generating units per cross-sectional area, or to decreased activation of those units. To rule out the later possibility, we studied the final step in activation of the contractile elements.

Myosin Light Chain Phosphorylation

Contraction of smooth muscle is proceeded by an increase in intracellular calcium, which subsequently binds to calmodulin.[48] The calcium-calmodulin complex activates the enzyme myosin light chain kinase which, in turn, phosphorylates the myosin light chain, leading to activation of myosin ATPase activity and crossbridge cycling which results in cell shortening and force-generation.[2,17,29] Decreased activation of the contractile elements within the cell could be due to down-regulation of cholinergic receptors, loss of cell-cell contact, or alterations in signal transduction. All of these abnormalities, however, should ultimately lead to impaired myosin light chair phosphorylation. Therefore to determine if upstream regulatory events were playing a role in obstruction-induced bladder hypertrophy, we studied this final step in smooth muscle cell activation. Myosin light chain phosphorylation, in response to a variety of different stimuli, was the same in smooth muscle isolated from obstructed rabbit bladder and control bladders.[15] These data suggest that upstream regulatory events are <u>not</u> the cause of altered contractility in this animal model. In our opinion, signal transduction abnormalities are likely to play a role in the other common clinical feature of obstruction, detrusor instability.

Contractile Protein Studies

Contraction of smooth muscle occurs when myosin ATPase activity is activated by calcium-calmodulin. The subsequent ATP-dependent interactions between myosin and actin produce a sliding motion along the thick and thin filament present within the cell, leading to shortening and force generation.

Myosin is a hexameric molecule consisting of two heavy chains (MHC), two phosphorylatable and two non-phosphorylatable light chains (MLC). Its role in muscle contraction is to structurally comprise the "thick filaments" and enzymatically convert chemical energy in the form of ATP into mechanical force in muscle[61,32]. The two myosin heavy chains (MHC) intertwine to form an alpha-helical rod at their carboxyl-terminal end, while the amino-terminal region of each heavy chain assumes a globular configuration. The globular head region is known to contain the myosin light chains and actin binding sites, as well as the site of ATPase activities. Previous studies have shown the existence of at least two MHC isoforms by SDS polyacrylamide gel electrophoresis of purified myosin from smooth muscle tissue.[51,66-68,70] SM1 and SM2, the designations for the isoforms of MHC, have approximate molecular weights of 204 kDa and 200 kDa, respectively. It has also been reported that the difference in size and structure of SM1 and SM2 might be due to heterogeneity in the carboxyl terminus as shown by peptide mapping analysis.[51] SM1 and SM2 are expressed in approximately equal proportion, have been found in several smooth muscle preparations from swine.[70] However, other studies from various smooth muscle tissues of human, monkey, rat, and turkey suggested that the proportion is unequal and varies[67,50] which may imply that smooth muscle MHC isoform expression is species and tissue-specific. SM1 and SM2 appear to form homodimers.[31] Furthermore, the ratio of SM1 and SM2 is altered by certain physiological and pathological conditions such as obstruction-induced hypertrophy and hypertension.[33,27,22] The ratio of MHC isoforms is known to be developmentally regulated.[59] Human MHC from SM has been recently cloned and mapped to chromosome 16q12.[46]

We used quantitative gel electrophoric techniques to study the total amount of myosin heavy chain protein in obstructed rabbit bladders as well as the ratio between the two isoforms. There is a decrease in the amount of total myosin heavy chain seen per cross-sectional area of cell.[1] These data agree with similar observations in the rat model of obstruction.[43] It appears that although increasing load (obstruction) produces hypertrophy of the smooth muscle cell, myosin heavy chain expression does not increase proportionately, leading to a decrease in the number of myosin thick filaments per cross-sectional area. The decrease in myosin content is in direct proportion to the decrease in force-generation per cross-sectional area.

In the normal adult rabbit bladder, there is a slight predominance of the SM2 (lower molecular weight) isoform of MHC. Two-to-three weeks after the onset of obstruction, however, there is a significant shift toward SM1 predominance at the protein level. [1] These data suggest that obstruction-induced hypertrophy not only leads to down-regulation of myosin heavy chain, but also an alteration in alternate splicing, leading to differences in isoform expression.

Myosin Heavy Chain Gene Expression

Molecular cloning of SM1 and SM2 complementary DNA has been recently reported from chicken[23,30], rabbit[11,83], and rat[69]. From the characterization of the rabbit clone SMHC-40 (SM1) and SMHC-29 (SM2) which both encode part of the light meromyosin, Nagai et al. have suggested that both mRNAs originate from a single smooth muscle MHC gene through an alternative splicing mechanism at the 3' end of the coding region. Comparison of the nucleotide sequences reveals that these two mRNAs share an identical sequence of at least 1424 nucleotides with divergence, in part of the 3' carboxyl coding region and in a portion of the 3'-untranslated region. They concluded that a unique 39 nucleotide insertion of SMHC-29 leads to the difference between SM1 and SM2 by having a unique stretch of 43 and 9 amino acids respectively on their carboxyl-termini.[83] Furthermore, they raised antibodies against two synthetic peptides deduced from these unique carboxyl-termini and demonstrated by immunoblotting that indeed SMHC-40 mRNA encodes for SM1 and SMHC-29 encodes for SM2.[83] Recently Kuro-o et al. reported that the expression of rabbit vascular MHC isoforms were subject to developmental regulation at the RNA splicing level. They also suggested that a third MHC isoform in the embryonic and perinatal vascular smooth muscle may exist.[18] Non-muscle myosin is also expressed in SM cells under stress conditions.[60,63,64,81]

Using cDNA probes to rabbit smooth muscle myosin heavy chain we studied the transcription of both total MHC as well as the individual isoforms by Northern Blot and S1 nuclease protection assays, respectively.[16] Northern Blot analysis of the normal rabbit bladder demonstrates a high constitutive level of MHC expression. Shortly after the onset of obstruction, MHC expression is significantly down-regulated. By the third week, MHC expression tapers down to barely detectable levels. Nuclease protection assays, capable of simultaneously quantitating the individual isoforms, demonstrate a significant shift towards SM1 expression.[16] Interestingly, the predominant SM1 expression pattern is also seen in the developing fetal rabbit bladder and in dedifferentiated smooth muscle cells in culture. These studies suggest that obstruction-induced hypertrophy in the rabbit model is associated with a dedifferentiation of the smooth muscle cell.

Caldesmon Expression

Caldesmon is a major smooth muscle protein that interacts in vitro with calmodulin in a Ca^{2+}-dependent manner and with actin, tropomyosin and myosin in a Ca^{2+}-independent manner.[31,34,79] The stoichiometry of Ca^{2+}-dependent binding of caldesmon to calmodulin was reported to be 1:1.[65] However, its binding to F-actin is Ca^{2+}-independent and saturable, with maximal binding of 1caldesmon : 6-26 actin monomers. [80,67,68, 50,33, 69,44, 77,10] In the presence of micromolar concentrations of Ca^{2+}, calmodulin dissociates caldesmon from F-actin. However, at lower levels of Ca^{2+}, caldesmon binds exclusively to F-actin.[66,10] Domain mapping studies of chicken caldesmon have localized the various binding sites on the molecule.[26] The actin-, calmodulin-, and tropomyosin-binding sites are all located within the C-terminal region of the polypeptide, whereas the myosin-binding site is near the N-terminus.

It has been reported that the inhibitory effect of caldesmon on smooth muscle actomyosin ATPase activity could not be overcome by Ca^{2+} and calmodulin alone, but that phosphorylation of caldesmon could reverse this inhibition.[50,74] Caldesmon appears to be an inhibiting factor to SM contraction.[8,14,21,24,30,71] Caldesmon can be phosphorylated by Ca^{2+}-calmodulin-dependent protein kinase II[78,62], and protein kinase C[74,78] in vitro. Phosphorylation on the C-terminus of caldesmon, where the actin- and calmodulin-binding sites are located, diminishes the caldesmon inhibition of actin-activated myosin Mg^{2+} ATPase activity. Thus, phosphorylation of caldesmon may also play an important role in regulating smooth muscle contraction.

Two caldesmon isoforms have been identified. The h-form from smooth muscle cells migrates on SDS PAGE gels with a apparent molecular weight of 140 kDa. The l-form from non-muscle cells (e.g. fibroblasts) is approximately 75 kDa. The tissue and cell distribution of the two isoforms are distinctively different, although the l-form has been observed in dedifferentiated smooth muscle cells seen in developing tissues and cell culture. cDNAs of chicken h-caldesmon have been cloned.[25,12] However, the sequences are discrepant, suggesting that two sub-isoforms of h-caldesmon may exist in chicken gizzard SM. The larger sub-isoform contains 771 amino acids and has a deduced molecular weight of 88,743[25]; the smaller one is identical in sequence but lacks a 15 amino acid segment and has a deduced molecular weight of 86,974[12]. The fact that the deduced molecular weight values are substantially lower than the values determined by SDS-PAGE may due to a high acidic amino acid content.[13] The l-caldesmon cDNAs from chicken brain[26], rat liver[82], and human platelet[28] have also been cloned. Sequence comparison of l- and h-caldesmon from these species indicates that the caldesmon gene is highly conserved. Major differences between the chicken h- and l-forms include: (1) a 247 amino acid section in the central region of the h-caldesmon is absent in l-form, (2) insertion of Ala-508 in l-caldesmon, (3) divergence in the N-terminus for first 25 amino acid residues in h-form and 16 amino acid residues in l-form. Based on southern and northern analysis, it is suggested that the two species of caldesmon mRNA, 4.8 kb for h- and 4.1 kb for l-form, may be generated from a single gene by alternative splicing.[26]

To date, the functional nature of the sequence differences between h- and l-caldesmon remains unclear. Although they share similar binding specificity and capacity, l-caldesmon may not allow the same degree of force maintenance as the h-isoform.[26] Moreover, change in expression of the two caldesmon isoforms is closely associated with the phenotypic modulation of the SM cell.[75]

Caldesmon from non-muscle[82] and SM cells[40] can be phosphorylated in vitro by $p34^{cdc2}$ kinase, resulting in the inhibition of caldesmon binding to F-actin and Ca^{2+}-calmodulin. Furthermore, the phosphorylation of the l-form by $p34^{cdc2}$ kinase is mitotic-specific.[82] These findings strongly suggest that phosphorylation of caldesmon by $p34^{cdc2}$ may play a significant role in smooth muscle proliferation and/or other functions of smooth muscle.

To improve our understanding of *h*- and *l*-caldesmon expression in our rabbit model of obstruction, we have isolated several caldesmon cDNA clones. Knowing that fibroblasts express only *l*-caldesmon, a pcDNAII plasmid expression library was constructed from rabbit dermal skin fibroblast mRNA and screened using a monoclonal anti-caldesmon antibody (antibody C21 from Sigma Chemical Co.). An initial positive clone (cd72) was characterized and its fusion protein found to react with the antibody C21. Further screening of the same library using cd72 as a probe allowed us to identify 5 additional clones which overlapped the full length mRNA. After sequencing all of these clones, we deduced a 4065 nucleotide long mRNA sequence for *l*-caldesmon with an open reading frame that encodes a 505 amino acid long *l*-caldesmon protein. Sequence homology analysis using an IntelliGenetics Suite Program reveals that rabbit *l*-caldesmon has significant homology to both *h*- and *l*-chicken caldesmon (Mean score of 479 and 360 respectively based on the algorithm of Wilbur and Lipman analysis). Comparison of the deduced peptide sequence of rabbit *l*-caldesmon to that of chicken shows a 85% homology at the C-terminal region (residues 292-493). When the JN5 clone, one of the 5 clones which covers the entire coding region, is *in vitro* transcribed and translated, the resulting product is a 75 kDa polypeptide which corresponds to the molecular weight of *l*-caldesmon. Western analysis further demonstrates that the *in vitro* translated JN5 protein reacts to the anti-caldesmon antibody and co-migrates with *l*-caldesmon from two week obstructed bladder and rabbit fibroblast. These data indicate that the cDNA clone we have isolated is indeed rabbit *l*-caldesmon.

Northern blot analysis of RNAs from hypertrophied bladder probed with random-primed fragments of this *l*-caldesmon cDNA (see Figure 9) revealed the existence of two hybridizing species of mRNAs (4.9 and 4.1 kb). The 4.1 kb message is *l*-caldesmon and is expressed in obstructed bladder, SM cells in culture and liver tissue. The 4.9 kb caldesmon specific message is apparently the *h*-form which shares a high degree of sequence homology with the *l*- caldesmon probe. Furthermore, the mRNA size of these two isoforms is consistent with those reported from the chicken.[26] *h*-caldesmon mRNA is expressed in normal and one week obstructed bladder but not in other tissues. By contrast, both forms are evident in the one week obstructed bladder.

These findings coincide with the specific expression of caldesmon protein isoforms in the corresponding tissues. When we examined the time course of caldesmon mRNA expression, Northern analysis reveals that induction of *l*-caldesmon mRNA is detectable 48 hours after the onset of obstruction and significantly expressed after only one week compared to that of the normal bladder. Up to two weeks of obstruction, *h*- and *l*-caldesmon mRNA expression are essentially equal. *l*-caldesmon mRNA becomes the predominant isoform after two weeks of obstruction. Preliminary studies in our laboratory have demonstrated a predominance.

Given the importance of caldesmon in regulating SM contractility, the induction of l-caldesmon in obstruction-induced bladder hypertrophy may contribute to the decreased SM contractility seen in obstruction-induced bladder dysfunction. Basic research in caldesmon gene expression in SM will provide insight into the molecular basis of normal and abnormal bladder function. In addition, these studies should contribute to our general understanding of smooth muscle pathophysiology in other disease states such as erectile impotency, hypertension and atherosclerosis. Moreover, characterization of detrusor smooth muscle cells at the cellular and molecular level is recognized as a major priority in bladder research.

SUMMARY

In an animal model of obstruction, increasing load induces significant smooth muscle hypertrophy which is associated with a down-regulation of myosin heavy chain expression. This undoubtedly contributes to the decreased smooth muscle contractility seen in this model. Moreover, obstruction-induced hypertrophy leads to the development of a dedifferentiated smooth muscle phenotype, as evidenced by a revision of the cell to fetal (of non-muscle) gene expression patterns. Similar alterations are seen in atherosclerotic vessels and other pathologic smooth muscle systems. In these systems, dedifferentiation is also associated with significant alterations in extracellular matrix expression. It seems likely that obstruction in the bladder induces dedifferentiation of the smooth muscle cell which alters contractility as well as extracellular matrix expression, leading to altered bladder performance and decreased compliance.

REFERENCES

1. Abernathy BB, Kadesky KT and McConnell JD: Myofilament protein alterations in the obstructed rabbit bladder. J Urol 139:195A, 1988.
2. Aksoy MO, Murphy RA and Kamm KE. (1982) Role of Ca^{2+} and myosin light chain phosphorylation in regulation of smooth muscle. Am. J. Physiol. 242:C109-C116.
3. Arner A and Uvelius B. (1982) Force-velocity characteristics and active tension in relation to content and orientation of smooth muscle cells in aortas from normotensive and spontaneously hypertensive rats. Circ. Res. 50:812-821.
4. Arner A, Malmqvist U and Uvelius B. (1985) Effects of Ca^{+2} on force-velocity characteristics of normal and hypertrophic smooth muscle of the rat portal vein. Acta. Physiol. Scand. 124:525-533.
5. Arner A, Malmqvist U and Uvelius B. (1990) Metabolism and force in hypertrophic smooth muscle from rat urinary bladder. Am. J. Physiol. 258:C923-C932.
6. Beckers-Bleukx G and Mar_chal G. (1985) Detection and distribution of myosin isozymes in vertebrate smooth muscle. Eur. J. Biochem. 152:207-211.

7. Berner PF, Somlyo AV and Somlyo AP. (1981) Hypertrophy-induced increase of intermediate filaments in vascular smooth muscle. J. Cell Biol. 88:96-101.
8. Brenner B, Yu LC and Chalovich JM: Parallel inhibition of active force and relaxed fiber stiffness in skeletal muscle by caldesmon: Implications for the pathway to force generation. Proc. Natl. Acad. Sci. 88:5739-5743, 1991.
9. Brent L and Stephens FD. (1975) A quantitative study of smooth muscle cells in reflux, obstructed and triad bladder. Invest. Urol. 12:503-508.
10. Bretscher A. (1984) Smooth muscle caldesmon. Rapid purification and F-actin cross linking properties. J. Biol. Chem. 259:12873-12880.
11. Brozovich FV, Walsh MP and Morgan KG. (1990) Regulation of force in skinned, single cells of ferret aortic smooth muscle. Pflugers Arch. 416:742-749.
12. Bryan J, Imai M, Lee R, Moore P, Cook RG and Lin WG. (1989) Cloning and expression of a smooth muscle caldesmon. J. Biol Chem. 264:13873-13879.
13. Bryan J. (1989) Smooth muscle thin filaments. caldesmon, acidic amino acids and molecular weight determinations. J. Muscle Res. Cell Motil. 10:95-96.
14. Chen Y and Chalovich JM: A mosaic multiple-binding model for the binding of caldesmon and myosin subfragment-1 to actin. Biophys J 63:1063-1070, 1992.
15. Cher ML, Kamm KE and McConnell JD: Stress generation and myosin phosphorylation in the obstructed bladder. J Urol 143:355A, 1990.
16. Cher ML, Lin VK and McConnell JD: Myosin gene expression in obstruction-induced bladder hypertrophy. J Urol 145:310A, 1991.
17. Dillon PF, Aksoy MO, Driska SP and Murphy RA. (1981) Myosin phosphorylation and crossbridge cycle in arterial smooth muscle. Science 211:495-497.
18. Dingus J. Hwo S and Bryan J. (1986) Identification by monoclonal antibodies and characterization of human platelet caldesmon. J. Cell Biol. 102:1748-1757.
19. Gabella G. (1979) Hypertrophic smooth muscle. Cell Tissue Res. 201:63-288.
20. Gosling JA and Dixon JS. (1980) Structure of trabeculated detrusor smooth muscle in cases of prostatic hypertrophy. Urol. Int. 35:351-372.
21. Haeberle JR, Trybus KM, Hemric ME and Warshaw DM: The effects of smooth muscle caldesmon on actin filament motility. J Biol Chem. 267:23001-23006, 1992.
22. Hai CM and Murphy RA. (1988) Cross-bridge phosphorylation and regulation of latch state in smooth muscle. Am. J. Physiol. 254:C99-C106.
23. Hamada Y, Yanagisawa M, Katsuragawa Y, Coleman JR, et al.: Distinct vascular and intestinal smooth muscle myosin heavy chain mRNAs are encoded by a single-copy gene in the chicken. Biochem Biophy Res Comm. 170:53-58, 1990.
24. Harricane M, Fabrizio E, Arpin C and Mornet D: Involvement of caldesmon at the actin-myosin interface. Biochem. J 287:633-637, 1992.
25. Hayashi K, Kanda K, Kimizuka F, Kato I and Sobue K. (1989) Primary structure and functional expression of *h*-caldesmon complementary DNA. Biochem. Biophys. Res. Commun. 164:503-511.
26. Hayashi K, Fujio Y, Kato I and Sobue K. (1991) Structural and functional relationships between *h*- and *l*-caldesmon. J. Biol. Chem. 266:355-361.
27. Hemric ME and Chalovich JM. (1988) Effect of caldesmon on the ATPase activity and the binding of smooth and skeletal myosin subgragments to actin. J. Biol Chem. 263:1878-1885.
28. Humphrey MB, Herrera-Sosa H, Gonzalez G, Lee L and Bryan J. (1992) Cloning of cDNA encoding human caldesmons. Gene 112:197-204.
29. Kamm KE and Stull JT. (1989) Regulation of smooth muscle contractile elements by second messengers. Annu. Rev. Physiol. 51:299-313.
30. Katsuyama H, Wang CLA and Morgan KG. (1992) Regulation of vascular smooth muscle tone by caldesmon. J. Biol. Chem. 267:14555-14558.
31. Kelley CA, Sellers JR, Goldsmith PK and Adelstein RS: Smooth muscle myosin is composed of homodimeric heavy chains. J Biol Chem. 267:2127-2130, 1992.
32. Kim YS, Levin RM and Chacko S. (1993) Alterations in the expression of _-cytoplasmic and aslpha-smooth muscle actin isoforms in obstruction-induced hypertrophy of urinary bladder smooth muscle. J. Urol. 147:854a.
33. Lash JA, Sellers JR and Hathaway DR. (1986) The effects of caldesmon on smooth muscle heavy actomeromyosin ATPase activity and binding of heavy meromyosin to actin. J. Biol. Chem. 262:16155-16160.
34. Lehman W, Moody C and Craig R: Caldesmon and the structure of vertebrate smooth muscle thin filaments. An NY Acad Sci. 75-83.

35. Levin MR, Longhurst PA, Monson FC, Kato K and Wein AJ. (1990) Effect of bladder outlet obstruction on the morphology, physiology, and pharmacology of the bladder. Prostate Suppl. 3:9-26.
36. Lin VK, Lee IL and McConnell JD. (1991) Expression of l-caldesmon in obstruction-induced bladder hypertrophy. J.Cell Biol. 115:1380a.
37. Litten RZ, Martin BJ, Low RB and Alpert NR. (1982) Altered myosin isoenzyme patterns from pressure-overloaded and thyrotoxic hypertrophied rabbit hearts. Circ. Res. 50:856-864.
38. Lutzeyer W, Hannapel J and Schafer W. (1983) Sequential events in prostatic obstruction. in "Benign prostatic hypertrophy". ed. Hinnan C. New York:Springer-Verlag, p693-700.
39. Mahdavi V, Lompre AM, Chambers AP and Nadal-Ginard B. (1984) Cardiac myosin heavy chain isozymic transitions during development and under pathological conditions are regulated at the level of mRNA availability. Eur. Heart J. 5(suppl F):181-191.
40. Mak AS, Watson MH, Litwin CME and Wang JH. (1991) Phosphorylation of caldesmon by cdc2 kinase. J. Biol. Chem. 266:6678-6681.
41. Malhotra A, Schaible TF, Capasso J and Scheuer J. (1984) Correlation of myosin isoenzyme alterations with myocardial function in physiologic and pathologic hypertrophy. Eur. Heart J. 5(suppl F):61-67.
42. Malkowicz SB, Wein AJ, Elbadawi A, Van Arsdalen K, Ruggieri MR and Levin RM. (1986) Acute biochemical and functional alterations in the partially obstructed rabbit urinary bladder. J. Urol. 136:1324-1329.
43. Malmqvist U, Arner A and Uvelius B. (1991) Cytoskeletal and contractile proteins in detrusor smooth muscle from bladder with outlet obstruction-A comparative study in rat and man. Scan. J. Urol. Nephrol. 25:261-267.
44. Marston SB. (1990) Stoichiometry and stability of caldesmon in native thin filaments from sheep aorta smooth muscle. Biochem. J. 272:305-310.
45. Matsumura F and Yamashiro S: Caldesmon. Curr Op in Cell Biol. 5:70-76, 1993.
46. Matsuoka R, Hoshida MC, Furutani Y, Imamura S, et al: Human smooth muscle myosin heavy chain gene mapped to chromosomal region 16q12. Am J Med Gene 46:61-67, 1993.
47. Murakami N, Mehta P and Elzinga M: Studies on the distribution of cellular myosin with antibodies to isoform-specific synthetic peptides. Fed Euro Biochem Soc. 278:23-25, 1991.
48. Murphy RA, Aksoy MO, Dillon PF, Gerthoffer WT and Kamm KE. (1983) The role of myosin light chain phosphorylation in regulation of the crossbridge cycle. Fed. Proc. 42:51-56.
49. Nagai R, Kuro-o M, Babij P and Periasamy M. (1989). Identification of two types of smooth muscle myosin heavy chain isoforms by cDNA cloning and immunoblot analysis. J. Biol. Chem. 264:9734-9737.
50. Ngai PK and Walsh MP. (1984) Inhibition of smooth muscle actin-activated myosin Mg^{++}-ATPase activity by caldesmon. J. Biol. Chem. 259:13656-13659.
51. Ngai PK and Walsh MP. (1985) Detection of caldesmon in muscle and non-muscle tissues of the chicken using polyclonal antibodies. Biochem. Biophys. Res. Commun. 127:533-539.
52. Olivetti G, Anversa P, Melissari M and Loud AV. (1980) Morphometry of medial hypertrophy in the rat thoracic aorta. Lab. Invest. 42:559-565.
53. Onji T, Takagi M and Shibata N. (1987) Caldesmon specifically inhibits the effect of tropomyosin on actomyosin system in platelet. Biochem. Biophys. Res. Commun. 143:475-481.
54. Ooshima A, Feller GC, Cardinal GJ, Spector S and Underfriend A. (1974) Increased collagen synthesis in blood vessels of hypertensive rats and its reversal by antihypertensive agents. Proc. Natl. Acad. Sci. U.S.A. 71:3019-3023.
55. Owens GK, Rabinovitch PS and Schwartz SM. (1981) Smooth muscle cell hypertrophy versus hyperplasia in hypertension. Proc. Natl. Acad. Sci. U.S.A. 78:7759-7763.
56. Owens GK and Schwartz SM. (1982) Alterations in vascular smooth muscle mass in the spontaneously hypertensive rat: Role of cellular hypertrophy, hyperploidy and hyperplasia. Circ. Res. 51:280-289.
57. Pettaway CA, Murphree S and McConnell JD: The response of detrusor smooth muscle nucleus to obstruction. J Urol 143:367A, 1990.

58. Radzinski C, McGuire EJ, Smith D, Wein A, Levin RM, Miller F and Elbadawi A. (1991) Creation of a feline model of obstructive uropathy. J. Urol. 145:859-863.
59. Rovner AS, Thompson MM and Murphy RA. (1986) Two different heavy chains are found in smooth muscle myosin. Am. J. Physiol. 250:C861-C870.
60. Saez CG, Myers JC, Shows TB and Leinwand LA: Human nonmuscle myosin heavy chain mRNA: Generation of diversity through alternative polyadenylation. Proc. Natl. Acad. Sci. 87:1164-1168, 1990.
61. Samuel M, Kim Y, Horiuchi KY, Levin RM and Chacko S. (1992) Smooth muscle myosin isoform distribution and myosin ATPase in hypertrophied urinary bladder. Biochem. Intern. 26:645-652.
62. Scott-Woo GC, Sutherland C and Walsh MP. (1990) Kinase activity associated with caldesmon is Ca^{2+}/calmodulin-dependent kinase II. Biochem. J. 268:367-370.
63. Shohet RV, Conti MA, Kawamoto S, Preston YA, et al: Cloning of the cDNA encoding the myosin heavy chain of a vertebrate cellular myosin. Proc. Natl. Acad. Sci. 86:7726-7730, 1989.
64. Simons M, Wang M, McBride OW, Kawamoto S, et al: Human nonmuscle myosin heavy chains are encoded by two genes located on different chromosomes. Circul Res. 69:530-537, 1991.
65. Sobue K, Muramoto Y, Fujita M and Kakiuchi S. (1981) Calmodulin-binding protein from chicken gizzard that interacts with F-actin. Biochem. Int. 2:469-476.
66. Sobue K, Muramoto Y, Fujita M and Kakiuchi S. (1981) Purification of a calmodulin-binding protein from chicken gizzard that interacts with F-actin. Proc. Natl. Acad. Sci. USA 78:5652-5655.
67. Sobue K, Morimoto K, Inui M, Kanda K and Kakiuchi S. (1982) Control of actin-myosin interaction of gizzard smooth muscle by calmodulin and caldesmon-linked flip-flop mechanism. Biomed. Res. 3:188-196.
68. Sobue K, Takahashi K and Wakabayashi I. (1985) $Caldesmon_{150}$ regulates the tropomyosin-enhanced actin-myosin interaction in gizzard smooth muscle. Biochem. Biophys. Res. Commun. 132:645-651.
69. Sobue K. Tanaka T, Kanda K, Ashino N and Kakiuchi. Purification and characterization of $caldesmon_{77}$: A calmodulin-binding protein that interacts with actin filaments from bovine adrenal medulla. (1985) Proc. Natl. Acad. Sci. U.S.A. 82:5025-5029.
70. Sobue K, Kanda K, Tanaka T and Ueki N. (1988) Caldesmon: A common actin-linked regulatory protein in the smooth muscle and non-muscle contractile system. J. Cellu. Biochem. 37:317-325.
71. Sobue K and Sellers JR: Caldesmon, a novel regulatory protein in smooth muscle and nonmuscle actomyosin systems. J Biolog Chem. 266:12115-12118, 1991.
72. Susset JG, Servot-Viguir D, Lamy F, Madernas P and Black R. (1978) Collagen in 155 human bladders. Invest. Urol. 16:204-215.
73. Susset JG. (1985) Effects of aging and prostatic obstruction on detrusor morphology and function. in "Benign prostatic hypertrophy". ed. Hinan C. New York:Springer-Verlag, P653-665.
74. Tanaka T, Ohta H, Kanda K, Hidaka H and Sobue K. (1990) Phosphorylation of high-M_r caldesmon by protein kinase C modulates the regulatory function of this protein on the interaction between actin and myosin. Eur. J. Biochem. 188:495-500.
75. Ueki N, Sobue K, Kanda K and Hada T. (1987) Expression of high and low molecular weight caldesmons during phenotypic modulation of smooth muscle cells. Proc. Natl. Acad. Sci. U.S.A. 84:9049-9053.
76. Uvelius B, Arner A and Johansson B. (1981) Structural and mechanical alterations in hypertrophic venous smooth muscle. Acta. Physiol. Scand. 112:463-471.
77. Velaz L, Hemric ME, Benson CE and Chalovich JM. (1989) The binding of caldesmon to actin and its effect on the ATPase activity of soluble myosin subfragments in the presence and absence of tropomyosin. J. Biol. Chem. 264:9602-9610.
78. Vorotnikov AV, Shirinsky VP and Gusev NB. (1988) Phosphorylation of smooth muscle caldesmon by three protein kinases: implication for domain mapping. FEBS Lett. 236:321-324.
79. Walsh MP. (1990) Smooth muscle caldesmon. in "Progress in clinical and biological research. Vol. 237. Frontier in smooth muscle research". Sperelakis N and Wood JD. eds. Wiley-Liss, New York. pp127-140.
80. Walsh MP. (1991) Calcium-dependent mechanisms of regulation of smooth muscle contraction. Biochem. Cell Biol. 69:771-800.

81. Weidong S and Chantler PD: A unique cellular myosin II exhibiting differential expression in the cerebral cortex. Biochem Biophy Res Comm. 175:244-249, 1991.
82. Yamashiro S, Yamakita Y, Hosoya H and Matsumura F. (1991) Phosphorylation of non-muscle caldesmon by $p34^{cdc2}$ kinase during mitosis. Nature 349:169-172.
83. Zhan Q, Wong SS and Wang CLA. (1991) A calmodulin-binding peptide of caldesmon. J. Biol. Chem. 266:21810-21814.

BLADDER OUTLET OBSTRUCTION - DISCUSSION

William DeGroat - The changes you described as created by obstruction were referred to as dedifferentiation, and this dedifferentitaion is really a matter of degree. I would think that in the developing bladder in utero or postnatally, that part of the maturation process is really due to stimulation by increased urine formation and the resulting volume work done by the bladder. We might expect that even in a mature bladder that if there was a mild degree of obstruction, and increased work load, that it might continue to develop in the way that it has early on in the neonate. Perhaps only when you push it on to the extreme limit, does it beign to dedifferentiate.

John McConnell -There are so many other factors going on here as well. It (work load) was probably playing a crucial role, and in development clearly changes in innervation are occurring. We do know that for striated muscle, innervation directly affects expression of myosin heavy chain genes and so innervation assumes quite a role in this so-called maturation process. This point is well taken; we freely admit that our primary interest is actually looking at the signals within the cell that cause the dedifferentiation or hypertrophy. We have not done a time course experiment that would be required to look at the role of innervation nor have we assessed the issue of reversibility of these changes. All of these experiments would be crucial to providing the best answer to your question.

Participant - Church and Krivella in one of their papers on hypertrophy had an interesting two sentences. They said that growth is clearly different than hypertrophy but sometimes it is very difficult to distinguish between them. What is your comment to that?

John McConnell - Growth of the bladder in utero and in the neonate, I would presume, results in increases in numbers of cells as well as increases in the sizes of these smooth muscles cells. It may be the same sort of thing that happens early on with obstruction as the cells undergo hypertrophy.

Participant - Is there a way to differentiate between normal growth and hypertrophy?

John McConnell - I hope we can find the answer to that. One of the interesting things that I think about is why don't these cells start proliferating. Certainly in atherosclerosis, in response to another pathologic circumstance, you start getting proliferation. Smooth muscle cells are capable of proliferation but we just don't see that, as if these cells are locked in G zero and aren't capable of going past that. So I think it clearly is different from growth. What I did not have time to talk about today is the observation that there are very bizarre change that occur in the nuclei of these cells. They increase their surface area by 13 fold, there are probably changes in nuclear matrix that occur that probably have some impact on gene expression, as well. No, it is not normal growth.

Participant - I have two questions. One is related to whether or not you have begun to look at the inotropic agents that impact the processing of SM1 and SM2 or the switching between h and l caldesmon, like fibroblast derived growth factor or PDGF?

John McConnell - No, we have not looked at any of that. The only thing that we have looked at early on when it sort of became intrigued with how innervation and thyroid hormone affects the myosin heavy chain expression in the heart. In other systems, David Ewalt when he was in our lab, looked at thyroid hormone. Based on his work it does not appear that thyroid hormone has any affect on the normal developmental transition that occurs. We have not looked at any other aspects of growth regulation.

Participant - There is an interesting analogous system to the switching between SM1 and SM2 and that is in the lipoprotein system. Between Apo B which is the major protein on LDL, there is a genetic switch from pre-birth to post and the silent isoform of Apo B which is seen on LDL. I was wondering if you have looked at this analogous processing; does it appear to be post-transcriptional regulation in both cases.

John McConnel - You are the third person besides Mike Brown and Joe Goldstein to point this out to me. We have not pursued that.

Participant - It is interesting that with regard to Apo B that thyroid hormone does in fact affect the processing and it actually induces the processing for one of the forms of Apo B.

John McConnell - Well the only thing against that though, is that in thyrotoxicosis there is very good evidence that the excess thyroid hormone in a clinical setting affects ventricular function, but as far as we can tell, there is no evidence that it can alter bladder function. There is no clinical evidence that thyroid hormone plays a clinical role in the function of the urinary bladder.

Samuel Chacko - You showed there was no difference in the degree of light chain phosphorylation. It would be interesting to look at the kinase and phosphatase activity in the muscle. There are all kinds of regulatory pathways that might be operating here. Have you looked at this?

John McConnell - No.

Stephen Zderic - This question is directed to both John McConnell and Sam Chacko. As I have done some reading on the literature regarding myosin light chain phosphorylation, I came across a paper by William Gerthoffer in the Journal of Pharmacology and Experimental Therapeutics in which he did a time course study using trachealis smooth muscle and he demonstrated that light chain phosphorylation remained constant over time in response to cholinergic stimulation. This is in contrast to what Sam showed in which the phosphorylation diminished over time as tension was maintained in what I believe was vascular smooth muscle. This brings to mind the possibility that all smooth muscle may not be created equally. So one question I have is did either of your groups ever do time course studies in bladder smooth muscle? The other question I wish to raise is about a series of experiments done by Gerthoffer in which he depleted muscle strips of their calcium and then repleted the calcium while simultaneously measuring stress generated and myosin light chain phosphorylation. What he found was most interesting in that there was a definite relationship between the stress generated and the calcium concentration of the bath, and yet the percentage phosphorylation remained relatively constant at some 70% with no major changes taking place although the stress actually was increasing. This leads me to question what other alternative pathways there might be to force generation in smooth muscle other than the step initiated by myosin light chain phosphorylation?

John McConnell - Well I can't answer all of those questions, Sam might be better equipped to do that, but even within the trachealis muscle there can be significant differences reported between the labs. Certainly Kris Kamm and Jim Stull working with bovine trachealis smooth muscle have clearly confirmed the observation of others that phosphorylation of light chain is correlated with the development of force, and that this phosphorylation reaches a peak that diminishes to a plateau phase. But you are quite right in that there have been a number of reports in certain organ systems and under certain conditions in which there has been a constant phosphorylation depending upon what the agonists and the conditions are. I really don't have an explanation for that. In terms of what other force generating regulatory processes may be occurring in the cell, you could conjure up all kinds of ideas. That is why I brought up the issue of Caldesmon phosphorylation because there are multiple phosphorylation sites on Caldesmon, protein kinase C and other kinases. We have evidence that myosin heavy chain phosphorylation occurs in vitro, but not in vivo. We have the whole phosphatase story as alluded to by Sam Chacko which is on the other side of the fence in terms of dephosphorylation. I think having said all that, I think that most people investigating this area feel pretty strongly that myosin light chain phosphorylation is still the major player. Sam, would you agree with that?

Sam Chacko - Yes, I think you have brought out a very interesting point. I would agree that among the investigators that there is a lot of debate about phosphorylation and dephosphorylation. Still I think that among these investigators there is general agreement that light chain kinase phosphorylation is the major and primary step that can be correlated with force generation in smooth muscle. That is not to say that it is the only regulatory point, only that it is accepted as the predominant one. Certainly there is agreement that phosphorylation is required for the initiation of smooth muscle contractility. I think the majority of the people believe that in smooth muscle, there are multiple regulation sites. Certainly smooth muscles from different parts of the body have different functions and so it is understandable that they may have multiple regulatory paths. Some of these experimental situations where they selectively manipulate the physiology are very nice, but it is difficult to manage these things and it is difficult to know what system they are activating. These experiments might result in phosphorylation of myosin light chain kinase, caldesmon, other kinases, and of course we must not forget the possible roles of phosphatases plus other factors as yet not discovered. In short there are so many variables that may influence contractility that an experimentalist must try and control for. It is important to keep in mind that, maybe this complexity is the best way to optimize and individualize smooth muscle function, because smooth muscles are not all the same.

Participant - It has been our observation that non-muscle myosins are seen in the presence of experimentally induced outlet obstruction, and yet this is not present in the fetal bladder. So we wonder if this represents a rearrangement of smooth muscle cells, or would this still be a dedifferentiation.

John McConnell - Do you know which non-muscle myosin isoform was expressed? This important because some people have described non muscle myosins developing in smooth muscle cells, and others have described them in hypertrophy too which is what you are saying? But your point is that if are using non-muscle myosins as a marker as one example where you don't see a fetal development pattern appear during the hypertrophy of obstruction. However I would stress that I don't think the whole developmental sequence may play backwards in these systems but rather that just certain selective changes may present as a consequence of obstruction.

Robert Levin - We have seen that fetal bladders are capable of generating high tension and high pressures so that the fetal forms of all the proteins within the fetal bladders are fully functional. We have to be careful about trying to relate the alterations in the fetal forms of protein within an obstructed bladder to the contractile dysfunction, because they don't necessarily have to go together.

John McConnell - That is an excellent point, and we have in all honesty, given up any hope of ever showing that these myosin isoform shifts really have any physiological significance because of Adelstein's work. But, as Sam pointed, if these there may be reasons why shifts in l-Caldesmon may play a role.

Robert Levin - On the other hand while these changes do not correlate with force generation, they may be very useful markers of obstruction. Thus these isoform shifts may serve as markers of obstructive hypertrophy.

Sam Chacko - The SM-1 and SM-2 content changes, although the enzymatic action is not different. I think people feel there is good data to suggest that these may play a role in filament activation. The filaments are organized within the cell in a certain pattern, and it would be very interesting for someone to use electron microscopy methods to see how these filaments are spatially arranged. Perhaps there is a changes that occurs in the spatial organization with maturation from fetal to adult smooth muscle where in the transition from SM1 to SM2 has a role.

John McConnell - You can't forget that there is a fundamental decrease in the expression of myosin. That certainly explains part these findings.

Alan Wein - People are beginning to get tired and hungry, but I wish to make sure we have exhausted all questions for our speakers.

Philip Ransley - My question is for all of them, some of them, or none of them. By the way my understanding of smooth muscle physiology as it relates inversely to my degree of bladder filling. In the course of this morning, there have been a large number of experimental studies that have been set up over many years, yet none of them appear to have gotten infected. Just because of its relation to the clinical scenario, I just wonder if anybody on the experimental side would make any comment as to how infection alters any of the changes which they see in relation to obstruction in the experimental studies. For the clinician this would be extremely relevant.

Alan Wein - I see Dr Steers sliding up to the microphone obviously knows the answer.

William Steers - I don't know any of the answers, just lots of questions. With regards to infection, we have been looking at the defenses of the bladder and the roles of cytokines as well as interleukin 1 and interleukin 6. Two things are apparent: 1) like the intestine you get hypertrophy of the bladder, and 2) things like nerve growth factor expression are increased. So it is easy to see why infection would alter function of bladder smooth muscle. Again, there are molecular similarities with cancer and normal growth and so it would not be a major surprise if some of these processes are activated during infection. But cytokines or a combination of TGF and interleukins are the most potent stimuli we have come across for nerve growth factor production. So I would not be surprised at all if infection produces changes in bladder smooth muscle function simply because in response to a stress there is only a limited repertoire of what the genome will do. These are some thoughts from our perspective, it is an area we are beginning to look into further.

Robert Levin - Certainly we can complicate the matter further, because in creating an obstruction in rabbits we increase the bacterial adherence of the surface so these animals will be more susceptible to infection. We don't routinely determine the state of infection in the animals. What we have determined that obstructed animals have bacterial adherence factors if you inject radioactive bacteria into them. The result of this increased binding of bacteria to the obstructed mucosa is a greater likelihood of cystitis developing. Unfortunately, we don't routinely determine the state of infection with the animals and that is something we should do.

John McConnell - Since we have Professor Uvelius, a prominent physiologist with us I would like to ask, what makes in these bladder less likely to empty after obstruction? We have only talked about pressure generation, tension, contractility, or velocity and many of these parameters recover to most of their baseline over time following obstruction. Yet these bladders even when they don't have extensive alterations in the extracellular matrix still have emptying problems. Has this been described by your group? Could these cells, because of alterations in filament assembly that we talked about or because of the cells being packed with desmin shorten as a normal smooth muscle cell would? Does this cell shortening have anything to do potentially with residual urine independent of the extracellular matrix? I would appreciate your comments.

Bengt Uvelius - Based on our data I would say whether that the length tension relations of such obstructed cells when properly normalized are exactly the same as for control smooth muscle cells. If you set the strip or the cells to a certain length and then stimulate the muscle to see if it will shorten to a certain length, I cannot say.

John McConnell - When one looks at some of the morphologic data one can see that sometimes the cells are 5 times longer than it used to be, and even if it undergoes a substantial shortening, it could still be twice the length of a normal cell. Given the billions of cells in a bladder could it be that this anatomic explanation in part accounts for the failure to fully empty that is observed?

Bengt Uvelius - It is interesting to note that when one compares cell length between obstructed and control animals at comparable bladder volumes, the lengths for the obstructed group have actually been shown to be longer.

DEVELOPMENTAL PHYSIOLOGY-OVERVIEW

William C. de Groat, Ph.D.

Department of Pharmacology
University of Pittsburgh
School of Medicine
Pittsburgh, PA. 15261

Urinary excretion is an essential aspect of body homeostasis and therefore appears at an early stage of fetal development. In the primitive lower urinary tract, storage and elimination of urine are presumably mediated by nonneural mechanisms; however at later stages of fetal and postnatal life these mechanisms are regulated by neural pathways that trigger involuntary or reflex voiding. As the nervous system matures during the postnatal period reflex voiding is eventually subject to voluntary control which originates in higher centers in the brain (1). Thus, the functions of the lower urinary tract can be grouped according increasing levels of complexity from the most primitive which are intrinsic to the target organs to the most complicated which require the participation of the cerebral cortex.

An analysis of the developmental changes in the bladder and urethra provides an opportunity to examine certain features of urinary tract function in isolation before the emergence of more complicated functions. Developmental studies can also provide insights into the mechanisms which govern the emergence and assembly of the hierarchical control system. This information is likely to be useful in understanding issues related to pediatric urology as well as to urologic disorders in adults since diseases and injuries to the urinary tract or the nervous system in some instances can lead to the reemergence in adults of primitive functions that were prominent during early stages of development (1).

Many of the morphological, biochemical and physiological changes that occur in the lower urinary tract during development are excellent examples of plasticity of muscle and nervous system to changing environmental demands. For example, the increased volume of urine produced as a child grows requires a larger more compliant bladder and presumably hypertrophy of the smooth muscle. Other changes, such as the emergence of voluntary micturition are likely to be programmed in the genes but may be modified by environmental factors. The mature lower urinary tract also retains a certain degree of plasticity throughout adult life and therefore is able to adapt to different conditions including disease and injury. Since plasticity is so prominent and predictable during development and since many of the biochemical mechanisms that control neuronal and smooth muscle plasticity are likely to be similar throughout life, studies of developmental plasticity may provide important insights into the adaptive mechanisms that accompany disorders of the lower urinary tract in adults.

The lower urinary tract exhibits a number of properties that distinguish it from other visceral organs. For example, micturition in infants and adults is completely dependent on central nervous control (1); whereas many other organs such as the heart and blood vessels maintain a certain level of activity even after elimination of extrinsic neural input. The lower urinary tract is also unusual in its pattern of activity exhibiting only two modes of operation; storage and elimination. Thus many of the neural circuits controlling the bladder and urethra exhibit phasic firing unlike the tonic activity characteristic of autonomic pathways to the cardiovascular organs. The parasympathetic excitatory control of the bladder is organized as a simple on-off switch. Later in postnatal development this switching circuitry is controlled voluntarily; whereas most other visceral functions are regulated involuntarily. Micturition also requires the central

Muscle, Matrix, and Bladder Function,
Edited by S. Zderic, Plenum Press, New York, 1995

integration of autonomic and somatic efferent mechanisms to coordinate the activity of visceral organs (the bladder and urethra) with that of the urethral striated muscles.

Since the nervous system plays a key role in lower urinary tract physiology it is important determine its contribution to urine storage and elimination at different stages of development. In addition since neural mechanisms and peripheral target organs are undergoing simultaneous maturation, it is also of interest to evaluate the influence of neural-target communication on this maturation. For example, nerves release chemicals including transmitters that may influence the development of the bladder and urethra. Conversely, the target organ can influence the nervous system via the release of neurotrophic substances. This two way communication (dialogue) between nerves and targets presumably plays a role in the coordinated development of neural-muscular activity in the fetus and neonate, as well as in the maintenance of normal function or the initiation functional plasticity in the adult (see chapter by W. Steers for a summary of the role of neurotrophic factors in neural plasticity in the adult lower urinary tract).

The section of the conference dealing with developmental physiology consisted of four presentations. Dr. Kogan discussed the physiology and functions of the lower urinary tract in the fetal lamb. He presented evidence that adrenergic and nitric oxide mechanisms regulate voiding in the fetus. Dr. Baskin discussed the changes in bladder compliance during fetal development and how compliance may be related to the types of collagen expressed in the bladder. Dr. Zderic focused on the changes in bladder contractile mechanisms during the neonatal period and the contribution of intracellular and extracellular calcium to contractions at different stages of development. Dr. Peters summarized his studies dealing with the effects of fetal outlet obstruction on the properties of the bladder. These presentations revisited issues that were introduced earlier in the conference, including structure-function relationships and cell-cell interactions within the target organ.

The four speakers provided important insights into the normal properties as well as pathology-induced changes in the developing lower urinary tract. The presentations also raised many questions that need to be addressed in future research. For example, are the biochemical differences between fetal, neonatal and adult bladder physiologically significant; do they alter function? The fetal and neonatal bladder utilize extracellular calcium to trigger contractions; whereas the adult bladder relies more on intracellular calcium stores. Why is it necessary that contractile mechanisms in the smooth muscle change during maturation of the bladder when the functions of the bladder to store and release urine are maintained throughout development? Is this change necessitated by hypertrophy of the smooth muscle cell, by a switch from intrinsic to neural control of the smooth muscle or is it a mechanism to increase compliance and allow the bladder to hold larger volumes of urine? Neurons at various sites in the peripheral and central nervous system are known to exhibit prominent developmental changes in the expression of ion channels as well as the synthesis of neurotransmitters and transmitter receptors (2-4) . These changes can regulate the firing properties and signalling mechanisms of neurons and in turn influence the transfer of information from the spinal cord to the peripheral organs. In future studies it will be important to determine if these changes also occur in the neural pathways to the lower urinary tract.

What are the factors that control developmental changes in the lower urinary tract? It was mentioned at the conference that increased urine formation and stretch of the bladder might be an important stimulus for the maturation of the bladder smooth muscle. In this regard it would be interesting to know if outlet obstruction in the fetus accelerates or accentuates the developmental changes. Neural-target organ interactions may also be important for normal maturation of the bladder and urethra since neural input to the smooth muscle may have a trophic influence on target cells. Conversely, trophic factors elaborated by the target can influence the chemical phenotype and the expression of ion channels in neurons (3,4). Thus the dialogue between nerves and target may have an important impact on the maturation of voiding function. This could be evaluated by studying patients or animals in which the neural pathways have been interrupted either surgically or by congenital neurological defects. The study of the developing urinary tract has been facilitated by the introduction of a variety of new techniques. In this session of the conference, methods ranging from sonogram recordings of voiding function in utero to the analysis of chemical and contractile properties of single muscle cells were discussed. Clearly, future progress in this field will depend on the availability of preparations and methods which will allow questions to be explored at different levels of complexity, ie., functions of individual molecules to integrated voiding behavior. Recently, our laboratory has developed in vitro preparations to study voiding reflex mechanisms in the neonatal rat (5). These preparations (the brain stem-spinal cord-bladder, spinal cord-bladder and spinal cord slice preparations) retain in varying degrees normal reflex connections and therefore represent model systems, intermediate between the intact behaving animal and the single cell. These preparations have been useful for studying the development of coordinated voiding. Data obtained in these and other experiments (1) indicate that mature micturition reflex circuits are anatomically connected within 2 to 3 days after birth but are functionally inactive until approximately 16-18 days of age due to tonic inhibition in the central nervous system. These findings indicate that certain developmental changes in lower urinary tract activity can occur in discrete steps and that the emergence of a particular physiological function may lag far behind the appearance of the molecular mechanisms that underlie that function.

The impact of pathology on the developing urinary tract is also an important issue. In the field of neurobiology it is generally accepted that the neonatal brain is more plastic than the adult brain and exhibits greater recovery of function following disease or injury. Is the fetal or neonatal urinary tract also more adaptable than the adult urinary tract and more able compensate for pathology? Also does voiding function recover to a different extent when treatment occurs at different stages of development? These issues could have an important clinical impact since they might determine the timing of surgical intervention or other types of treatment of disorders of the developing urinary tract.

In summary, analysis of the maturation of lower urinary tract function has provided insights into structure-function relationships in the bladder and also has revealed prominent examples of neural and target organ plasticity. The research described in the following papers indicates that developmental issues are very attractive topics for future investigation.

REFERENCES

1. de Groat, W. C., A. M. Booth, and N. Yoshimura. Neurophysiology of micturition and its modification in animal models of human disease. In: *The Autonomic Nervous System, Vol. 3, Nervous Control of the Urogenital System*, edited by C. A. Maggi. London: Harwood Academic Publishers, p. 227-290, 1993.
2. Spitzer, N. C. A developmental handshake: Neuronal control of ionic currents and their control of neuronal differentiation. *J. Neurobiol.* 22: 659-673, 1991.
3. Dourado, M. M. and S. E. Dryer. Regulation of A-currents by cell-cell interactions and neurotrophic factors in developing chick parasympathetic neurones. *J. Physiol.* 474: 367-377, 1994. 4. Rao, M. S. and S. C. Landis. Cell interactions that determine sympathetic neuron transmitter phenotype and the neurokines that mediate them. *J. Neurobiol.* 24(2): 215-232, 1993.
5. Sugaya, K. and W. C. de Groat. Micturition reflexes in the in vitro neonatal rat brainstem-spinal cord-urinary bladder preparation. *Am. J. Physiol.* 266: R658-R667, 1994.

FETAL LOWER URINARY TRACT PHYSIOLOGY: IN VIVO STUDIES

Robert A. Mevorach and Barry A. Kogan

Department of Urology
University of California School of Medicine
San Francisco, CA 94143-0738

INTRODUCTION

Fetal urine is an important contributor to amniotic fluid, and hence there is no question that the fetus voids in utero. However, the maturation of the fetal lower urinary tract function has not been studied extensively. Nonetheless, defining the developmental changes is critical to understanding the pathophysiology of abnormal voiding, both pre- and post-natally. For example, the long-term bladder problems in patients with posterior urethral valves, which can cause renal insufficiency, are directly related to developmental changes owing to the obstruction. Furthermore, abnormalities of lower urinary tract function are seen prenatally and result in problems with urine storage and elimination even before birth. As a consequence, some instances of prenatal hydronephrosis may be caused by in utero vesico-sphincteric dysfunction (Kopp and Greenfield 1993). Furthermore, the advent of many new technologies in the evaluation and treatment of fetal abnormalities now allows for intervention, in the form of decompression of the urinary tract, in cases of in utero hydronephrosis. Because these interventions are technically feasible, it behooves us to understand better the physiology of fetal lower urinary tract function in to direct treatment more effectively. As we improve our understanding of normal bladder and sphincteric function in the fetus, we will be able to distinguish better between innocent urinary tract changes worthy of observation and significant lower urinary tract disorders for which therapeutic intervention would be advantageous.

The only methods reported at this time for studying the function of the fetal lower urinary tract in vivo are those in which the human fetus is observed by ultrasound and those in which physiologic measurements are made in an animal preparation. This information can then be correlated with histologic studies of the anatomic development of the urinary tract and with other physiologic studies of bladder function, including whole organ preparations and muscle strip studies. Because the latter studies are covered in other chapters, we will concentrate on the findings of ultrasound observations and animal preparations.

PRENATAL SONOGRAPHY: THE WINDOW

Since the introduction of prenatal ultrasound in the mid 1970s, technological advances have permitted in utero diagnosis of pathologic conditions in many different organ systems. Visualization of the urinary tract is particularly easy because the fluid-filled bladder, as well as the collecting system when dilated, is readily identified against the more echogenic tissue of the fetal abdomen. Further, fetal voiding is so linked to amniotic fluid status that, as a matter of routine, considerable emphasis is given to observing the bladder, as well as its filling and emptying.

Muscle, Matrix, and Bladder Function,
Edited by S. Zderic, Plenum Press, New York, 1995

It is merely an extension of this standard technique that allowed for characterization of the normal appearance, size, and behavior of the fetal bladder as a function of gestational age (Hata and Deter 1992). The normal bladder may be seen as early as 15 weeks' gestation, as only a few mL of urine is required for visualization. Bladder emptying occurs frequently in the human fetus, varying from every 30 minutes at 28 weeks to once an hour at 40 weeks' gestation. In a different study, evaluation of bladder size by ultrasound every 2-5 minutes allowed for calculations of hourly fetal urine production, maximal bladder volume, and effectiveness of bladder emptying in 101 fetuses (Rabinowitz et al. 1989). Hourly fetal urine production increased from 5 mL/hr at 20 weeks' gestation to 51 mL/hr at 40 weeks. It would be anticipated that this increase would necessitate an increase in bladder size and, indeed, maximal bladder volume was approximately 1 mL at 20 weeks (0.5-1 mL/kg) and 36 to 54 mL at 41 weeks (6 mL/kg). Voiding lasted a mean of 9.5 seconds, but emptying was complete in only 28%. Residual volume was less than 25% of capacity in 55 of the others, but in the remaining 18 of these normal fetuses the residual volume varied from 26 to 65% of capacity. In summary, although the human fetus does void as early as 20 weeks' gestational age, it is apparent that this function is not fully developed in all, as compared with the neonate, in whom voiding to completion is the rule rather than the exception. A similar study, in which ultrasound was also used to monitor bladder function, in fetuses at 36-40 weeks' gestation, demonstrated that vibroacoustic stimulation could induce fetal voiding at significantly lower volumes than normal capacity (Zimmer et al. 1993). This suggests intact neural pathways between the central nervous system and the bladder in the late gestation human fetus.

A number of disorders of bladder function have also been studied, to contrast with the data seen in normals. For example, in fetuses with spina bifida, a variety of bladder dysfunctions have been observed. Despite the lack of any anatomic obstruction, incomplete, delayed, and/or abnormal emptying of the bladder occurs, often accompanied by hydronephrosis (Nicolaides et al. 1988). In a series of 13 of these fetuses, residual urine volume was greater than 75% of bladder capacity in more than half, in contrast to the findings in normals noted above. Additionally, observed voiding revealed more delayed emptying, often exceeding 1-2 minutes' duration. In contrast, other fetuses displayed apparently weak sphincteric resistance with virtually constant urinary drainage and no bladder filling at any time.

Other studies have been done looking at bladder emptying in fetuses with prune belly syndrome and posterior urethral valves. Emptying in these cases is poor, although it may improve markedly after even a single bladder aspiration. In fact, in utero punctures of the bladder in fetuses with posterior urethral valves have revealed an intravesical pressure in excess of 80 mmHg (normal values are unknown, but would be anticipated to be much less), which suggests excellent bladder function, despite poor emptying (Cullen et al. 1989). From these preliminary investigations it is quite apparent that the in utero effects of voiding dysfunction are measurable, progressive, and likely to influence the structure and function of the entire urinary tract and the overall well being of the newborn child.

Animal Preparations To Study Fetal Lower Urinary Tract Function

Anatomic and functional information determined from sonography has provided insight into the in utero function of the lower urinary tract; however, to understand the physiology of lower urinary tract function in the fetus better, we developed a fetal lamb model that allows for in vivo urodynamic studies and pharmacologic manipulation. To evaluate normal voiding, its neural control, and the effects of in vivo pharmacologic manipulation, fetal bladder function was studied in time-dated pregnant sheep of 118-125 days' gestation (term=147 days) (Kogan and Iwamoto 1989). The surgical preparation consisted of placing catheters in the fetal bladder, aorta and inferior vena cava (Figure 1). After recovery, and after assuring good maternal and fetal health, cystometrogram (CMG) studies were performed. In contrast with other functional studies (in which whole organ baths or muscle strips are used), this technique allows for investigations of the lower urinary tract with the normal interactions among the bladder, sphincter and central nervous system intact and thereby uniquely simulates the normal situation.

Pressure changes in the bladder during normal filling were observed to determine fetal bladder compliance and to attempt to distinguish bladder contractions. Because of the anatomic position of the fetus, it is essential to record intraabdominal and intrauterine

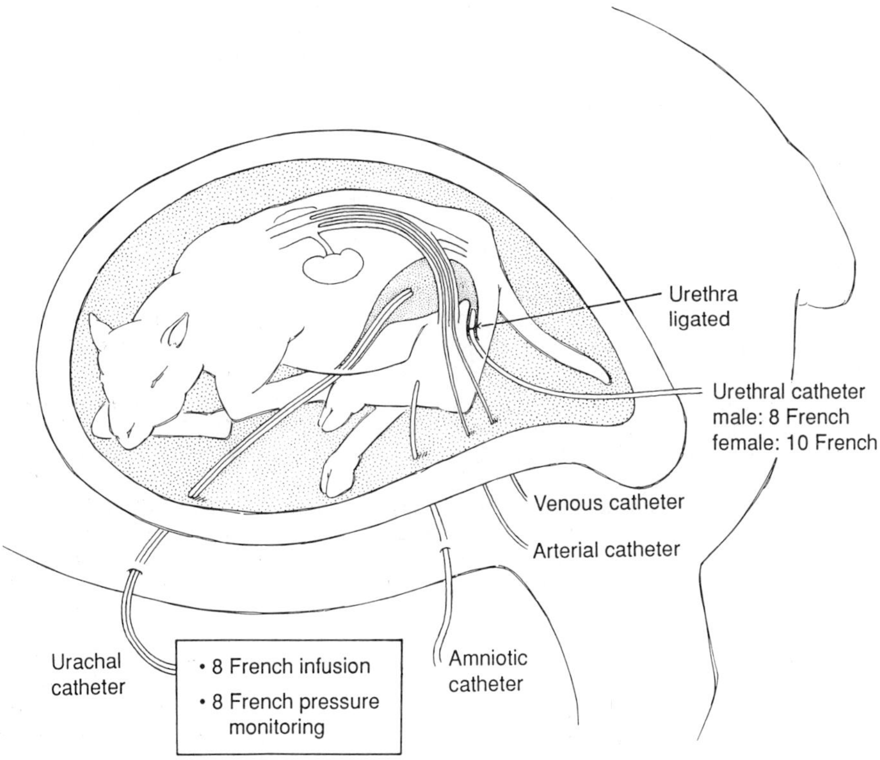

Figure 1. Monitoring set-up for in vivo urodynamics in the fetal lamb.

pressures simultaneously (via catheters in the inferior vena cava and amniotic cavity, respectively) to rule out fetal or maternal movement or uterine contractions as the cause of the bladder pressure changes. We found that the fetal bladder was compliant and that fetal bladder contractions could be discerned (Figure 2). In fact, two distinct types of contractions could be discerned, one typical of those seen in mature animals and one phasic. Although we believe the phasic contractions are the result of immaturity, this has never been substantiated (Figure 3). Intravesical pressure at the time of contraction varied greatly (10-24 mmHg), but residual urine was invariably less than 1 mL.

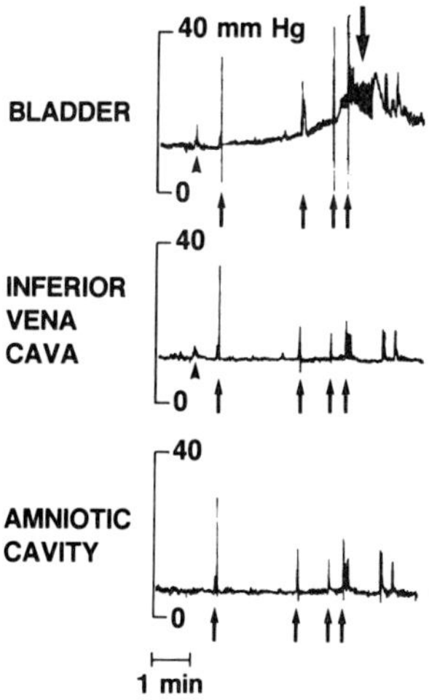

Figure 2. Tracing of intravesical, inferior vena cava, and amniotic pressures after furosemide infusion. Bladder contraction (large arrow) is distinct from low level intraabdominal pressure spikes (small arrows). (Reprinted with permission from Kogan and Iwamoto, 1989).

Numerous drugs were found to alter fetal bladder function. We first tested cholinergic and anticholinergic medications as they are known to alter bladder function in the mature animal. In fact, bethanechol decreased bladder capacity and, conversely, atropine markedly inhibited bladder contractions, increased bladder capacity and caused a marked residual urine. We also tested α-adrenergic drugs and found little alteration in bladder function, which was not surprising on the basis of adult studies and the fact that, even in the neonate, the α-adrenergic system is incompletely developed. In contrast, β-sympathomimetic stimulation (with ritodrine) resulted in unexpected and marked inhibition of bladder contractions, again causing a significant residual urine. This difference in response from that seen in the mature animal is due most likely to the sensitization of β-adrenergic receptors by progesterone, the same mechanism that exists in the uterus. Because of this finding, we tested other drugs known to inhibit uterine contractions, magnesium sulfate and diltiazem. Both also inhibited bladder contractions and resulted in an increase in both bladder capacity and residual urine (Figure 4). These findings suggest that bladder function in utero is under cholinergic and β-adrenergic control and can be influenced by systemic modulation of the activity of these neurotransmitters. Additionally, we found that medications commonly used to inhibit preterm labor also inhibit fetal bladder contractions. This has practical significance in that it may explain partially the high incidence of fetal hydronephrosis that resolves spontaneously after birth and also the occasional finding of urinary retention in the neonate in whom preeclampsia is treated with magnesium sulfate.

Nitric oxide (NO) is a non-adrenergic/non-cholinergic (NANC) neurotransmitter that has been shown to be active throughout the central nervous system. It is critical in penile erection and abnormalities of its function have been shown to be involved in numerous diseases, including pyloric stenosis and Hirschsprung's disease. We wondered whether it might be involved in lower urinary tract function also. To test its role, we used the preparation described above and gave infusions of N^{ω}–nitro–L–arginine to inhibit NO synthesis and L-arginine to stimulate NO activity (Mevorach et al. 1994). N^{ω}–nitro–L–arginine resulted in increased numbers of ineffective contractions at relatively low bladder volumes, a larger overall capacity and poorer emptying in comparison with controls (Figure 5). It was particularly interesting that the poor emptying occurred in spite of normal bladder pressures during micturition. We performed two additional studies to confirm these findings. First, we demonstrated that we could reverse the changes with L-arginine, the substrate for NO synthesis, but not with D-arginine, an inactive form. Second, because the infusion of N^{ω}–nitro–L–arginine resulted in systemic hypertension, the studies were repeated with mechanically induced hypertension via an aortic constrictor; no significant alterations in voiding occurred.

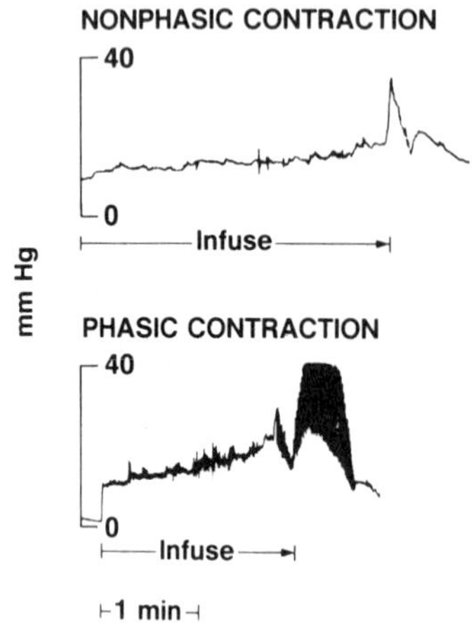

Figure 3. Comparison of non-phasic and phasic bladder contractions. Phasic contractions, which we suspect to be consistent with bladder immaturity, occurred in 7 animals and "mature" contractions were observed in 6. (Reprinted with permission from Kogan and Iwamoto, 1989).

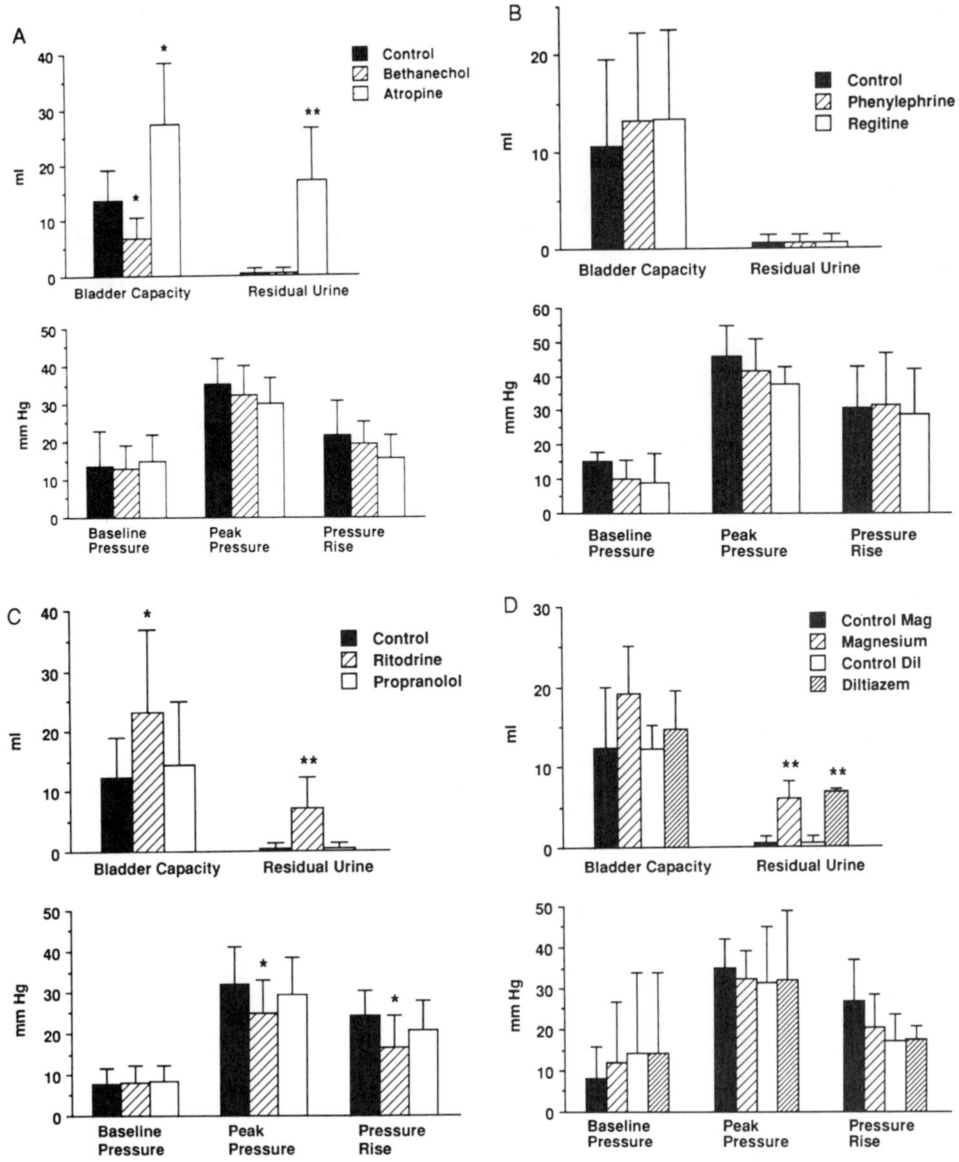

Figure 4. Effects of various medications on fetal bladder functional parameters. The increased post-void residual after atropine demonstrates the cholinergic dependence of fetal voiding. (Reprinted with permission from Kogan and Iwamoto, 1989)

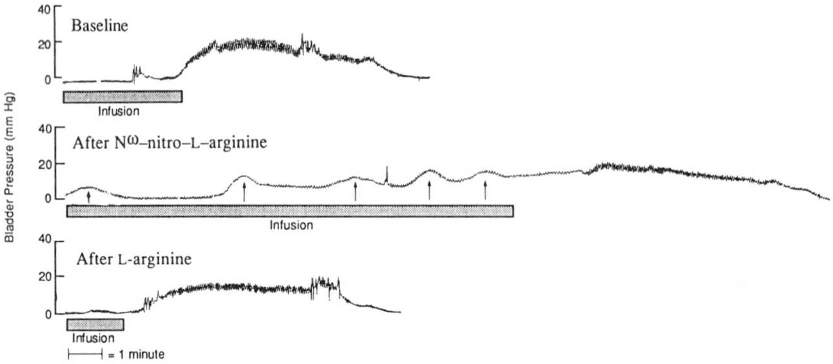

Figure 5. Urodynamic tracings show an increase in detrusor instability and increased bladder capacity with inhibition of nitric oxide synthase by Nω–nitro–L–arginine. (Reprinted with permission from Mevorach et al. 1994)

L-arginine, when given as the first drug, resulted in constant urinary drainage of the cystometric infusion from the urethra. This striking finding lasted for two hours after discontinuing the infusion and prevented any urodynamic studies of bladder function with this agent. Therefore, we modified the preparation by ligating the bladder neck to test the effects of L-arginine on the bladder. Nω–nitro–L–arginine still increased bladder capacity, but L-arginine had no effect on bladder function with this technique (Figure 6).

The finding that L-arginine induces constant urinary drainage despite its failure to cause any significant alteration in bladder function suggests that stimulation of the NO system causes marked sphincteric relaxation (in this instance, both bladder neck and striated external sphincter). Conversely, the finding that Nω–nitro–L–arginine causes poor bladder emptying despite failing to alter the pressure generated during bladder contractions suggests that a failure of sphincteric relaxation is the cause of the poor emptying. Hence, modulation of the NO system suggests that NO has a major role in determining lower urinary tract sphincteric function. Because our preparation depends on systemic infusions of the drugs used, we cannot ascertain whether the effect is local or central and further studies are necessary to clarify this point. Nonetheless, studies in the gastrointestinal tract both in vivo and in isolated muscle strips suggest the actions are at least local in the sphincteric musculature.

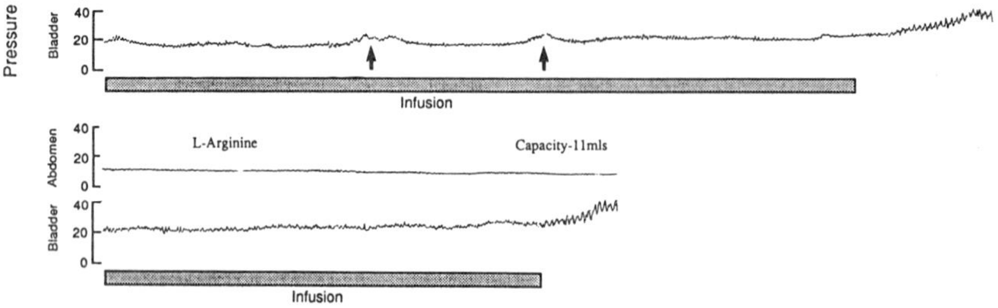

Figure 6. With bladder neck occluded, no additional effect of L-arginine is noted. Nω–nitro–L–arginine infusion again induces detrusor instability and increases bladder capacity. (Reprinted with permission from Mevorach et al. 1994)

In the only other in vivo fetal bladder study published at this time, Wlodek et al studied the relationship between micturition and electrical activity in the brain in the late gestation ovine fetus (Wlodek et al. 1989). It was found that micturition in utero occurs coincidentally with a change to low voltage electrocortical activity, suggesting some central nervous system control of the fetal bladder. Interestingly, continuous catheter drainage of the bladder failed to alter the previously noted electrocortical activity, suggesting that development of sensory neural pathways and their interconnections with motor pathways are incomplete. This finding suggests that, at least by late gestation, spontaneous voiding is regulated by descending information from the brain.

SUMMARY

Fetal lower urinary tract function is under continuous maturation throughout gestation and is an integrated neurophysiologic event by late gestation. Preliminary in vivo study suggests that peripheral and central nervous system regulation of micturition occurs in utero, and that these events can be modulated pharmacologically and through external stimulation. Normal bladder function in utero is essential to normal development of the entire fetus, and in utero modulation of bladder dysfunction may be feasible as our understanding and diagnostic acumen increase.

REFERENCES

Cullen, M. T., Athanassiadis, A. P., Grannum, P., Green, J. J., and Hobins, J. C., 1989, In utero intravesicular pressure and the prune belly syndrome, *Fetal Ther* 4: 73-7.

Hata, T., and R. L. Deter, 1992, A review of fetal organ measurements obtained with ultrasound: normal growth, *J Clin Ultrasound* 20: 155-74.

Kogan, B. A., and Iwamoto H. S., 1989, Lower urinary tract function in the sheep fetus: studies of autonomic control and pharmacologic responses of the fetal bladder, *J Urol* 141: 1019-24.

Kopp, C., and Greenfield S. P., 1993, Effects of neurogenic bladder dysfunction in utero seen in neonates with myelodysplasia, *Br J Urol* 71: 739-42.

Mevorach, R., Bogaert, G., Kogan, B., 1994, Role of nitric oxide in fetal urinary tract function, *J Urol* 152: 510-4

Nicolaides, K. H., Rosen, D., Rabinowitz, R., and Campbell, S., 1988, Urine production and bladder function in fetuses with open spina bifida. *Fetal Ther* 3: 135-40.

Rabinowitz, R., Peters, M. T., Vyas, S., Campbell, S., and Nicolaides, K. H., 1989, Measurement of fetal urine production in normal pregnancy by real-time ultrasonography, *Am J Obstet Gynecol* 161: 1264-6.

Wlodek, M. E., G. D. Thorburn, and Harding, R., 1989, Bladder contractions and micturition in fetal sheep: their relation to behavioral states, *Am J Physiol* 257: R1526-32.

Zimmer, E. Z., Chao, C. R., Guy, G. P., Marks, F., and Fifer, W. P., 1993, Vibroacoustic stimulation evokes human fetal micturition, *Obstet Gynecol* 81: 178-80.

FETAL BOVINE COMPLIANCE CHANGES

Laurence Baskin,[1] Dave Meaney,[2] Adam Landsman, Ph.D.,[2]
Serban Constantinescu,[3] and Edward Macarak[3]

[1]Division of Urology (Present address: Department of Urology, U575, University of California School of Medicine, San Francisco, CA 94143
[2]Department of Bioengineering,
[3]Connective Tissue Research Institute, University City Science Center
Children's Hospital of Philadelphia, University of Pennsylvania, Philadelphia, PA 19104

INTRODUCTION

The elastic nature of the urinary bladder allows for the accumulation of urine without a marked increased in intravesical pressure. Indirectly, the elasticity of the bladder can be measured by calculating compliance using pressure volume measurements. Changes in bladder compliance are clinically relevant because of the direct association of a poorly compliant bladder with urinary tract infections, vesico-ureteral reflux, calculi, hydronephrosis, renal scarring and urinary incontinence. To date bladder compliance in normal fetal development has not been defined. In this study, we have characterized the elastic properties (compliance) of the normal bovine bladder throughout fetal life, the newborn period and into adulthood using a novel circularly clamped bladder testing system. We hypothesize that the mechanical strength of the bovine bladder changes during these periods, with early changes triggered by developmental events such as the initial production of urine and later changes due to the normal aging process.

To account for these changes in bladder compliance, we have quantitatively analyzed the major structural collagens, type I and III, that are found in the bladder.

MATERIAL AND METHODS

Experimental Apparatus

The circularly clamped bladder testing system (Figure 1) used to apply a uniform biaxial strain to segments of bovine bladder tissue was based on the technique developed by Thibault and Fry to study the mechanical properties of arterial samples.[1,2] The bladder tissue specimen is clamped between two circular plates, where the central portion of each plate has been removed to expose a circular portion (r = 0.3445") of tissue. For smaller, thinner animal specimens the radius of the test sample can be reduced accordingly. The base plate assembly is configured to function as a reservoir, and is connected in turn to a manometer. The range of pressures used in this study was selected to fall within the range of pressures considered in the normal physiologic range (0-40 cm H_2O).

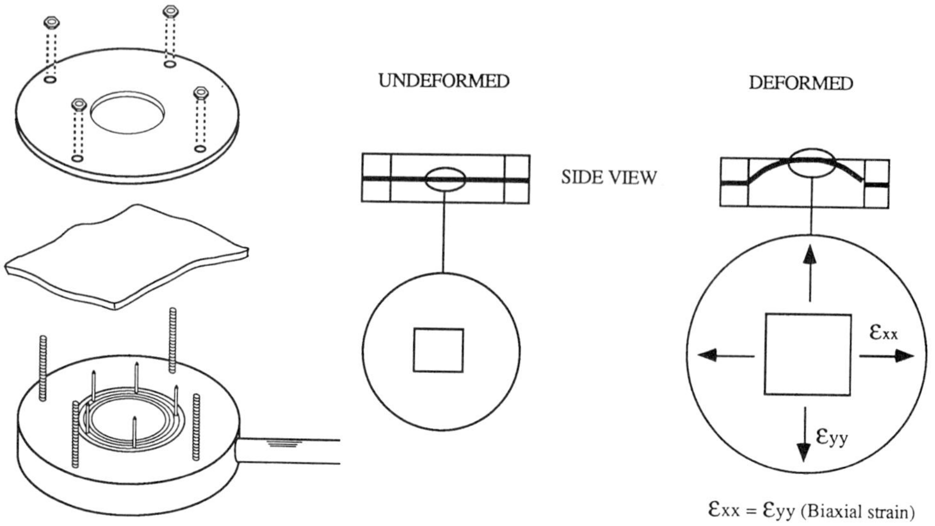

Figure 1. Uniform biaxial testing device. The bladder tissue is clamped firmly between two circular plates, where a central circular portion of each plate has been removed. Pins and grooves machined into the device are used to prevent tissue slippage. Upon pressurization from beneath, the bladder tissue deforms in the shape of a spherical cap. If the tissue is assumed homogeneous and isotropic, then a uniform biaxial strain state is created in the bladder sample. The elastic modulus, interpreted as an inverse measure of compliance, can be calculated from the centerline tissue deflection and the pressure.

If the tissue is assumed to be isotropic, an increase in the bottom reservoir pressure will cause the tissue to uniformly deflect upwards in the shape of a spherical cap. A fiber optic displacement transducer probe (MTI Instruments, Latham, NY, USA) is positioned on the exposed surface of the tissue to measure the centerline displacement, w_o, caused by pressurization from underneath (Figure 2). Since a creep response was observed in the tests, the tissue was allowed to achieve an equilibrium position before recording the centerline displacement. To assess the time dependent response of the bladder tissue, the centerline displacement was continuously recorded (Dell Computer Corp) before and immediately following the pressure change.

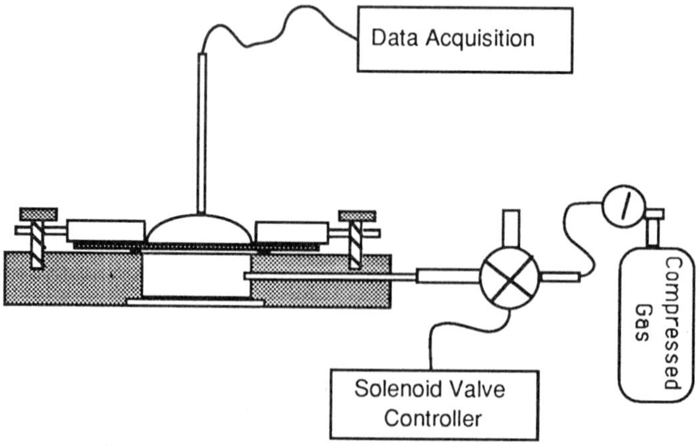

Figure 2. Control and data acquisition. A solenoid valve controller is used to control the flow from the pressure source to the testing device. A fiber optic displacement probe centered on the tissue sample is used to monitor the response of the tissue to pressurization. Data are recorded on an automated data acquisition system and later transferred to an analysis work station.

Total Collagen Quantitation

To determine the total collagen content per unit weight of five representative bovine bladder specimens quantitative collagen amino acid determination was performed.[10-12] Dried bovine bladder protein specimens were subjected to vapor phase hydrolysis by 6 N HCL at 110° C for 48 hours under reduced pressure and a nitrogen atmosphere. Phenol (1%) was added as a free-radical scavenger. The resulting amino acids were reacted with Edman reagent, phenyl isothiocarbamyl under alkaline conditions to form the respective phenylthiocarbamyl amino acid derivative (PTC-amino acid). Chromatography was then performed on a reverse phase column using high performance liquid chromatography. Amino acids are identified and quantified by comparing the observed retention times with the areas generated by known amounts of standards eluting with defined retention times. The standards consisted of a commercial amino acid mix (Sigma, St. Louis, MI) supplemented with hydroxyproline, extra proline and hydroxylysine. The retention time of each amino acid typically varied less than 3-5 seconds from run to run with baseline separation usually obtained for ~90% of the standard amino acids. Total hydroxyproline concentration was then multiplied by a factor of 7.46 as determined by Neuman and Logan to give the total collagen concentration.[13]

RESULTS

Initially experiments were performed with air to test the reproducibility and accuracy of the circularly clamped bladder testing system. In all cases the bladder tissue reacted as an elastic membrane deforming reproducibly into a hemisphere and thereby fulfilling the criteria to calculate the elastic modulus. Additionally, repeated pressurizations of a specimen produced the same tissue deflection behavior and therefore the same calculated elastic modulus. Figure 3A illustrates a typical experiment where the vertical displacement of the bladder specimen is plotted versus time for a constant pressure. The vertical displacement caused by the pressurization remains constant until the pressure is removed, upon which the bladder specimen moves to a position slightly below its initial, unpressurized position. Within less than 10 seconds, however, the initial resting position of the bladder is restored. Additionally, the vertical displacement caused by pressurization was not sensitive to specimen conditioning, since repetitive pressurization's resulted in identical peak vertical displacements (Figure 3B).

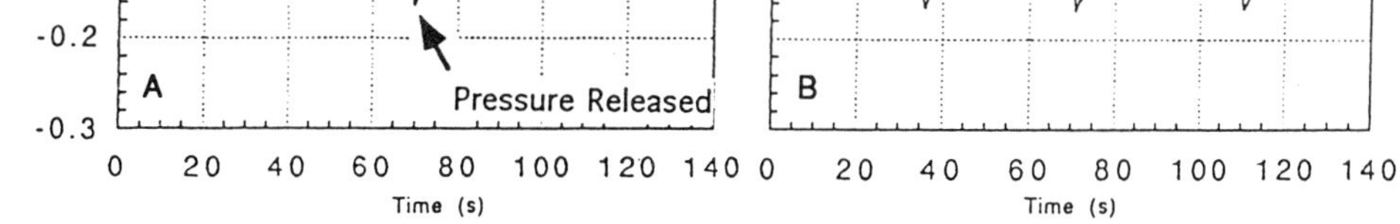

Figure 3. (A). Repeatability and time dependent behavior. Testing of bladder tissue exhibited little long term residual deformation from a given pressure (B). Furthermore, repetitive pressurizations produce little variation in maximum tissue deflection.

To determine the elastic modulus, pressure-displacement relationships were obtained for each sample over a range of 0 to 40 cm of H_2O. Figure 4 illustrates that the elastic modulus was calculated beginning with points on the pressure/vertical displacement curve only where the bladder specimen assumes membrane behavior. This point begins where the deflection is significantly larger than the thickness of the membrane ($w_o/h > 2$).

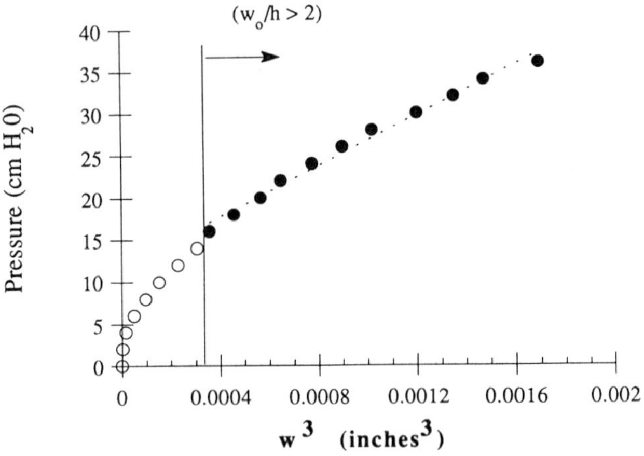

Figure 4. Calculation of elastic modulus. If the bladder tissue is modeled as homogeneous and isotropic, then the bladder behaves as a membrane for large centerline deflections. Elastic modulus of the sample can be determined by formula (1) described in the text.

Bladder thickness increased with fetal age (Figure 5). The bladder wall thickness was the same for fetuses of the same age irrespective of sex. The adult bladders were all thicker then the fetal and new born bladders.

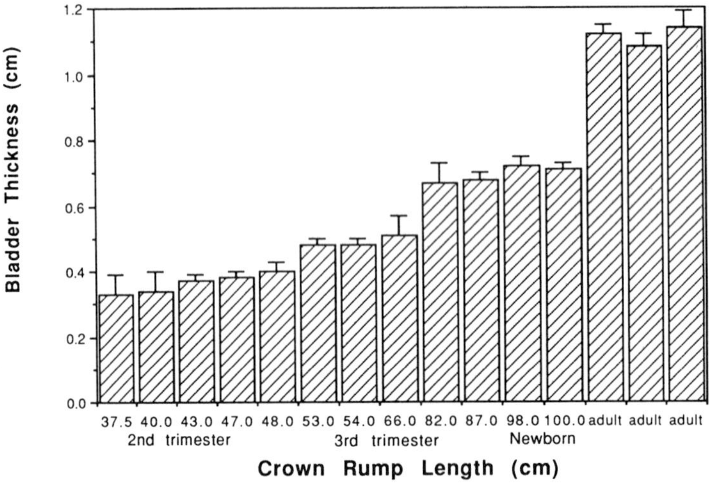

Figure 5. Fetal and adult bovine bladder wall thickness illustrating the increase in bladder wall size with increasing age.

The elastic modulus, which can be considered an inverse measure of bladder compliance, decreased with fetal age with the youngest fetus studied (end of 1st trimester/beginning of the 2nd trimester) having the largest elastic modulus and the newborn bovine bladders having the smallest modulus (Figure 6). Comparing the newborn bladders with the adult bladders showed an increase in the elastic modulus for the adult bovine bladders.

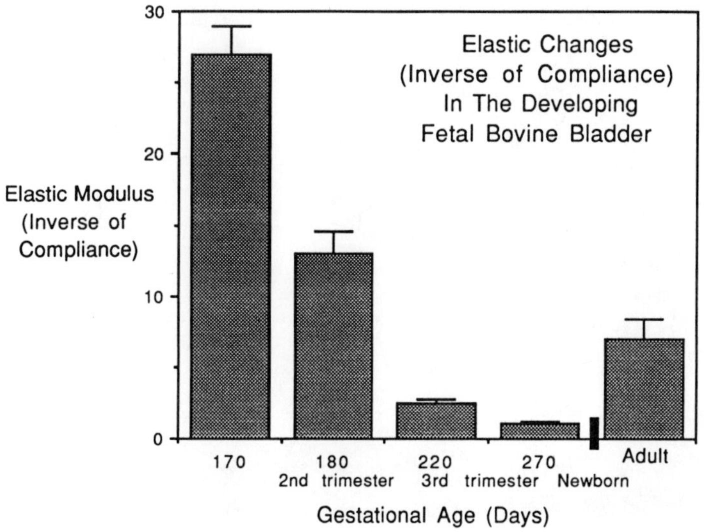

Figure 6. Elastic modulus changes with gestational age. Samples from across a broad range of gestational ages were tested using the system described in this report, with results showing modulus through the gestational period. A decrease in the elastic modulus, interpreted as an increase in the compliance, was found to coincide with the initial urine production stage in fetal development. These observations were attributed to the volume work of the bladder and its effect on the bladder behavior.

The percent dry weight of the processed insoluble proteins after extraction and purification of which the major component was collagen was not different when comparing the fetal bovine bladder specimens and the adult specimens (Table 1).

Table 1. Percentage dry weight of insoluble proteins, collagen and elastin after extraction and purification. Note that there is no difference between the fetal bladders and the adult bladders.

		Mean	S.D.
Fetal Bovine Bladders	(N = 13)	45.3%	7.6%
Adult Bovine Bladders	(N = 4)	43.9%	7.4%
			P = 0.746

CNBR peptides of the different age bovine bladders were separated by two dimensional gel electrophoresis. The peptide spots for type I collagen (a1[I]-CB 8) and type III collagen (a1[III]-CB 5) were adequately separated to allow densitometric quantitation (Figure 7). The percentage of type III/I collagen in the bovine bladder varied as a function of age (Figure 8). The youngest fetal animals from the end of the 1st trimester/beginning of the 2nd trimester had the highest percentage (~40%) of type III collagen. The percentage of type III collagen decreased with increasing fetal development to a low of 19% in the newborn bovine bladders. From the newborn period to the mature adult bladder their was a modest increase in the percentage of type III/I collagen (25%).

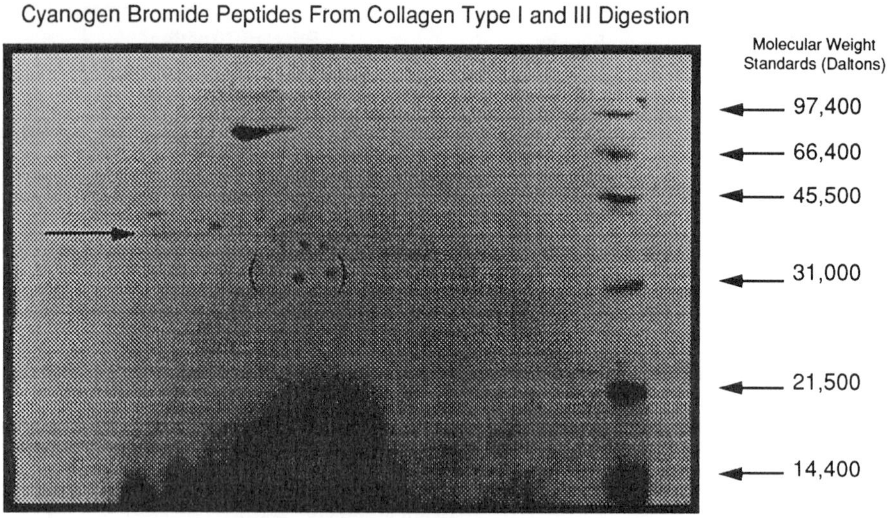

Figure 7. Quantitation of type I and III collagen in fetal, newborn and adult bovine bladder tissue was performed by collagen extraction and purification, cyanogen bromide digestion, peptide separation by two dimensional gel electrophoresis and collagen quantitation using scanning densitometer. This representative gel of a third trimester fetal bovine bladder shows the peptide spots for type I collagen a1[I] - CB8 (arrow) and type III collagen a1[III] -CB5 (parenthesis). Note molecular weight standards.

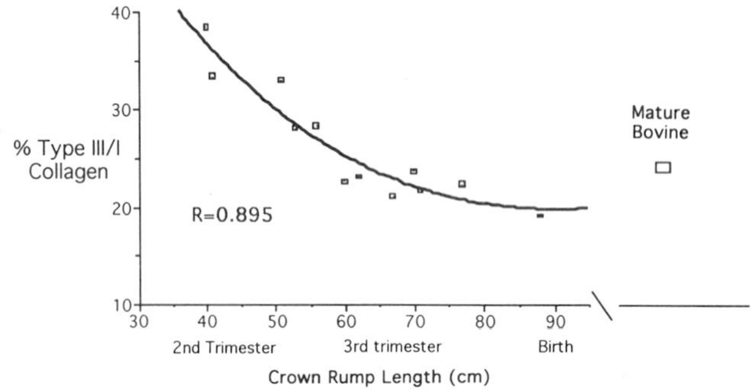

Figure 8. Quantitation of the ratio of type III/I collagen as a function of age (crown-rump length) in the fetal, newborn and adult bovine bladder. Note that the percent of type III collagen decreases in the developing bladder from the end of the 1st trimester until the newborn period. From the newborn period to the adult bladder note the modest increase in the ratio of type III/I collagen.

Hydroxyproline and total collagen analysis of five representative bladder bovine specimens also varied as a function of fetal age (Table 2). Using hydroxyproline as a marker for total collagen the youngest fetus had a greatest amount of hydroxyproline per dry bladder weight with the amount decreasing as a function of age. From the newborn period to the adult period their was an increase in total collagen content as determined by hydroxyproline analysis.

Table 2. Hydroxyproline and total collagen concentration in the fetal bovine and adult bladder.

Crown-Rump cm fetus	Gestational Age	Hydroxyproline μg/mg dry weight	Total Collagen μg/mg dry weight
48	Early 2nd trimester	58.2	434
53	Middle 2nd trimester	57.5	429
66	Early 3rd trimester	48.5	361
87	Late 3rd trimester	44.6	333
Adult	Mature	65.4	487

The percent of type I and III collagen and the total amount per unit dry bladder weight of type I and III collagen is given in Table 3. This data shows that the absolute amount of type III collagen decreases by greater then one-half (134 to 63 μg/mg dry bladder weight) from the beginning of the 2nd trimester until the newborn period. In contrast the amount of type I collagen remained within the same range (Table 3). From the newborn period to the mature bladder their was also a modest increase in the absolute amount of type III collagen.

Table 3. Percent type I and III collagen and total type I and III collagen in the fetal bovine and adult bladder.

Crown-Rump cm fetus	% Type I Collagen	% Type III Collagen	Type I Collagen μg/mg dry weight	Type III Collagen μg/mg dry weight
48	69%	31%	299	134
53	74%	26%	317	112
66	79%	21%	285	76
87	81%	19%	270	63
Adult	75%	25%	365	121

DISCUSSION

Previous investigators have characterized the compliance or mechanical properties of the bladder wall utilizing either whole bladder or isolated bladder strip preparations.[14-21] These studies have described both static and time dependent relationships for the distension or extension of bladder tissue in response to a pressure or force.[14-21] Studies employing isolated strips have been particularly useful in documenting the regional variations in the mechanical properties of human and animal bladders. In contrast, whole bladder preparations are presently not capable of providing such detailed information on specific anatomic regions in the bladder, but do represent the most accurate means of simulating the in vivo state in the laboratory to assess bladder function during both normal and pathologic states, as well as in response to pharmacologic agents.[22,23]

It is important to note, however, that the biomechanical relationship between the whole bladder and isolated strip testing has not been clearly stated. Certainly, the non-ideal shape and anatomic variability in bladder wall composition presents a challenge in properly describing the deformation experienced in regions of the bladder during normal physiologic function. However, if the bladder is idealized as a thin walled spherical shell, it has been shown that internal pressurization of this shell results in a biaxial strain within the shell material.[1,2] This state of deformation, termed equibiaxial strain, is distinct from the uniaxial tests currently conducted for isolated bladder strips. Therefore, it is reasonable to suggest that a method to apply an equibiaxial strain, in contrast with a uniaxial strain, to isolated bladder tissue samples may be a more accurate simulation of the normal physiologic state in the bladder.

We have used an equibiaxial strain device to determine the elastic modulus and hence a measure of the elasticity or compliance of the developing fetal bovine bladder, the newborn bladder, and the adult bovine bladder. We have found that the elastic modulus is highest for the youngest fetal bladders and then decreases with age, reaching its nadir in the newborn period. One possible explanation for this observation is that bladder compliance increases in conjunction with the production of urine in the developing kidneys. In the human fetus, urine production begins at 8-10 weeks, increases to 5 ml/hr at 20 weeks, and is 56 ml/hr at term. Interestingly, the duration of bladder filling and emptying cycle remains unchanged at ~ 25 minutes during gestation.[24] This increase in urine production or volume work may account for the increase in elasticity or compliance that we have observed in the developing fetal bovine bladders. Although the urine production in the developing fetal bovine is not documented like in the human, we hypothesize that their embryogenesis is similar.

Comparing the elastic modulus of the newborn with the adult bovine shows an increase in the elastic modulus in the adult or a decrease in the compliance of the adult bladders. The magnitude of this change was not nearly as large as the change from the youngest fetus to the newborn period (Figure 9). We speculate that this decrease in bladder compliance may be the natural aging process secondary to changes in the connective tissue and/or bladder smooth muscle.

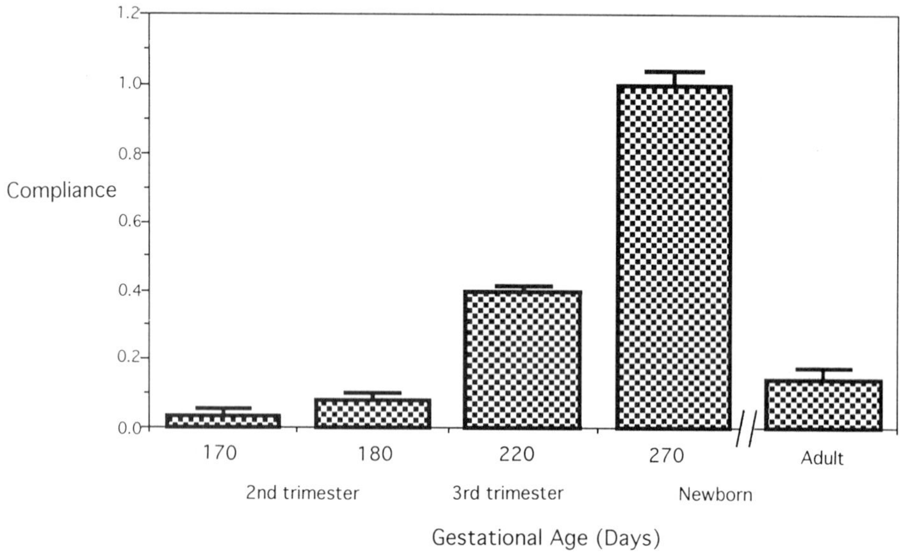

Figure 9. Compliance changes in the developing fetal, newborn and adult bovine bladder.

We compared this system against the standard uniaxial isolated bladder smooth muscle strip preparation. The findings were similar in the smooth muscle strips with the newborn bovine bladder tissue exhibiting the lowest passive tension (stiffness) compared to the youngest fetus and the adult bovine bladder (data not shown). Next a comparison was made to the whole bladder model performing pressure-volume cystograms.[25] The results were the same as the equibiaxial strain circularly clamped bladder testing system with the compliance of the bovine bladders increasing as a function of fetal age and the adult bladder showing a relatively modest increase in compliance.

A potential source of error in our system is the assumption that the thickness of the bladder wall is uniform throughout the sample. In reality, there are variations in the cell population in the bladder wall which can certainly introduce some microheterogeneity. This difference can give misleading information if pathologic changes occur in a focal area which may be missed by the testing system. To avoid this situation in the abnormal state multiple samples can be taken so that the whole bladder is actually randomly sampled. In the normal developing bovine bladder we found the data quite reproducible thus validating our assumption that the microheterogeneity of the normal bladder wall was not significant.

From this study, we learned that the bladder wall has excellent elastic properties well suited for its function as an expandable and contractible storage container. Figure 8 illustrates the repetitive displacement that the bladder tissue underwent with a return to the baseline. Analysis of the graph shows that after releasing the pressure under the bladder tissue, the bladder recoils below the baseline but quickly recovers back to the baseline. Performing this experiment over a wide array of pressures (0-100 cm of H_20) and in a repetitive fashion (> 10 times over a three minute period), revealed that the bladder was resilient and always returned to the baseline position. The reproducible recovery back to the baseline may be interpreted as a reversible stretching or unfolding of either connective tissue elements such as collagen and elastin or reversible changes in smooth muscle stretching.

To account for these differences in bladder compliance, we first looked at the thickness of the bladder wall which increased as a function of age,[26] the thinnest bladders were, as expected, found in the youngest fetuses while the adult bovine bladders were the thickest (Figure 5). This relationship did not correlate with the compliance graph that was determined by our mechanical studies (Figure 9).

Next we choose to analyze the collagenous content of the bladder wall connective tissue. The bladder wall is composed of an outer layer of fat, a smooth muscle layer with connective tissue dispersed in and around the smooth muscle bundles, the lamina propria which is composed mostly of connective tissue with a small amount of smooth muscle, the muscular mucaris and the urothelium. Presently, it is not known which component is the most critical in determining bladder compliance. Previous studies by a number of authors have shown changes both in connective tissue,[27-33] and in the detrusor in bladders with abnormal compliance.[34-36] Specifically, Shapiro et al. showed an increase in connective tissue in myelomeningocele stillborn fetuses when compared to controls using quantitative morphometric analysis.[28] Similar results were also obtained from older patients with myelomeningocele who were having augmentation cystoplasty for a decreased bladder compliance. An electron microscopic analysis of neuropathic, outlet obstruction and stretch injured bladder, showed an increase in the intercellular collagen deposition between the muscle cells by Mayo et al.[36] These authors concluded that the formation of intercellular collagen may be a very non-specific response to a variety of bladder insults. In a quantitative hydroxyproline analysis of normal human bladders obtained from autopsy specimens, Susset et al. showed no significant difference between pre and post puberty age groups in relative collagen content.[33] In contrast, in females greater then 50 years of age they found an increase in the percent of collagen which is consistent with the presence of prostatic obstruction in males. They postulated that the reaction to detrusor outlet obstruction is proportionally greater for detrusor hypertrophy then connective tissue synthesis which would cause a proportional reduction in percent collagen in the male patients although its absolute value may increase. Finally, Ewalt et al. looked at the percent change in the major structural collagen, type I and III, in the noncompliant bladder and showed an increase in the percent of type III collagen when compared with controls.[27] In all these studies, there was a change in the connective tissue content in the non-complaint bladder.

Taking advantage of our developmental bovine bladder model, we analyzed the percent change of type I and III collagen as a function of age and then the total collagen content as determined by hydroxyproline assay. The unique aspect of this study is the direct correlation that can be made to the biomechanical compliance studies of the bladder wall, Baskin et al.[26] We have demonstrated that the ratio of type III/I collagen (Figure 8) decreases with increasing fetal development. Determination of the elastic modulus for these tissue samples shows that a similar relationship exists, i.e., elasticity increases with increasing fetal age (Figure 4).[26]

To determine total collagen per unit weight of bladder tissue we assayed for the total hydroxyproline content. This amino acid is a marker for collagen since it is not found in significant levels in other proteins. The results of the hydroxyproline determination and absolute collagen quantitation show that type III collagen decreases with fetal development. These results inversely correlate with bladder wall compliance.[26]

In summary, in the developmental bladder model, total collagen by hydroxyproline amino acid analysis correlates with elasticity. When the bladder wall collagen is analyzed by type, the percent of type III collagen also correlates with bladder elasticity as a function of age.

Whether in the developing fetus type III collagen synthesis decreases with age or degradation increases cannot be determined from these set of experiments. Conversely, type I collagen synthesis may increase with fetal development or degradation may increase.

This study adds structural and physiologic information to the complicated puzzle of what determines bladder compliance. We have started with the major structural collagens type I and III of the bladder wall. Many other variables may be involved such as collagen cross linking, collagen cellular interactions, elastin and other connective tissue proteins. Certainly the bladder smooth muscle is a critical factor in determining bladder compliance. Clinically, the work of Cartwright and Snow has shown that bladder compliance may be significantly increased by removing the detrusor layer and allowing the lamina propria to balloon out as a large bladder diverticulum.[27]

In fetal bovine bladder development we hypothesize that with the initiation of urine production the bladder wall begins to undergo mechanical forces or pressure/volume work. As gestation progresses the amount of urine production increases with a parallel increase in bladder wall compliance.[26] This study shows that the ratio of type III/I collagen and the absolute amount of type III collagen correlates with this increase in bladder compliance in the developmental process. In the human noncompliant bladder, Ewalt et al., has shown that type III collagen increases.[27] A similar situation exist in patients with ischemic cardiomyopathy.[28] In this situation, a hypertrophic process is accompanied by a pressure overload inducing an increase in collagen synthesis and structural and biomechanical remodeling of the extracellular matrix.[29] Mukherjee and Sen have shown that the increased collagen concentration in the myocardium of patients with ischemic cardiomyopathy is type III collagen with type I collagen remaining unchanged.[28] Further work by Weber concluded that the amount of collagen present in the heart is not the only factor in myocardial fibrosis but that the arrangement of the collagen fiber with respect to the myocyte is also critical.[29] Finally Mukherjee and Sen also showed that the antihypertensive medication captopril normalized the ratio of type III/I collagen as well as reducing the total amount of collagen synthesis.[28] The mechanism of captopril action maybe owing to an improvement in the hemodynamic status of the heart or captopril may directly effect the biosynthetic pathway of the myocardial collagen.

Additional evidence that the bladder responds to mechanical forces has been documented in our laboratory where we have shown that isolated cultures of both bladder smooth muscle cells and urothelial cells respond to mechanical forces by changing the rate of synthesis of collagens.[30] Further work has also shown that mechanical force increases the synthesis of collagen type I from urethral fibroblasts which may be important in the ongoing process of urethral stricture formation.[31]

In conclusion, during normal development of the bovine fetal bladder, there is a progressive change from a rather stiff noncompliant bladder characterized by a high elastic modulus to a compliant bladder with a lower elastic modulus. The increase in compliance appears to correlate with the developmental period during which urine production first occurs, suggesting that volume work may be required for the normal development of bladder compliance. The poorly compliant fetal bladder may explain some of the transient dilations of the upper urinary tract which have been documented in utero.

From the newborn period to the mature adult bovine, we documented a relatively modest increase in the elastic modulus or decrease in bladder compliance which may be explained by the normal aging process.

Finally, type III collagen correlates with bladder wall compliance in the bovine model. Although other factors such as the smooth muscle, elastin, collagen cross linking and connective tissue myocyte interaction are certainly important in determining compliance. This study adds further evidence that type III collagen is a marker for bladder wall compliance.

Material in this chapter has been drawn, in part, from two articles published by Laurence S. Baskin, M.D., in the Journal of Urology Pediatric Supplement, *August, 1994: "Bovine bladder compliance increases with normal fetal development," and "Type III collagen decreases in normal fetal bovine bladder development."*

REFERENCES

1. L.E. Thibault, Mechanical characterization of membrane like biological tissue, *J Biomech. Eng.* 105:31 (1993).
2. F.K. Winston, E.J. Macarak, S.F. Gorfien, and L.E. Thibault, A system to reproduce and quantify the biomechanical environment of the cell, *J Appl Physiol.*. 67(1):397 (1989).
3. P. Eckstein, and W.A. Kelly, Implantation and development of the conceptus, *in*: Reproduction of Domestic Animals, H.H. Cole and P.T. Cupps, eds., Academic Press, New York (1977).
4. D. Cheung, P. Benya, N. Perelman, P.E. DiCesare, and M.E. Nimni, A highly specific and quantitative method for determining type III/I collagen ratios in tissues, *Matrix*. 10:164 (1990).
5. A. Nordwig, and Y.P. Dick, Cleavage with cyanogen bromide of methionyl bonds in collagen, *Biochem Biophys Acta.*. 97:179, (1965).
6. G. Bellon, Quantification and specific detection of collagenous proteins using an enzyme-linked immunosorbent assay and an immunoblotting for cyanogen bromide peptides, *Anal Biochem.*. 150:188, (1985).
7. W.G. Cole, D. Chan, and J.F. Bateman, Two-dimensional assays of peptide fragments, *Methods Enzymol.*. 145:183 (1987).
8. L.S. Baskin, S. Constantinescu, P.S. Howard, J.W. McAninch, D. Ewalt, J.W. Duckett, H.M. Snyder, and E.J. Macarak, Biochemical characterization and quantitation of the collagenous components of urethral stricture tissue, *J Urol* . 150:642 (1993).
9. W.G. Cole, and D. Chan, Analysis of the heterogeneity of human collagens by two dimensional polyacrylamide-gel electrophoresis, *Biochem J.* 197:377 (1981).
10. E. Adams, and L. Frank, Metabolism of proline and the hydoxyprolines, *Annu Rev Biochem.*. 49:1005 (1980).
11. B.A. Bidlingmeyer, S.A. Cohen, and T.L. Tarvin, Rapid analysis of amino acids using pre-column derivatization, *J Chromatogr.* 336:93 (1984).
12. T. M. Penning, W. R. Abrams, and J.E. Pawlowski, Affinity labeling of 3 alpha-hydroxysteroid dehydrogenase with 3 alpha-bromoacetoxyandrosterone and 11 alpha-bromoacetoxyprogesterone. Isolation and sequence of active site peptides containing reactive cysteines; sequence confirmation using nucleotide sequence from a cDNA clone. *J Biol Chem.*. 266:8826 (1991).
13. R.E. Neuman, and M. Logan, The determination of collagen and elastin in tissues, *J Biol Chem.*. 186:549 (1950).
14. R.S. Alexander, Mechanical properties of urinary bladder, *Am J Physiol.*. 220:1413 (1971).
15. J.T. Apter, P. Mason, and G. Lang, Urinary bladder wall dynamics, *Invest Urol.*. 9:520 (1972).
16. A. Kondo, J.G. Susset, and J. Lefaivre, Viscoelastic properties of bladder. I. Mechanical model and its mathematical analysis, *Invest Urol.*. 10:154 (1972).
17. A. Kondo, and J.G. Susset, Physical properties of the urinary detrusor muscle. A mechanical model based on the analysis of stress relaxation curve, *J Biomech.*. 6:141 (1973).
18. R.S. Alexander, Viscoplasticity of smooth muscle of urinary bladder, *Am J Physiol.*. 224:618, 1973.
19. B.L. Coolsaet, W.A. van Duyl, R. van Mastrigt, and A. van der Zwart, Visco-elastic properties of the bladder wall, *Urol Int.*. 30:16 (1975).
20. B.L. Coolsaet, R. van Mastrigt, W.A. van Duyl, and R.E. Huygen, Viscoelastic properties of bladder wall strips at constant elongation, *Invest Urol.*. 13:435 (1976).
21. J.G. Susset, and C.H. Regnier, Viscoelastic properties of bladder strips: standardization of a technique, *Invest Urol* . 18:445 (1981).
22. R.M. Levin, K. Brendler, and A.J. Wein, Comparative pharmacological response of an in vitro whole bladder preparation (rabbit) with response of isolated smooth muscle strips, *J Urol.*. 130:377 (1983).
23. R.M. Levin, and A.J. Wein, Response of the in vitro whole bladder (rabbit) preparation to autonomic agonists, *J Urol.* 128:1087 (1982).

24. R. Rabinowitz, M.T. Peters, S. Vyas, S. Campbell, and K.H. Nicolaides, Measurement of fetal urine production in normal pregnancy by real-time ultrasonography, *Am J Obstet Gynecol.*. 161:1264 (1989).
25. D.E. Coplen, E.J. Macarak, and R.M. Levin, Developmental changes in normal fetal bovine whole bladder physiology, *J Urol.*. 151:1391 (1994).
26. L.S. Baskin, D. Meany, A. Lansman, S.A. Zderic, and E. Macarak, Bladder compliance increases with normal fetal development, *J Urol* . (in press).
27. D.H. Ewalt, S.C. Constantinescu, P.S. Howard, B. Blyth, H.M. Snyder, J.W. Duckett, and E.J. Macarak, Increased collagen type III in the neurogenic non-compliant bladder (AUA abstract, # 904 Washington D.C., 1992).
28. E. Shapiro, M.J. Becich, E. Perlman, and H. Lepor, Bladder wall abnormalities in myelodysplastic bladders: a computer assisted morphometric analysis, *J Urol.*. 145:1024 (1991).
29. S.A. Gilpin, J.A. Gosling, and R.J. Barnard, Morphological and morphometric studies of the human obstructed, trabeculated urinary bladder, *Br J Urol.*. 57:525, (1985).
30. M.E. Mayo, and F. Hinman Jr., Structure and function of the rabbit bladder altered by chronic obstruction or cystitis, *Invest Urol.*. 14:6 (1976).
31. B. Uvelius, and A. Mattiasson, Collagen content in the rat urinary bladder subjected to infravesical outflow obstruction, *J Urol.*. 132:587 (1984).
32. R. Cortivo, F. Pagano, G. Passerini, G. Abatangelo, and I. Castellani, Elastin and collagen in the normal and obstructed urinary bladder, *Br J Urol.*. 53:134 (1981).
33. J.G. Susset, D. Servot-Viguier, F. Lamy, P. Madernas, and R. Black, Collagen in 155 human bladders, *Invest Urol.*. 16:204 (1978).
34. E.J. McGuire, Detrusor response to outlet obstruction, *World J Urol.*. 2:208 (1984).
35. R.M. Levin, P.A. Longhurst, F.C. Monson, K. Kato, and A.J. Wein, Effect of bladder outlet obstruction on the morphology, physiology, and pharmacology of the bladder, *Prostate.* 3:9 (1990).
36. M.E. Mayo, R.W. Lloyd-Davies, K.E. Shuttleworth, and J.R. Tighe, The damaged human detrusor: functional and electron microscopic changes in disease, *Br J Urol.*. 45:116 (1973).
37. P.C. Cartwright, and B.W. Snow, Bladder autoaugmentation: early clinical experience, *J Urol.*. 142:505 (1989).
38. Mukherjee, D. and Sen, S., Alteration of collagen phenotypes in ischemic cardiomyopathy, *J Clin Invest.*. 88:1141 (1991).
39. K. T. Weber, Cardiac interstitium in health and disease: the fibrillar collagen network, *J Am Coll Cardiol.*. 13:1637 (1989).
40. L. Baskin, P.S. Howard, J.W. Duckett, H.M. Snyder, and E. Macarak, Effect of physical forces on bladder smooth muscle and urothelium, *J Urol.* 150:601 (1993).
41. L.B. Baskin, P.S. Howard, and E. Macarak, Effect of mechanical forces on extracellular matrix synthesis by bovine urethral fibroblasts in vitro, *J Urol.* 150:637 (1993).

DEVELOPMENTAL ASPECTS OF EXCITATION CONTRACTION COUPLING IN URINARY BLADDER SMOOTH MUSCLE

Stephen A. Zderic MD, Chaoliang Gong PhD, Joseph Hypolite BS, Robert M. Levin PhD

Urology Research Laboratories
Children's Hospital of Philadelphia
Hospital of the University of Pennsylvania
University of Pennsylvania School of Medicine
Philadelphia, PA 19104

Over the past 5 years, this laboratory has been studying the normal developmental changes that take place in the active and passive properties of urinary bladder smooth muscle. The majority of our experimental work has been carried out using the New Zealand White rabbit as the model of choice because of our extensive experience with it. Exact age groups are available allowing for accurate comparisons. If obstructions are to be performed, the rabbit is a hardy survivor with the added benefit of lower housing costs. The smaller bladder size is easier to manipulate in physiology experiments, but this benefit is offset by the fact that more tissue is required when one is studying protein expression. Because so many of our assays looking at protein expression require sizable amounts of tissue, we have now begun to use the fetal bovine bladder in our research efforts. Large amounts of tissue can be collected cheaply allowing for both physiologic and molecular studies at the protein and nucleic acid levels. This system has the disadvantage of requiring the investigators to take whatever specimens are available at the abattoir that day. One cannot be guaranteed that the proper age groups or sexes will be available. In addition to losing convenience, the bovine model would be extremely expensive to manipulate experimentally requiring the use of a large animal veterinary facility. Despite these limitations each model has something to offer in the study of normal bladder development.

Keating et al using a whole organ model showed that bladders from neonatal rabbits (1 day & 1 week) generated higher intravesical pressures than those from mature animals (8 weeks) when subjected to field stimulation, cholinergic, or purinergic agonists (1). These differences in pressure generating capacity were maintained over a wide range of volumes which would seem to rule out geometric artifacts as an explanation. Despite some limitations, most investigators agree that the whole bladder model offers unique insight into the whole organ properties (2). It is the best in vitro method for determining the emptying properties of the bladder, a parameter which cannot be obtained from muscle strips. No differences were noted in the fraction of expelled urine (95%) along the course of normal bladder development.

A series of experiments were then designed using muscle strip techniques to assess developmental changes in the normal developing rabbit urinary bladder (3,4). Length tension curves were carried out for all of the age groups to determine the length (Lo) were maximal active tension was developed. We reported that the optimal active tension generated for each strip was at a point that was equal to 2.5 times the slack length (Ls) (3). This compared very favorably with the work of Uvealis who showed that optimal force was developed at a length that was 3.0 times Ls (5). These data were also very comparable with those reported by Longhurst who found optimal force generation at 2.5 x Ls (6). Should all smooth muscle strip experiments be carried out at their exact Lo, or can one use an average value and set all

Muscle, Matrix, and Bladder Function,
Edited by S. Zderic, Plenum Press, New York, 1995

strips to this length? This has the benefit of simplifying the experiment and subsequent data analysis, but some have argued that it might induce experimental artifact. We tested this by carrying out a series of length tension curves using bethanechol challenges (100 μM) and comparing the stress generated at the absolute Lo with the stress generated at 2.5 x Ls (7). We found an even bell shaped distribution of Lo values around 2.5 x Ls. A propagation of errors analysis found that the error incurred by setting all the strips at 2.5 x Ls versus studying each strip at its exact Lo gave a 10% underestimate of the maximal stress generated (7). Thus if one is studying two sets of bladder smooth muscle strips whose contractile properties differ by 50% or more, the error induced by studying all strips at 2.5 x Ls can't explain the difference. This study also assessed the effect of using a fixed 2 grams of passive tension for each strip as the point where all experimental work is to be done. Again a statistical analysis of the errors showed that the error was approximately 10%.

These studies also measured the contractile properties of the bladder smooth muscle (3,4). In response to field stimulation the neonatal strips generated a greater force when normalized for strip mass (3). Neonatal strips also generated greater forces than mature strips in response to KCl, ATP, histamine, and bethanechol (4). These data were expressed as grams tension generated per 100 mg tissue weight, but since all the strips were studied at the same length, this correlates directly with stress (g/cm2). Controversy has arisen over how to express the active forces generated by smooth muscle when one compares one strip to another. Does one normalize based on total cross sectional area, total muscle area, or should one factor into account fiber orientation? It has been suggested by Stephens et al that normalizing to the total cross sectional area occupied by muscle is the most accurate when comparing tracheal to bronchial smooth muscle (8). The active forces generated differed substantially but could best be explained by a 50% variation in alterations in muscle fraction (8). We do not feel that the physiologic differences we reported could be explained by differences in smooth muscle area. Furthermore Stephens' group has also reported on developmental changes in rabbit airway smooth muscle and reported a greater force generating capacity in the neonates (9).

Prior work had already shown that the cholinergic receptor density and dissociation constants did not change with development (10). These developmental changes in contractility had to be accounted for by other molecular mechanisms. We considered the possibility that these changes might be explained by developmental changes in the way the smooth muscle cells handle cytosolic free calcium. In one series of experiments, 1x1x8 mm muscle strips were stripped of their mucosal layer and incubated over night in oxygenated Tyrode's solution containing the FURA-2 methyl ester dissolved in 2.5 % DMSO. Overnight incubation allows FURA-2 to diffuse into the cytosol where the methyl ester is cleaved away. This allowed for measurements of intracellular calcium through the use of dual emission fluorescence imaging. The excitation wavelengths of FURA-2 change from 380 nm to 340 nm once a calcium ion is bound to the molecule (11). By monitoring the fluorescent emissions at 510 nm following alternating excitations at 340 & 380 nm, one can determine the 340/380 ratio which correlates with intracellular free calcium. It is possible to calibrate this kind of system so as to know the exact cytosolic calcium concentration. Alternatively one may just compare the signals at these 2 wavelengths and determine a ratio which can then be used to determine the relative changes in free cytosolic calcium. In our lab, it is possible to simultaneously determine the FURA-2 ratios and measure the force generated. In these experiments strips from neonatal and mature groups were set at 2 grams of passive tension and equilibrated for 1 hour. Intracellular calcium was depleted by incubating in calcium free buffer containing EGTA. Then in the presence of 200 μM bethanechol, extracellular calcium was added back to the baths. Tension and the FURA-2 ratio were recorded once a stable plateau was reached. The neonatal strips were far more sensitive to the effects of extracellular calcium when one considered both the force and the fluorescence signal that developed (figure 1) (12). These experiments were repeated in strips with intact mucosa, and the same findings were recorded.

With this preliminary data we proposed several possible mechanisms to account for the observed physiology. 1) Maturational changes in the membrane bound calcium channels, i.e. a neonatal membrane that would be "leakier" for the calcium ion. 2) Enhanced sensitivity of the filaments in the neonatal group. 3) Altered binding of the calcium ion to intracellular sites such as storage sites. 4) Alterations in the intermediary regulatory proteins that transduce the calcium signal into force such as calmodulin. With these ideas in mind we began to consider the first and fourth possibilities. Using the radioligand H^3-PN-200 and a Scatchard analysis, we were unable to demonstrate any maturational changes in the binding or dissociation constants of the voltage operated calcium channel (figure 2) (13). With RIA techniques we could not detect any maturational differences in calmodulin expressed as nmol/mg total protein (14).

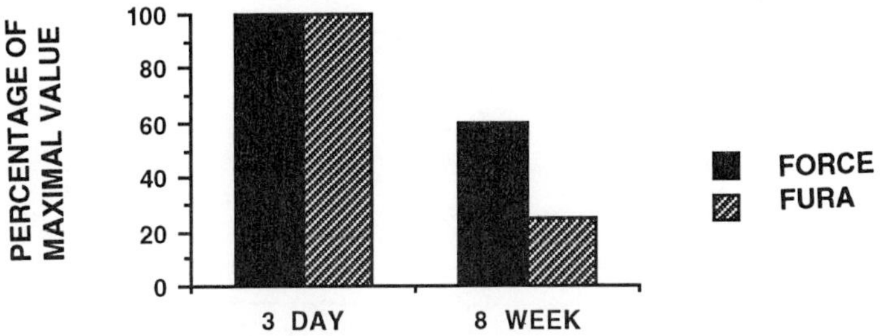

Figure 1 - This figure shows the relative ratios of the FURA-2 ratios at 340 & 380 nm which correlate with cytosolic calcium, and the relative stress generated between neonatal and adult bladder smooth muscle strips. The differences are statistically significant.

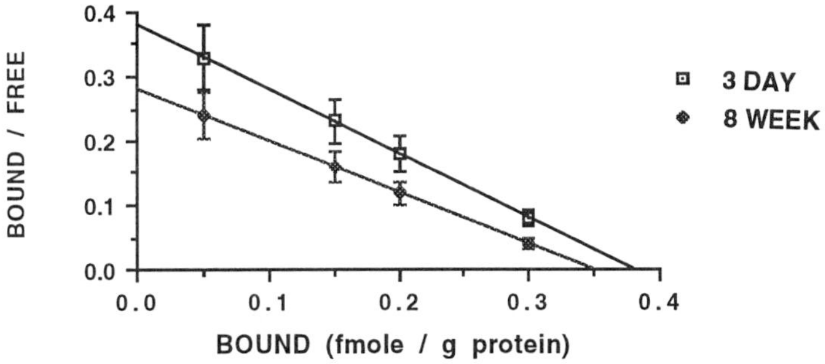

Figure 2 - This figure shows that there are no developmental shifts in the voltage operated calcium channel as measured with PN-200.

We then decided to do some further physiology experiments using muscle strip methods to test for the third possible explanation. Muscle strips from the 2 age groups were set at Lo. The optimal response to field stimulation was measured in grams of tension, and this starting point was defined as 100%. The buffer system was changed over to calcium free Tyrode's with 2 mM EGTA and repeat field stimulation was performed at 2 minute intervals. The time at which the strips generated less than 50% of their starting force was recorded. This differed substantially between the neonatal (2.1 min) and mature groups (4.2 min) ($P<0.05$), and seemed to imply that the mature smooth muscle might have a greater pool of stored intracellular calcium ions. However this result by itself would not be sufficient to prove this point because differences in strip size might induce diffusion effects and other artifacts. Additional experiments were designed to consider if there might be maturational differences in calcium sequestration. With the strips depleted of their intracellular calcium, the Tyrode's solution was changed over to a calcium free version with potassium having been substituted for sodium. At this point this calcium was added back to the baths and a dose response curve was determined in the presence of membrane depolarization. Despite the absence of changes in the voltage operated calcium channel, there were still substantial changes in the calcium dose response curves (figure 3)(13).

This data lead us to consider the real possibility that the major physiologic differences we were seeing could be explained by alterations in intracellular calcium storage and release mechanisms. This would explain why the differences in calcium sensitivity could be seen with no changes in the cholinergic receptor or voltage operated calcium channel densities. Experimental evidence for intracellular calcium storage sites in bladder smooth muscle was first proposed by Mostwin (15). In vascular smooth muscle Somlyo had already demonstrated the existence of intracellular calcium storage sites which might account for up to

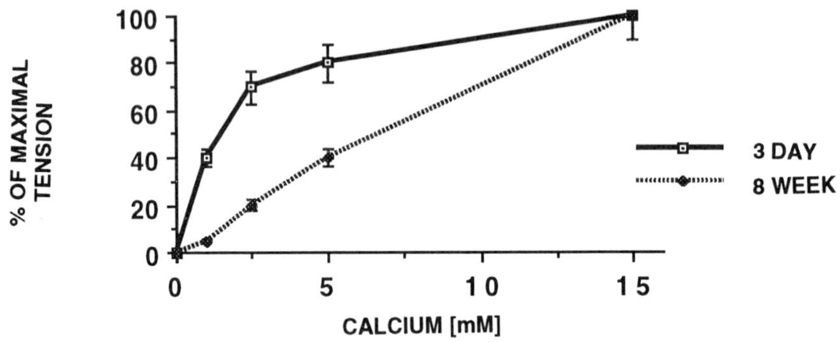

Figure 3 - Despite an absence of changes in the voltage operated calcium channels, there is a marked difference in the calcium dose response curves of neonatal and mature rabbit urinary bladder smooth muscle (after preliminary depletion of the intracellular pools).

5-10% of cytosolic volume (16). However no one had investigated the possibility of developmental changes in the system that regulated intracellular calcium storage and release in a smooth muscle system.

Cardiac physiologists, however, have been extremely interested in this topic and have accumulated substantial evidence to suggest that the sarcoplasmic reticulum (SR) undergoes substantial changes with normal development. In early morphologic studies using electron microscopy, Friedman and Sheldon demonstrated increases in the SR of sheep myocardium associated with normal development (17). In the rabbit Klitzner and Friedman have shown that there are major differences in the calcium pools that are activated during cardiac contraction with the neonatal myocardium being far more dependent upon extracellular calcium to power its contraction (18). With normal maturation, there is a substantial increase in the role of intracellular calcium stores in the generation of contractile force (19). They used ryanodine as a pharmacologic probe of muscle function. This alkaloid binds to an ion channel associated with the SR that releases intracellular calcium stores into the cytosol (19). In their system ryanodine had a minimal effect in neonatal rabbit myocardium, and a pronounced

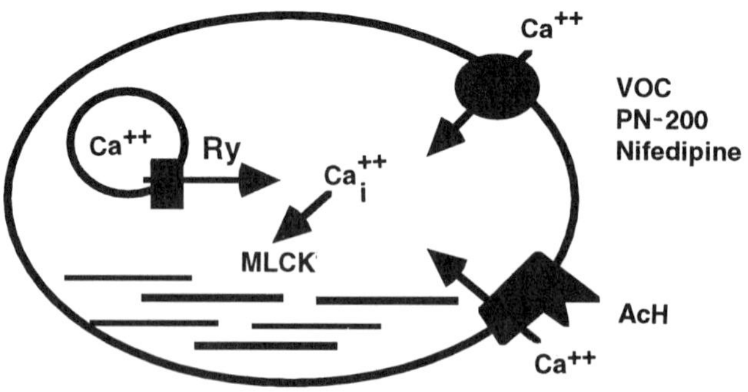

Figure 4 - A schematic simplification of excitation contraction coupling. Ry = ryanodine sensitive calcium release channel. VOC = voltage operated calcium channel that is site of action of PN-200 or nifedipine. AcH = acetylcoline which binds to a muscarinic receptor to trigger calcium influx. Free cytosolic calcium (Cai) activates myosin light chain kinase which is felt to initiate contractile force generation.

inhibitory effect in the mature myocardium (19). The role of the ryanodine sensitive ion channel is shown schematically in our simplified schematic representation of excitation contraction coupling (figure 4). At doses of 1 μM or less ryanodine binds to this ion channel and enhances calcium release from the SR. However at higher concentrations (> 10 μM) ryanodine inhibits the release of calcium from storage sites into the cytosol. These ion channels have been described in numerous tissue types, and at least 3 differing isoforms have been described (20).

We tested ryanodine in rabbit bladder smooth muscle strips of differing age groups to see if there was a physiologically significant difference. In addition we tested another drug with a completely different mechanism of action. Nifedipine works by primarily blocking the influx of extracellular calcium by the voltage operated channels (identified by PN-200). These experiments were done by comparing each strip to itself, and calculating the degree of inhibition incurred by the addition of these drugs. This analysis obviated any of the controversy about normalization of data, for it in effect identifies qualitative ways in which muscle function is altered during development. When these 2 drugs were tested in the neonatal and mature age groups, the following observations were made. Neonatal strips were significantly less sensitive to ryanodine, but very sensitive to nifedipine relative to the mature group (figures 5 & 6)(21). Taken together these physiologic data would suggest thatneonatal smooth muscle relies more heavily on extracellular calcium influx, and that with maturation the intracellular calcium pool increases.

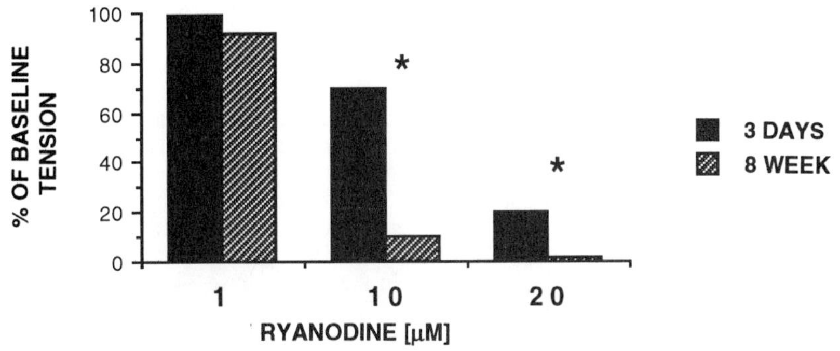

Figure 5 - The effects of ryanodine on contractile force in response to field stimulation are more pronounced in the mature group (* = P<0.05).

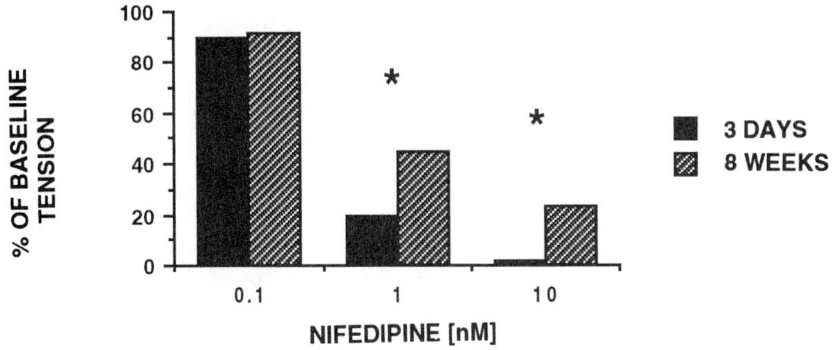

Figure 6 - The effects of nifedipine on contractile force in response to field stimulation are more pronounced in the neonatal group (* = P<0.05).

Having made these observations, we began to search for differences in protein expression that could be correlated with this physiology. The ryanodine ion channel is unique to the SR, and if one is speculating that there are developmentally associated increases in calcium storage sites, the density of this protein should increase. Wibo et al demonstrated this very nicely in the developing rat myocardium where a clear cut increase in receptor density independent of any changes in the dissociation constant was observed during normal development (22). We used H^3-Ryanodine to perform Scatchard analyses, and found dramatic increases in the density of the ryanodine sensitive ion channel with no changes in the dissociation constant (Bmax expressed as fmol ryanodine/mg membrane protein) (figure 7) (21,23). We concluded that with normal development there is an upregulation of the ryanodine sensitive ion channel, and this correlated with physiologic evidence suggesting that the total pool of intracellular calcium stores increases.

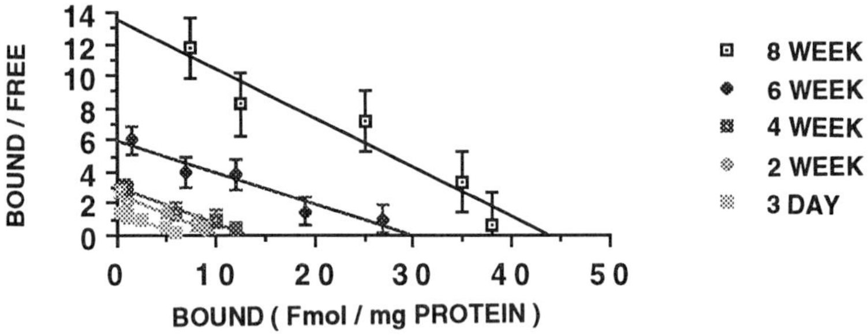

Figure 7 - The effects of normal maturation result in a dramatic increase in the density (Bmax) of the ryanodine sensitive ion channel with no change in the dissociation constant (Kd).

This research group has already begun to accumulate evidence that these mechanisms of calcium storage and release are also affected during obstruction. During the acute phase of bladder overdistension following acute obstruction, there is a decrease in the ryanodine receptor density which correlates with a phase of poor contractility. However as this acute period resolves there is partial recovery of bladder function, and receptor density begins to approach normal levels (24). We have interpreted these findings as a part of the hypertrophic response of a bladder subjected to pressure work. In the acute phase 24-48 hours the bladder is acutely decompensated, and smooth muscle is in a "stunned state". However as it becomes clear that the bladder must perform pressure work, the smooth muscle begins to undergo hypertrophy with an increase in cell volume. If cell volume is to be increased, then the cell must be able to increase its intracellular calcium stores. Failure to do so would mean it would have to rely on the inefficient diffusion of extracellular calcium ions to reach the contractile apparatus. We feel that the initial insult in these acute models is followed a period of compensatory hypertrophy, and a major component of this is upregulation of the calcium sequestration system. More will be said about this in our closing paragraphs.

The SR must take up, store, and then release free cytosolic calcium as dictated by the needs of the cell. As shown in figure 4, this is a complex process which requires the expenditure of energy in the form of ATP. Langer has estimated that up to 25% of the cardiac ATP expenditure goes towards maintaining the proper intracellular calcium gradients (25). The free cytosolic calcium concentration is maintained at about 0.1-1 μM, and the storage sites have concentrations in the 1 mM range (26). These concentration gradients require significant energy reserves to maintain. If one hypothesizes that there are changes in the SR component of bladder smooth muscle during normal development, then one should expect to see increases in some of the other membrane components that play a role in the regulation of cytosolic calcium. One such enzyme is the $Ca^{++}Mg^{++}$ATP-ase which is found on the SR and serves to transport cytosolic calcium into the storage sites. This enzyme was initially characterized in skeletal muscle by MacLennan (27) who subsequently identified it in cardiac muscle, brain, and kidney (28). His group has also cloned the gene for this enzyme, and demonstrated that at least 2 isoforms of this enzyme exist: a fast type (skeletal & cardiac), and a slow form (brain and kidney) (28). The base sequences are highly conserved in this setting with up to 85% or more homology between species suggesting that this enzyme has enough physiologic significance that it has not been amenable to evolutionary shifts. Monoclonal antibodies have been raised and characterized against these $Ca^{++}Mg^{++}$ATP-ases by Jorgenson and Campbell et al (29).

We began our study of the $Ca^{++}Mg^{++}$ATP-ases of the SR by developing a functional assay for calcium ion translocation so as to compare transport properties across the age groups we have been studying. Preliminary results in the rabbit model were disappointing because were trying to do the assays with minimal amounts of membrane protein in an effort to conserve tissue. We then tried these experiments using fetal bovine bladder because the tissues are plentiful, and inexpensive. Since each assay consumes about 0.3 g of tissue, working out the experimental protocol in the bovine model seemed quite logical. There are several possible assay techniques for the $Ca^{++}Mg^{++}$ATP-ase of the SR as reviewed by Tate and Phillips (30). Methods that measure the breakdown of ATP, liberation of PO4, or the trapping of a phosphorylated intermediate have been described (30), and have their respective advantages and disadvantages. The method we selected to use called for the trapping of radioactive Ca-45 in membrane vesicles using a Millipore filter system (figure 8) (31). The technique called for the isolation of a membrane fraction from the smooth muscle fraction of the fetal bovine bladders (the mucosa and lamina propia were stripped away). The assays are carried out at 37 C in a 20 mM TRIS Buffer system in 100 mM KCl at a pH of 6.8 to which 1 μM $Ca^{45}Cl2$ is added for a 5 minute preincubation. After 5 minutes, 2 mM ATP is added and 100 μl aliquots are withdrawn at 2 minute intervals, and filtered over a Millipore HA-45 filter (0.45 μm) followed by a 5 cc wash with chilled 20 mM TRIS and 2 mM EGTA. The actual uptake is enhanced by trapping the pumped calcium within the microsomes by using potassium oxalate. Failure to use oxalate means that labeled calcium pumped into the vesicles is able to diffuse out, and hence the counts retained (active uptake) on the filters will plateau. Use of oxalate trapping maintains a linear accumulation of the isotope from which a rate of uptake may be determined with linear regression analysis (velocity). This method is ideal for comparing the capacities of calcium transport between the various age groups. The Ca^{45} method has the following limitations: 1) it requires an intact vesicle (membrane fragments may contain enzyme activity that won't be detected) 2) the vesicles must seal properly to give the correct orientation for ion pumping. 3) it detects net calcium accumulation, and one cannot rule out diffusion from the vesicle 4) a radioisotope method is more expensive. On the other hand it is quite capable of detecting low levels of activity because the Ca^{45} used can be given a very high specific activity (10,000 - 20,000 DPM / nMole).

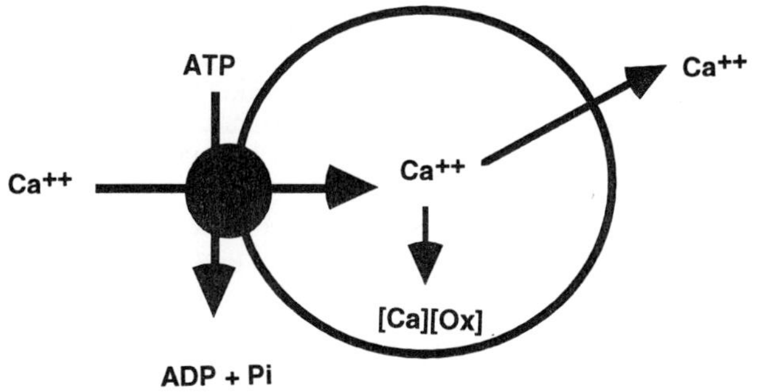

Figure 8 - A schematic representation of the Ca-45 uptake assay into membrane vesicles illustrating the principle of oxalate trapping. Failure to use oxalate will result in transported calcium leaking out of the vesicles.

Bladders were harvested from freshly killed bovine fetuses were segregated into trimesters (gestational ages were determined by their crown rump lengths). Adult bladders were obtained for comparison. Bladder smooth muscle was stripped away from the mucosa, weighed, and frozen at -80C until ready for use. After thawing muscle was minced and homogenized in a 20 mM TRIS-maleate buffer at pH=6.8 and subjected to centrifugations of 5000 x G and 10,000 x G which served to remove the insoluble matrix proteins, nuclei, and mitochondria. The resulting supernatant was subjected to ultracentrifugation at 63,000 x G for 1 hour at 4C, and the resulting membrane pellet was saved. The pellet was resuspended in 20 mM TRIS-maleate and protein content was determined by the method of Lowry. Aliquots of 60 μl were stored at -80C, with 10-20 vials being prepared per 3 to 6 grams wet weight tissue. Assays were run by adding 50 μl of membrane protein (vesicles) in to the assay system described in the preceding paragraph. Our results are shown below comparing the data for 2nd trimester versus adult bladder smooth muscle (figure 9). There is a clear cut increase in the ability of the adult membrane vesicles to transport Ca45. Furthermore the activity is increased no matter one expresses the data (i.e. nmol Ca/min/mg protein or nmol Ca/min/gr wet tissue weight).

These differences in calcium transport using the functional assay described above were then confirmed with a different technique. We used a Western Blot method with a monoclonal antibody kindly supplied to the slow form of the $Ca^{++}Mg^{++}$ATP-ase by Dr Kevin Campbell at the University of Iowa School of Medicine. The identical membrane fractions studied with the active calcium transport assay were then subjected to Western Blot assay. In addition to confirming the findings of our assay, the use of this technology with its

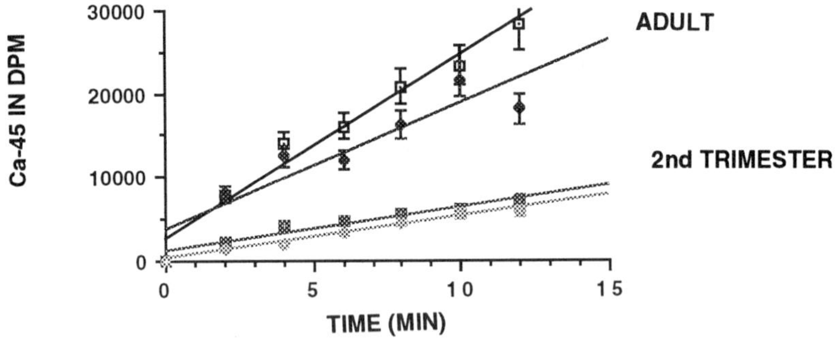

Figure 9 - The results of Ca-45 uptake assays in microsomal fractions from second trimester fetal versus adult bovine bladders. The Ca-45 taken up by the vesicles is plotted as DPM on the y axis as a function of time (min).

enhanced sensitivity allows for the use of smaller amounts of tissue. Membrane fractions were thawed and solubilized using SDS to free up the bound proteins which were then separated on a 5% SDS electrophoresis mini gel. After the separation is complete the proteins from the gel are electro-blotted onto a nitrocellulose filter. Using the monoclonal antibody at a 1:2000 dilution at 4C, the filter is equilibrated for 24 hours (after adsorbing non-specific sites with a powdered milk suspension). After 24 hours the filter is washed, and a second commercial antibody coupled to horseradish peroxidase is applied for another 24 period at 4C. The peroxidase is imaged by the application of peroxide and chloronapthol which results in a colorimetric detection of the protein bands to which antibody is bound (110 Kd). A representative blot is shown in figure 10. There is a greater intensity of stain in the membrane protein corresponding to the adult group as opposed to the fetal group. This result correlates very nicely with our functional assay using the Ca^{45} uptake method. These blots were done by using equal amounts of membrane protein (100 μg) from each group for the assays. In additional assays the protein loaded has been easily dropped to 10 μg which results in even greater differentiation between the fetal and adult ATPase signals. Signal amplification techniques may lower the total protein load required even further.

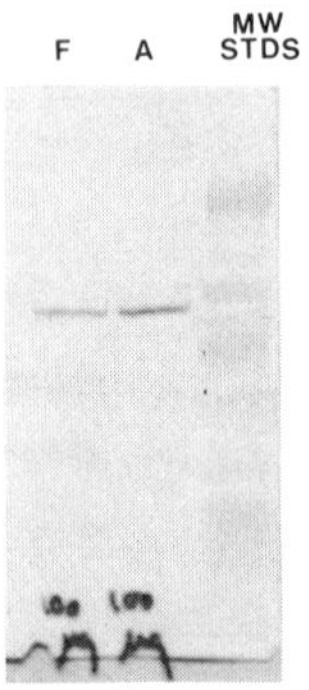

Figure 10 - A Western Blot assay using a monoclonal antibody against the slow form of the Ca++Mg++ ATPase shows an increased expression in the adult group versus the fetal. These immunologic findings correlate nicely with the functional Ca-45 assay. The primary antibody was a generous gift from Dr Kevin Campbell from the Howard Hughes Institute at the University of Iowa School of Medicine.

The expression of myocardial $Ca^{++}Mg^{++}$ATP-ase is influenced by the stage of development as well pathologic conditions such as compensatory hypertrophy and congestive heart failure. Fisher et al have shown a dramatic increase in the expression of the fast isoform of the $Ca^{++}Mg^{++}$ATP-ase associated with the transition from fetal to neonatal to mature rabbit myocardium (32). They proved this by using a functional Ca^{45} assay, a Western blot technique, and a Northern Blot method to probe for mRNA. Their Northern Blot data showed clear cut increases mRNA expression coding for the $Ca^{++}Mg^{++}$ATP-ase as normal development proceeded. In cardiac models abnormalities in calcium translocation correlate well with the clinical picture of obstruction both in human studies and in laboratory models making aortic stenosis a valuable model for urologists to consider. After the induction of aortic stenosis in a rat model there may be compensatory hypotrophy followed by congestive failure (at some point in time). In this model compensatory hypertrophy has been associated with an increased expression of the $Ca^{++}Mg^{++}$ATP-ase, whereas failure and a non-compliant ventricle are associated with diminished expression of this protein (33,34). These findings have been correlated nicely with human findings in studies of dilated cardiomyopathies due to outflow obstruction in hearts removed at the time of transplantation (35). In skeletal muscle hind limb unweighting (produces a neurologically intact muscle that develops disuse atrophy) is also associated with shifts in the expression of the $Ca^{++}Mg^{++}$ATP-ases (36). There is good evidence in both skeletal and cardiac muscle that mechanical stresses sensed by the muscle cells are transduced to the nucleus which influences the expression of proteins in the calcium regulatory system.

In terms of the fetal bovine bladder these findings can be interpreted in terms of the compliance changes that have been described by Baskin (37) and Coplen (38). These authors have shown that fetal bladder compliance changes dramatically with normal development. The bladders of the first trimester bovine fetus are much stiffer than those of the second and third trimester. Coplen has also shown that in first and second trimester bladders, this tone is highly dependent on extracellular calcium (38). In my own experience the adult muscle strips are even more distensible. The passive properties of the bovine bladder change dramatically with gestational age, and part of these normal developmental changes may be accounted for by maturation in the calcium handling properties of the smooth muscle cell. However we are the first to point out that one single explanation for the molecular basis of compliance is unlikely to be accurate. The developmental changes in compliance can also be related to changes in the collagen type III and type ratios (39), and will probably be influenced by elastin and the degree of cross linking between these complex molecules.

Our current work continues to stress the role of calcium homeostasis in the regulation of smooth muscle tone and contraction. We are seeking a better understanding of what binding proteins exist within the sarcoplasmic reticulum and what mechanisms may lead to release of this calcium pool to generate passive tone and active force. Our long term goal is to understand how these fine points of bladder smooth muscle cytosolic calcium homeostasis may be altered during normal development and in pathologic states such as bladder outlet obstruction.

REFERENCES

1. Keating, M.A., Duckett, J.W, Snyder, H.M., Wein, A.J., Potter, L., Levin, R.M.: Ontogeny of bladder function in the rabbit. J. Urol. 144:766,1990.
2. Levin, R.M., Wein, A.J.: Response of the in-vitro whole bladder preparation to autonomic agonists. J. Urol. 128:1087,1982.
3. Zderic, S.A., Duckett, J.W., Snyder, H.M., Wein, A.J., Levin, R.M.: The ontogeny of bladder compliance. Neurourol. & Urodyn. 9:595,1990.
4. Zderic, S.A., Duckett, J.W., Wein, A.J., Snyder, H.M., Levin, R.M.: Developmental factors in the contractile response of the rabbit bladder to both autonomic and non-autonomic agents. Pharmacology 41:119,1990.
5. Uvelius, B.: Isometric and isotonic length tension relations and variations in cell length in longitudinal smooth muscle from rabbit urinary bladder. Acta Physiol. Scand. 97:1,1976.
6. Longhurst, P.A., Kang. J., Wein, A.J., Levin, R.M.: Comparative length tension relationship of urinary bladder stirps from hamsters, rats, guinea pigs, rabbits, and cats. Comp. Biochem. Physiol. 96:221,1990.

7. Zderic, S.A., Liu, G.H., Haab, J.P., Monson, F.C., Gong, C., Levin, R.M.: What is the most accurate way to study the active properties of bladder smooth muscle? Pharmacology 48:380,1994.
8. Jiang, H., Halayko, A.J., Rao, K., Cunningham, P., Stephens, N.L.: Normalization of forces generated by canine airway smooth muscle. Am. J. Physiol. 260:L522,1991.
9. Sparrow, M.P., Mithcell, H.W.: Contraction of smooth muscle of pig airway tissues from before birth to maturity. J. Appl. Physiol. 68:468,1990.
10. Levin, R.M., Malkowicz, S.B., Jacobowitz, D., Wein, A.J.: The ontogeny of the autonomic innervation and contractile response of the rabbit urinary bladder. J. Pharm. Exptl. Ther. 219:250,1981.
11.Thomas, A.P., Delaville, F.: The use of fluorescent indicators for measurement of cytosolic free calcium concentration in cell populations and single cells. in McCormack, J.G., Cobbold, P.H. eds., Cell Calcium, IRL Press, Oxford, First Edition, 1991, pages 1-54.
12. Zderic, S.A., Hypolite, J.A., Duckett, J.W., Snyder, H.M., Wein, A.J., Levin, R.M.: Developmental aspects of bladder contractile function: the sensitivity to extracellular calcium. Pharmacology 43:61,1991.
13. Zderic, S.A., Sillen, U., Liu, Guo-Hua, Snyder, H.M., Duckett, J.W., Wein, A.J., Levin, R.M.: Developmental aspects of urinary bladder contractile function: evidence for an intracellular calcium pool. J. Urol. 150:623,1993.
14. Zderic, S.A.: unpublished data.
15. Mostwin, J.L.: Receptor operated intracellular calcium stores in the smooth muscle of the guinea pig bladder. J. Urol. 133:900,1985.
16. Somlyo, A.P.: Excitation contraction coupling and the ultrastructure of smooth muscle. Circ. Res. 57:497,1985.
17. Sheldon, C., Friedman, W.F.: Scanning electron microscopy of fetal and neonatal lamb cardiac cells. J. Mol. Cell. Cardiol. 8:853,1976.
18. Klitzner, T.S., Friedman, W.F.: Excitation contraction coupling in developing mammalian myocardium: evidence from voltage clamp studies. Ped. Res. 23:428,1988.
19. Klitzner, T.S., Friedman, W.F.: A diminished role for the sarcoplasmic reticulum in newborn myocardial contraction: effects of Ryanodine. Pediatric Research 26:98,1989.
20. Sorrentino, V., Volpe, P.: Ryanodine receptors: how many, where, and why? Trends in Pharm. Sci. 14:98,1993.
21. Zderic, S.A., Gong, C., Sillen, U., Liu, G.H., Duckett, J.W., Snyder, H.M., Levin, R.M.: Developmental aspects of excitation contraction coupling in rabbit bladder smooth muscle. J. Urol. 152:679,1994.
22. Wibo, M., Bravo, G., Godfraind, T.: Postnatal maturation of the excitation contraction coupling in the rat ventricle in relation to the subcellular localization and surface density of 1,4-Dihydropyridine and ryanodine receptors. Circ. Res. 68:662,1991.
23. Gong, C., Zderic S.A., Levin, R.M.: Ontogeny of the ryanodine receptor in rabbit urinary bladder smooth muscle. Mol. & Cell. Biochem. 137:169-172, 1994.
24. Levin, R.M., Levin, S., Zderic, S.A., Yoon, J.L., Wein, A.J.: Effect of partial outlet obstruction of the rabbit urinary bladder on the ryanodine inhibition of the contractile response to field stimulation and ryanodine binding to microsomal membranes. Pharmacology 1994 (in press).
25. Langer, G.A.: Calcium and the heart: exchange at the tissue, cell, and organelle levels. FASEB Journal 6:893,1992.
26. Ruegg, J.C.: Vertebrate smooth muscle. in Ruegg, J.C., author: Calcium and muscle contraction. Springer Verlag, New York, Second edition, 1992, pages 201-238.
27. Maclennan, D.: Purification and properties of an adenosine triphosphatase from sarcoplasmic reticulum. J. Biol. Chem. 245:4508,1970.
28. Lytton, J., Zarian-Herzberg, A., Periasamy, M., MacLennan, D.H.: Molecular cloning of the mammalian smooth muscle sarco(endo)plasmic reticulum Ca2+-ATPase. J. Biol. Chem. 264:7059,1989.
29. Jorgenson, A.D., Arnold, W., Pepper,D.R., Kahl, S.D., Mandel, F., Campbell, K.P.: A monoclonal antibody to the Ca-ATPase of cardiac sarcoplasmic reticulum cross reacts with slow type I but not with fast type II canine skeletal muscle fibers on immunocytochemical and immunochemical study. Cell Motil. Cytoskel. 9:164,1988.

30. Hamilton, S.L., Tate, C.A.: Proteins involved in the uptake and release of calcium from the sarcoplasmic reticulum. in McCormack, J.G., Cobbold, P.H. Eds. Cell Calcium, IRL Press, Oxford, 1991, First edition, pages 313-344.
31. Raymaekers, L., Wuytack, F.: Calcium pumps in smooth muscle cells. J. Muscle Research and Cell Motility 14:141,1993.
32. Fisher, D., Tate, C., Phillips, S.: Developmental regulation of the sarcoplasmic reticulum calcium pump in the rabbit heart. Ped. Res. 31:474,1992.
33. Feldman, A.M., Weinberg, E.O., Ray, P.E., Lorell, B.H.: Selective changes in cardiac gene expression during compensated hypertrophy and the transition to cardiac decompensation in rats with chronic aortic banding. Circ. Res. 73:184,1993.
34. Komuro, I., Kurabayashi, M., Shibazaki, Y., Takaku, F., Yazaki, Y.: Molecular cloning and characterization of a Ca Mg dependent ATP ase from rat cardiac sarcoplasmic reticulum. Regulation of its expression by pressure overload and developmental stage. J. Clin. Invest. 83:1102,1989.
35. Meuse, A.J., Perralut, C.L., Mogan, J.P.: Pathophysiology of cardiac hypertrophy and failure of human myocardium: abnormalities of calcium handling. in Hasenfuss, G., Holubarsch, C., Just, H., Alpert, N.R., editors, Cellular and Molecular Alterations in the Failing Human Heart. First edition, Springer Verlag, New York, 1992, 223-233.
36. Schulte, L.M., Navarro, J., Kandarian, S.C.: Regulation of sarcoplasmic reticulum pump gene expression by hind limb unweighting. Am. J. Physiol. 264:C1308, 1993.
37. Baskin, L., Meaney, D., Landsman, A., Zderic, S.A., Macarak, E.: Bovine bladder compliance increases with normal fetal development. J. Urol. 152:692, 1994.
38. Coplen, D., Macarak, E., Levin, R.M.: Developmental changes in normal fetal bovine whole bladder physiology. J. Urol. 151:1391,1994.
39. Baskin, L.S., Constantinescu, S., Duckett, J.W., Snyder, H.M., Macarak, E.: Type III collagen decreases in normal fetal bovine bladder development. J. Urol. 152:688, 1994.

CONGENITAL BLADDER OBSTRUCTION: RESEARCH STRATEGIES AND DIRECTIONS

Craig A. Peters M.D.

Children's Hospital
Harvard Medical School
Boston, Massachusetts

CLINICAL CONTEXT AND RELEVANCE

Congenital bladder obstruction (CBO) continues to challenge the clinician and scientist in terms of its health impact on patients, as well as its wide spectrum of biologic effect. Bladder dysfunction due to congenital obstruction can produce significant and often insidious injury to the upper urinary tract in addition to causing incontinence and urinary tract infection. The biological changes of congenitally obstructed bladder are not well described, yet represent alterations in growth regulation, connective tissue remodelling, and changes in developmental patterns, and therefore reflect fundamental principles of biological response.

Causes of Congenital Obstruction

Posterior Urethral Valves. Posterior urethral valves (PUV) represent the most pure form of CBO and continue to cause significant health effects long after the obstruction is removed. The detectable effects of PUV may be seen ultrasonographically as early as thirteen weeks of gestation, but most are detected at 16 to 18 weeks. Bladder distention is one sign, but this is not consistent, and severe valves may be present with a small, thick-walled bladder and hydroureteronephrosis. The renal effects are variable also, ranging from normal kidneys to markedly dysplastic, cystic kidneys with little functional potential. A wide spectrum of severity may be demonstrated; with a correspondingly wide variation in clinical manifestations post-natally.

Neurogenic bladder dysfunction. A form of obstructive uropathy in the fetus may be seen with neurogenic bladder dysfunction associated with myelomeningocele, a severe abnormality of neural tube closure. This is complicated by changes in bladder innervation, which may produce some or all of the changes seen. As early as twenty weeks, Shapiro and co-workers have demonstrated abnormal bladder wall histology in fetuses with myelodysplasia.[1] A dilated bladder and hydroureteronephrosis have been present in some fetuses, and this is presumed to correlate with functional bladder obstruction due to detrusor-sphincter dyssynergy. While not present in all children with myelodysplasia, these children are at the most risk for renal injury from their bladder dysfunction. This group of children require the most aggressive therapies in the newborn period, yet it seems apparent that changes in bladder structure and function have been ongoing well before birth.

Clinical Consequences of Congenital Obstruction

Bladder dysfunction. The effects of bladder dysfunction due to CBO include immediate renal and systemic consequences in the most severe forms, but may be more insidious in the long-term. The time of presentation is largely due to the severity of the obstruction. This produces variation in the severity and character of bladder dysfunction. As a result there may be a prolonged interval from valve treatment to clinical manifestations of bladder dysfunction. In recognized cases of severe CBO, anticipation of bladder dysfunction will serve to avoid delays in treatment.

Bladder dysfunction may be evident by urinary retention or in older children by incontinence. The patterns are variable and range from clinically obvious to more subtle voiding dysfunction associated with moderately obstructive valves.

Therapy for CBO induced bladder dysfunction is complex, associated with a variety of side effects and is not always effective in controlling the various manifestations of bladder dysfunction.[2] Pharmacologic measures may be useful but are often limited by side effects and seem to have a limited duration of action. These agents only reduce the manifestations of bladder dysfunction, and do not alter the underlying pathophysiology. Bladder augmentation using segments of the intestinal tract is effective in reversing the pathophysiology but represents a major surgical endeavor with recognized short and long term consequences.

Renal Injury. Two levels of renal injury are associated with CBO. The developmental injury produced by in utero obstruction includes renal dysplasia and severe hydronephrosis with functional impairment. The degree of renal injury is also variable and may be as severe as non-function of the kidney, or may be as minor as impaired concentrating capacity of the kidney. The second level of injury is a slowly progressive postnatal injury due to bladder dysfunction. This may occur after the valvular obstruction has been removed, or when the obstruction is less severe and has not been previously detected. Loss of functional renal mass in severe cases may require renal replacement therapy including dialysis and renal transplantation.

Pulmonary hypoplasia. The most lethal effect of CBO due to PUV is that of pulmonary hypoplasia. While the specific mechanisms are unclear, it is known that early gestation severe urinary tract obstruction will impair pulmonary growth and development.[3-5] This is usually associated with severe renal injury and oligohydramnios. Most of these infants die in the first days of life due to respiratory insufficiency. In utero bladder decompression into the amniotic space has preserved the lives of a few such children, but the precise indications for this intervention remain undefined.

The manifestations of CBO are therefore wide ranging in severity and systems, including bladder, renal and pulmonary development and postnatal function.

PRINCIPAL QUESTIONS

Review of the clinical patterns of CBO induced bladder dysfunction raises several basic questions that must be answered in order to provide for a useful understanding of this condition:

1. What are the fundamental bladder changes induced by CBO that produce the pattern recognized as the "valve bladder"? A precise description of the cellular and tissue alterations of CBO remains to be made, and yet is essential as a basis for interpreting subsequent mechanistic observations.

2. What is the source of the variability of the severity and character of the bladder dysfunction? The wide range of clincial manifestations may be simply the result of variation in severity, but may be due to more complex interactions of fetal bladder development and function.

3. What is the nature of the interdependence of prenatal bladder function and development? Alterations in fetal bladder function affect bladder development and this in turn alters function; the specific mechanisms of this relationship need to be understood.

4. How can long-term bladder dysfunction be predicted in the young child with valves and can the basic pathophysiologic mechanisms of CBO be manipulated?

BLADDER HYPERTONICITY

Bladder hypertonicity is seen in about one-third of all patients with post-valve bladder dysfunction.[6] The clinical entity of the "valve bladder" is best described as a hypertonic bladder, usually with small capacity. (Figure 1) Rapid increases in intravesical pressure with filling are indicative of the lack of compliance that underlies this pattern of dysfunction; a defect in the storage function of the bladder. Voiding pressures may be elevated, although that observation should also prompt an evaluation for possible residual obstruction.

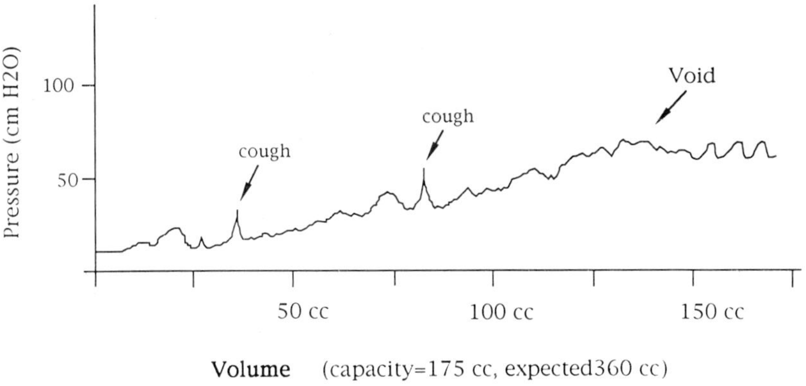

Figure 1. Cystometrogram in boy 10 years following valve resection experiencing urinary frequency and wetting. The rapid increase in bladder pressure with filling is indicative of a poorly compliant, hypertonic bladder. There was no evidence of residual obstruction. (Reprinted with permission from Peters, C.A., Bolkier, M., Bauer, S.B., et al. [6])

Bladder hypertonicity can produce upper urinary tract deterioration due to chronic high bladder pressures. Hydroureteronephrosis is the radiographic hallmark of this aspect of the hypertonic bladder. End stage renal impairment may be the ultimate outcome.

Treatment often requires chronic pharmacologic agents or bladder reconstruction, neither of which is entirely satisfactory. Both have potential problems, including inadequate long term efficacy as well as side effects.

Causes of Hypertonicity

Passive compliance of the bladder is largely determined by connective tissue properties.[7] The most important clinical factor of CBO is very likely to be due to alterations in the connective tissue compartment. Active compliance is more a product of smooth muscle activity and properties. Bladder contractility may be impaired by CBO as in the myogenic failure pattern of bladder dysfunction. Bladder instability may also be identified, and this is likely based in alterations in neuromuscular integration. Each has specific effects on bladder function with clinical consequences. In contrast to impaired compliance which produces bladder hypertonia with the potential for renal damage, impaired contractility or instability produce incontinence and infection. Clinical management of these patterns is usually more straightforward than that of the hypertonic pattern of bladder dysfunction.[8]

SMOOTH MUSCLE CHANGES WITH CONGENITAL OBSTRUCTION

Increased smooth muscle cell mass has been shown in several forms of bladder outlet obstruction in postnatal animals.[9-11] Controversy remains as to whether this represents hypertrophy or hyperplasia or a combination of the two. The underlying mechanisms of these two responses would likely be distinct and are not mutually exclusive. Up-regulation of TGF-ß in obstructed bladder has been demonstrated.[12] Similar findings have been reported in the renal pelvis of human congenitally obstructed kidneys, which may be similar to the bladder smooth musculature. It is unclear what effect this would have upon the functional properties of the smooth muscle, either contractility or compliance.

Observations on human fetal bladders with obstruction has indicated increased muscle thickness associated with increased bladder wall thickness. Smooth muscle bundles were larger than normal for gestational age, but no formal quantitation could be performed on those

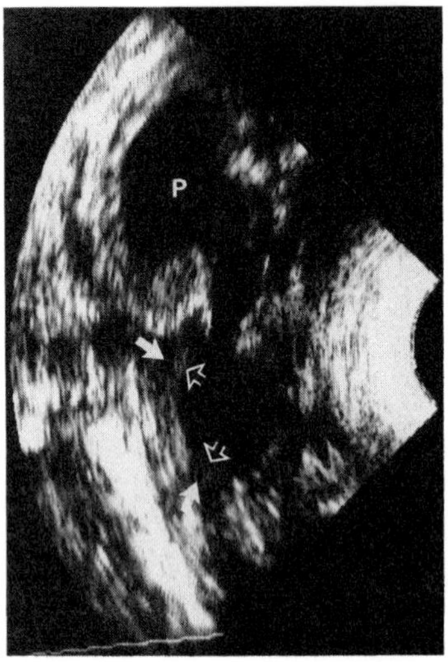

Figure 2. Prenatal ultrasound in a fetal sheep at term following 10 weeks of experimental bladder obstruction. Arrows mark the thickened bladder wall. "P" indicates the hydronephrotic renal pelvis. This appearance is similar to that seen in a human fetus with posterior urethral valves.

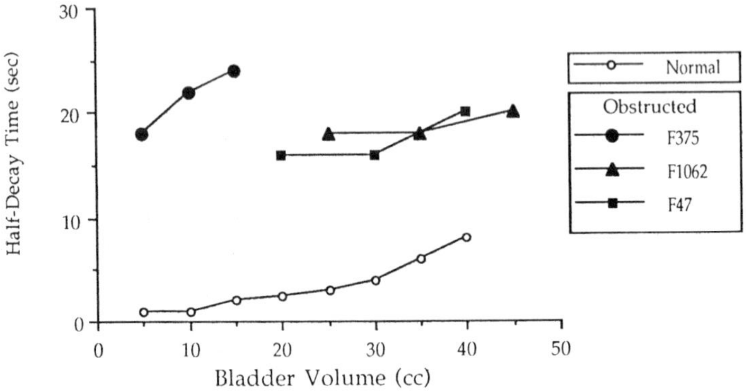

Figure 3. Graph showing estimates of stress relaxation estimates using half-decay times in normal term and three representative obstructed fetal sheep bladders at term (halothane anesthesia;135 days). Half-decay time was estimated following bolus injection of saline andtaken as half the time interval until new baseline intravesical pressure was reached (decay time). The markedly prolonged half-decay times indicate impaired compliance in the previously obstructed fetal sheep bladders relative to the normal. This parameter estimates the passive compliance characteristics of the bladder, indicative of the connective tissue elements of the bladder. *(Reprinted with permission from Peters, C.A., Vasavada, S., Dator, D., et al.* [15]*)*

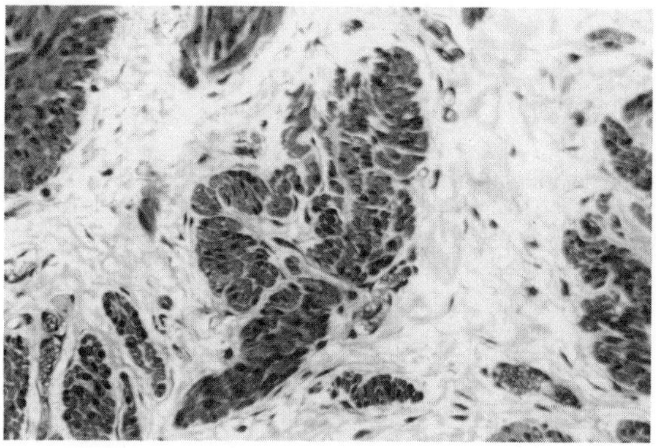

Figure 4. Histologic section of full term normal sheep bladder in which the dark staining tissues are the smooth muscle and the connective tissues are light. *(Reprinted with permission from Peters, C.A., Vasavada, S., Dator, D., et al.* [15]*)*

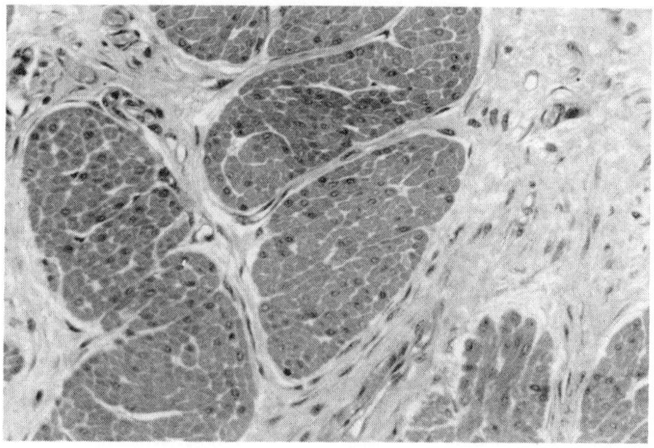

Figure 5. Histologic section of obstructed full term sheep after 10 weeks of fetal obstruction. The thickened smooth muscle bundles and denser areas of connective tissue are apparent in contrast to the normal appearance in Figure 4. Cross section of smooth muscle cells appear larger and fewer nuclei are seen relative to cytoplasmic area. (Methacrylate embedded, 3 micron section, toluidine blue stain, 400 x original magnification) *(Reprinted with permission from Peters, C.A., Vasavada, S., Dator, D., et al.* [15]*)*

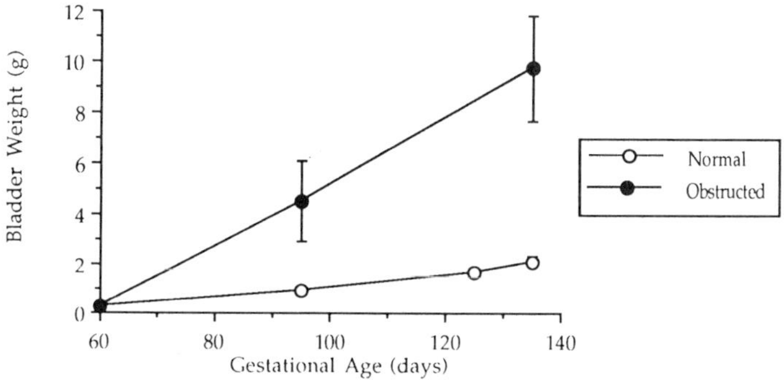

Figure 6. Graph indicating increased growth rate of obstructed fetal sheep bladder from 60 to 135 days. Values expressed as mean of more than 3 determinations (+/- 1 sem; error bars of normal bladders are too small to be resolved). *(Reprinted with permission from Peters, C.A., Vasavada, S., Dator, D., et al.* [15]*)*

specimens.[13,14] It is likely that human CBO produces a degree of smooth muscle hypertrophy and or hyperplasia.

Fetal animal experimental studies have demonstrated changes comparable to those seen in human cases of posterior urethral valves.[15] This model system involved creation of complete bladder obstruction at one-half gestation in the fetal sheep (term = 135 to 140 days). Ultrasonographic images demonstrated a thickened bladder wall as is often seen in humans (Figure 2). Functional (Figure 3) and histologic (Figures 4 & 5) evaluation of these congenitally obstructed bladders have been reported. Total bladder growth increases markedly with obstruction (Figure 6). Total bladder DNA increased with fetal obstruction, indicating a cellular hyperplasia. Total DNA increased to a lesser extent than total mass, and the DNA-to-protein ratio decreased with obstruction, which together indicate cellular hypertrophy. This observation is corroborated by morphometric data which indicate increased smooth muscle cell size (Figure 7). The fetal bladder responds to obstruction with both hypertrophy and hyperplasia. In a similar model system, flow cytometric analysis did not demonstrate a non-diploid cell population; cellular polyploidy is therefore not a likely explanation for the increased total DNA, as has been shown in postnatal non-bladder smooth muscle systems.[16]

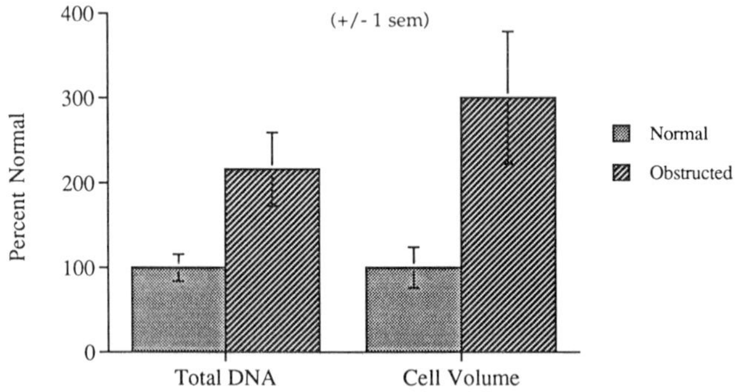

Figure 7. Graph indicating the response of the fetal sheep bladder to obstruction in terms of total bladder DNA and detrusor smooth muscle cell volume. Normal expressed as 100% and obstructed values represent mean of more than three animals (+/- 1 sem). Increased total DNA indicates tissue hyperplasia (more cells), while the increase in cell volume suggests tissue hypertrophy (larger cells). Both processes appear to be part of the fetal sheep bladder response to fetal obstruction. *(Reprinted with permission from Peters, C.A., Vasavada, S., Dator, D., et al.* [15]*)*

Smooth muscle innervation

In a model of late gestation moderate partial bladder obstruction, Karim, et al. reported increased total innervation of the bladder (as assessed by Choline Acetyl Transferase levels) in proportion to increases in bladder weight and cell number.[16] This suggests maintenance of normal levels of innervation, in contrast to observations in postnatal bladder obstruction in which decreased innervation is suggested.

Peters, et al.,[15] in a model of severe fetal bladder obstruction demonstrated an increase in muscarinic cholinergic receptors (MCR) per smooth muscle cell. When normalized to the contractile protein myosin, however, there was no increase (Figure 8). The functional significance of these alterations remains to be determined.

These alterations might also contribute to chronic hypertonicity. This is likely to be a component of the clinical picture, since some degree of hypertonicity is responsive to anticholinergic pharmacotherapy. These observations might also indicate a basis for bladder instability after CBO. Hyperreflexia is a frequent observation, occurring in about 25 to 30% of boys with bladder dysfunction after PUV.[6]

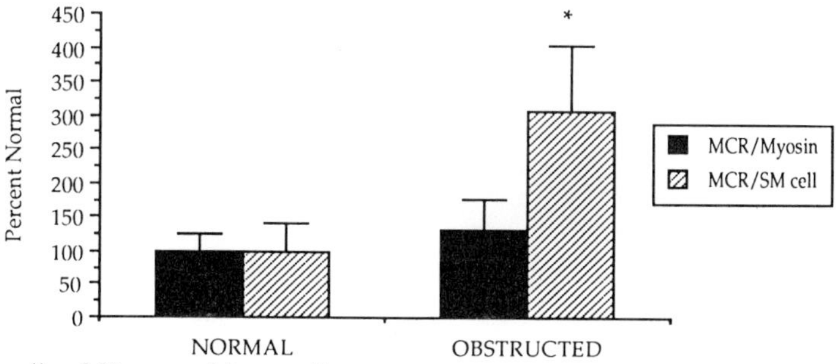

(* $p<0.05$ compared to normal)

Figure 8. Graph showing changes in muscarinic cholinergic receptors (MCR) in bladder smooth muscle cells following prenatal obstruction in the sheep. When normalized to each smooth muscle cell (estimated morphometrically), MCR increased to 300% of normal, yet when normalized to the contractile protein myosin, MCR concentration remained unchanged from normal. The functional significance of these alterations remains to be shown. Congenital obstruction produces altered patterns of neurotransmitter receptor protein distribution. *(Reprinted with permission from Peters, C.A., Vasavada, S., Dator, D., et al.* [15]*)*

Smooth muscle contractile proteins

Congenital obstruction produces alterations in the content of myosin, but not actin (Figure 9). This alteration in the balance of actin and myosin is likely to have functional consequences, which remain to be identified. An indication of the effect of CBO on developmental regulation of functional bladder proteins is demonstrated in the response of the myosin heavy chain isoforms to fetal obstruction. Two principle isoforms of the myosin heavy chain have been identified (SM1 and SM2). These proteins are the myosin-ATPase enzyme that catalyzes the contractile interaction of myosin and actin. Postnatal obstruction has been shown to alter the distribution of these isoforms.[9] The relative amount of these isoforms undergoes a developmentally regulated shift, with increasing relative amounts of SM1 during late gestation. Fetal obstruction accelerates this [15] (Figure 10). The functional consequences of these shifts are unknown, but they serve to demonstrate the effect of altered bladder function on bladder developmental regulation. These changes may contribute to myogenic failure as one of the clinical patterns of bladder dysfunction associated with congenital obstruction.

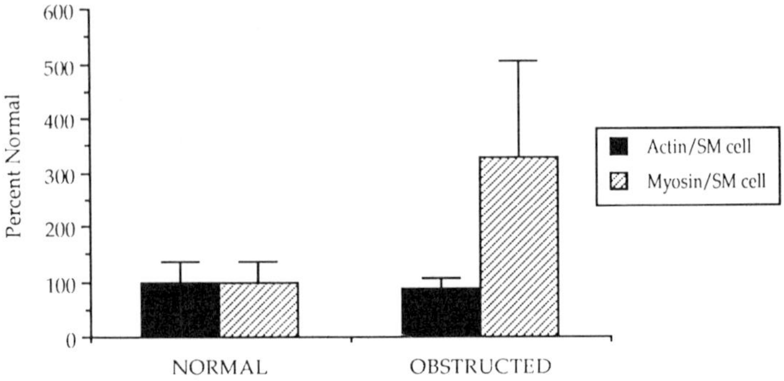

Figure 9. Graph showing changes in the amount of actin and myosin per smooth muscle cell in response to prenatal bladder obstruction in the fetal sheep. Normal equals 100 percent. Values are expressed as mean at term (135 days) of more than 3 animals (+/- 1 sem). Actin per cell does not increase, while myosin increases significantly. *(Reprinted with permission from Peters, C.A., Vasavada, S., Dator, D., et al.*[15]*)*

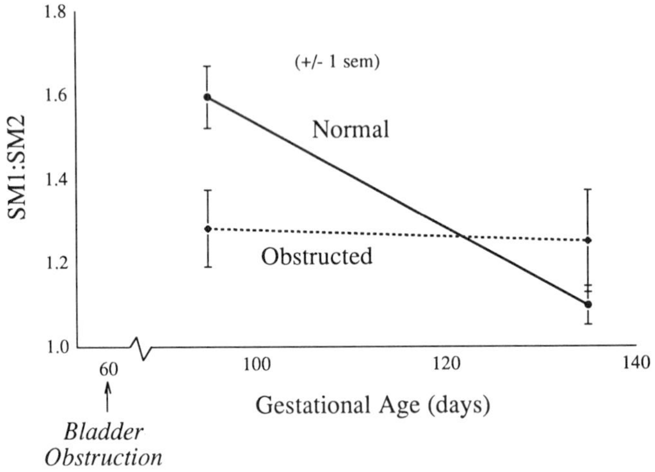

Figure 10. Changes in the contractile protein myosin heavy chain isoform ratios are shown in response to fetal bladder obstruction. The normal line indicates a developmental shift of the isoform ratios with maturation. Fetal obstruction produces a transient acceleration of that developmental shift. The functional significance of these observations remains to be determined, but indicate altered developmental regulation of contractile proteins in response to fetal obstruction. *(Reprinted with permission from Peters, C.A., Vasavada, S., Dator, D., et al.* [15]*)*

CONNECTIVE TISSUE AND CONGENITAL BLADDER OBSTRUCTION

Connective tissue mass

There is little information regarding quantitative assessment of bladder connective tissue in the setting of CBO. Collagen concentration was reported to be increased in one case of a young child with PUV.[17] Kogan and co-workers observed increases in connective tissues in human fetuses with obstruction. These changes were in conjunction with increased smooth muscle.[14] Experimental preparations in the sheep indicate increased connective tissue mass, about in proportion with smooth muscle increases.[15] (Figure 11)

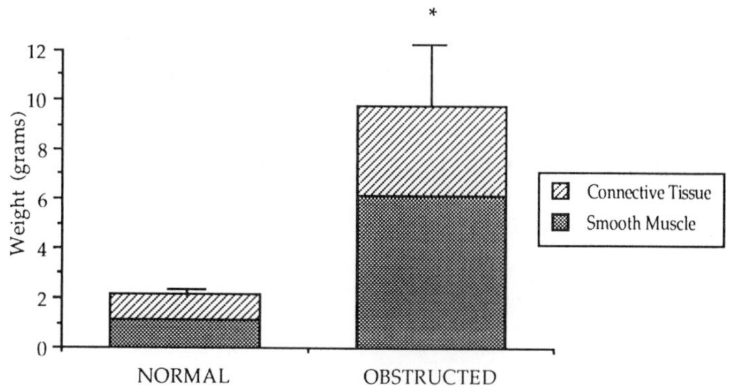

*(*significant difference from normal, p<0.002)*

Figure 11. Graph shows changes in the amount of smooth muscle and connective tissue fractions estimated morphometrically in fetal sheep bladders following prenatal obstruction for 10 weeks (60 days to 135 days). Both smooth muscle and connective tissue fractions increased in proportion. Expressed as mean +/- 1 sem. *(Reprinted with permission from Peters, C.A., Vasavada, S., Dator, D., et al.* [15]*)*

Bladder Collagen

In most postnatal experimental models of obstruction, increases in collagen are demonstrated and constitute the largest component of the increase in connective tissue mass.[18,19] Collagens type I and III are significant components of the fetal bladder in humans as well as in the bovine fetus. Type IV, or basement membrane, collagen has been identified in the human bladder and is presumably in the fetal bladder.

The spatial distribution of connective tissue elements is likely to influence functional characteristics of the tissue and may be altered with CBO. Connective tissue infiltration into the smooth muscle has been demonstrated in human fetuses with obstruction[14], and with myelodysplasia.[1] Connective tissue infiltration into the smooth muscle bundles has been shown in the detrusor of post-natal, non-compliant, neurogenic bladders, particularly with increased type III within the muscle fascicles.[20]

These patterns have not been confirmed in the fetal sheep models, and are best performed using immunohistochemical techniques to specifically locate collagen fibers, and define their position within the detrusor. Alterations in spatial distribution suggest altered regulation of collagen remodeling.

Collagen Subtypes

Collagen subtype distribution is another indication of altered connective tissue remodeling and may have significant functional consequences due to the specific physical characteristics of the collagen subtypes.

The ratio of collagen type III:I appears to decrease with increasing gestational age in the bovine bladder and this is associated with increasing bladder compliance[21,22]. Type III collagen may be "fetal" collagen with physical characteristics different from type I collagen. Alterations in this ratio due to obstruction may be the basis for altered bladder compliance in congenital obstruction. It is unclear at what level these developmental shifts are regulated (i.e.: gene expression, post-transcriptional or translational processing, procollagen hydroxylation, or collagen breakdown); it is also unknown at what level obstruction may have an effect.

Type I collagen is considered to be a "stiffer" collagen, and therefore changes in relative content may alter the compliance characteristics of the bladder. This may not be a correct assumption in that the degree of cross-linking, and other conformational characteristics contribute to the physical properties of the collagens.[23] An example of the relationship of between variations in collagen structure and mechanical properties may be found in a study of mammalian pericardia. Animals with thin, stiff pericardia were found to have greater relative amounts of type III collagen with a higher degree of cross-linking than animals with thicker pericardia. They had equivalent viscoelasticity.[24]

Using human fetal bladder tissues from both normal and obstructed bladders, Kim and Kogan examined the constituents of the bladder connective tissues.[14] Thick and thin collagen filaments correspond to collagen I (thick) and III (thin) and may be distinguished from each other using birefringent microscopy. This technique has been used to study rabbit hearts. Quantitation is difficult, but obvious alterations in the normal developmental shift were seen in the obstructed fetal bladders, indicating an increase in type I collagen as a factor in the response of the fetal bladder to obstruction.

Contrary findings were reported by Ewalt and associates in a heterogeneous group of patients with non-compliant neurogenic bladders.[20] The basis for these patients' bladder dysfunction was largely myelodysplasia. It was observed that the ratio of type III to type I collagen increases in the detrusor. These studies also made the observation that the lamina propria normally contained large amounts of collagens I and III, suggesting that the lamina propria determines compliance and provides the tensile strength of the bladder. Collagen deposition within the lamina propria was unchanged in the non-compliant bladders. It was progressively increased (type III) in the muscularis, in a radial fashion with increasing severity of the non-compliant state.

Elastin

The specific role of elastin in bladder function remains to be defined, yet changes in elastin concentration and deposition have been demonstrated by several investigators.[14,17,20] It is likely that concentration and patterns of deposition of elastin fibers contribute to the functional characteristics of the bladder, particularly with regard to compliance. The influence of elastin on the functional characteristics of the non-compliant bladder has not been studied.

Procollagen Gene Expression

The mechanisms responsible for the marked increases in total connective tissue in the congenitally obstructed bladder are important to understand for three reasons. In order to evaluate and prognosticate as to the long-term functional consequences of CBO, an understanding of these mechanisms is essential. This is particularly true to identify those patients with the potential for bladder dysfunction due to prior obstruction. More specific therapeutic strategies to prevent or correct bladder dysfunction due to congenital obstruction would be of significant benefit in terms of bladder and renal function. Finally, an understanding of CBO and its effects on bladder development and function would likely be of general applicability to other forms of bladder dysfunction and may be relevant to fibrotic diseases in other organs.

Increased levels of collagen synthesis may be the basis for increased content, and this should be reflected in the relative levels of gene expression. In a fetal sheep model we have examined the level of procollagen III gene expression using Northern blot analysis of mRNA. As shown in figure 12, fetal bladder obstruction does not appear to produce a greater level of procollagen III gene expression. The level of gene activity, represented by the intensity of the hybridization signals on the autoradiogram, is comparable between obstructed and normals at both 5 and 10 weeks following in utero obstruction.[25] At those time points, the total amount of connective tissue has increased more than 3-fold above normal.

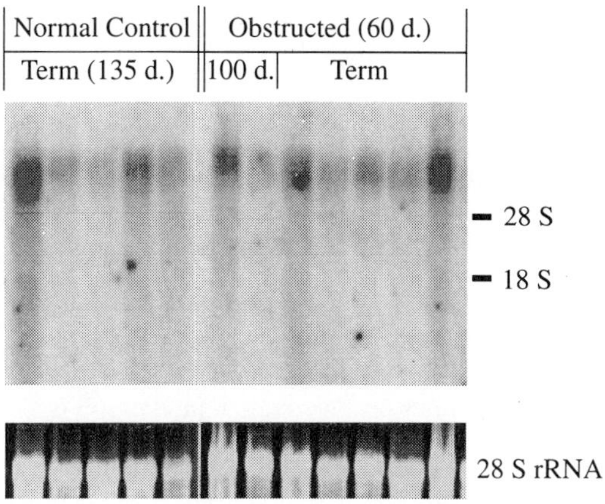

Figure 12. Northern blot analysis of mRNA for procollagen III in fetal sheep bladders. Obstruction of the bladder at 60 days does not produce an increased level of procollagen III mRNA signal at 95 or 135 days (5 and 10 weeks, respectively), despite an increase in connective tissue content.

This would suggest that other mechanisms are the basis for the remarkable alterations in the connective tissue compartment of the developing bladder. Collagen synthesis is complex, with several potential levels of regulation, including post-transcriptional or post-translational processing, processing of pro-collagen to collagen, and secretion and incorporation of collagen fibrils into the extracellular matrix.

Collagen Synthesis and Breakdown

An important level of regulation of collagen content is the balance between synthesis and breakdown.[26] A net increase in collagen would occur if breakdown activity were decreased while synthesis remained constant.

Regulation of collagen degradation has been examined in other disease states where it has been suggested to play an important role. These include cardiac hypertrophy and fibrosis[27], pulmonary fibrosis[28] and liver fibrosis[29]. Rheumatoid arthritis has been investigated from the view of the balance of synthesis and degradation, although in that disease, there is a net increase in degradation.[30] Collagen turnover rates have also been examined in the context of cardiac hypertrophy with outflow resistance, or in hyperdynamic states.[31] Total increased collagen turnover, with possible decreased breakdown has been suggested in several model systems. Collagen breakdown systems have been investigated in embryogenesis and tumor invasion.[32,33]

The principle mechanism of collagen degradation is by way of matrix metalloproteinases (MMPs).[32,34,35] These enzymes are synthesized locally, secreted as zymogens, and require activation as well as a metallic co-factor such as zinc. They may be inhibited by chelating agents. Their activity in vitro is specific for particular types of extracellular matrix components. The primary structural collagens (interstitial collagen), types I and III, are degraded preferentially by MMP 1 (type I collagenase, interstitial collagenase), a 52 kDa protein. Other important MMPs include MMP 2 and 9 (72 and 92 kDa gelatinases), specific for collagen type IV (basement membrane collagen).

Reduced activity of MMP 1 may be a possible explanation for the increased amounts of collagen in the congenitally obstructed bladder. Absolute levels of the MMP protein, however, may not represent activity levels. These enzymes require activation for in vivo activity and regulation of activation state is a likely mechanism of control of the action of MMPs.[35] It is therefore necessary to examine the MMPs at several levels to determine their possible role in bladder connective tissue remodeling. MMP gene expression, protein level, and most importantly, enzyme activity levels in experimental tissues are all important parameters to be studied.

Determination of MMP activity is best made using substrate gel electrophoresis (zymography).[36] Tissue extracts are electrophoresed on an agarose gel impregnated with denatured collagen (gelatin). Each MMP will migrate a specific distance determined by its molecular weight and at that point, active MMP will degrade the collagen impregnated in the gel. Intact collagen is stained with Coomassie blue and any areas with collagen degradation show as clear bands and indicate presence of MMP activity. Their relative position permits identification of the specific MMP. Parallel lanes on the gel include extracts incubated with an activating agent (APMA= aminophenylmercuric acetate). If the location of the MMP band seen on the unactivated lane moves downward with incubation with APMA, this suggests that an inactive MMP was present and has been activated. (activation is a cleavage process).[34]

An alternative means of assessing interstitial collagenase, MMP 1, activity is a radiometric collagenase assay. Tissue extracts are incubated with a known quantity of radiolabelled collagen type I. Release of radioactivity indicates the rate of collagenolytic activity when compared to a baseline level and to a standard bacterial collagenase which degrades collagen at a rate much higher than mammalian tissues. A unit of collagenase activity is defined as producing 10% cleavage of collagen in 2.5 hours at 37°C.

Regulation of Collagen Degradation

MMP activity is regulated at several levels as noted above. An important mechanism of control is with a group of enzymes that inhibit the activity of MMPs.[34,35] These are the Tissue Inhibitors of Metallo Proteinases (TIMPs), a family of proteins that inhibit the matrix degrading activity of MMPs. At least two TIMPs are known[37] (TIMP-1: 28.5 kDa, TIMP-2: 20 kDa), although a third and likely fourth have been reported. They demonstrate less specificity against the MMPs in vitro and are thought to act on the activated form of MMPs. It has been suggested that the TIMPs also bind the inactive MMPs, possibly with more specificity, and may therefore regulate activation.[38]

Assessment of activity is performed in a collagenase radiometric assay, described above. A standard amount of mammalian collagenase activity is added to the radiolabeled collagen with and without the experimental tissue extract. The reduction in activity due to the experimental tissue is the level of inhibitory activity present and is expressed as standard inhibitory units. One inhibitory unit (IC_{50}) is defined as the amount of protein required to inhibit one unit of corneal collagenase by 50%.

Imbalance of MMP and TIMP activity has been implicated in articular degeneration associated with rheumatoid arthritis.[30,34]

Collagenase Activity in Bladder Obstruction

In the congenitally obstructed sheep bladder, preliminary data has suggested that there are alterations in the activity levels of MMP 1 and the TIMPs (Peters, et al., unpublished data). If these observations are confirmed, this would indicate a potentially useful avenue of investigation to develop an understanding of the mechanisms of bladder dysfunction in the congenitally obstructed bladder. This basic mechanism may also have potential relevance to other fibrotic diseases characterized by altered remodeling and increased deposition of connective tissues.

Questions

If the basic hypothesis proves to be valid, and increased inhibition of MMPs is a factor leading to increased bladder connective tissue deposition, it will be important to consider several further questions.

1. What regulates collagenase and TIMP activity in the developing bladder?
2. Is this observation relevant to the human fetal obstructed bladder?
3. Is this observation relevant to other forms of bladder disease and dysfunction?

The answers to these questions will provide further insight into the potential for more specific therapeutic strategies in dealing with various forms of bladder dysfunction.

SUMMARY

The clinical significance of congenital bladder obstruction is evident to those who deal with children suffering from severe bladder dysfunction and renal failure. It is a lifelong effect often requiring complex surgical reconstruction and renal transplantation. CBO also represents an important pathophysiologic process with relevance to other organ systems in which hypertrophic and fibrotic processes occur. CBO occurs in a dynamic and developing organ in which function and development are interdependent. Alterations of one will induce alterations in the other which then affect the first. Multiple levels of organization are involved and relevant, from the cell to the whole bladder as a part of the organism. Functional and developmental changes have been demonstrated in models of CBO. Much attention has been focused on the connective tissue compartment of the bladder due to its important role in determining bladder compliance characteristics. Preliminary findings suggest that connective tissue remodeling may be affected by alterations in the balance of the regulatory mechanisms of collagen degradation. Further studies should be able to define these alterations and may permit specific modulation of these mechanisms.

ACKNOWLEDGEMENTS

This work has been supported in part by NIDDK Grant RO1 DK47582, and by the efforts of Dr. Michael R. Freeman and Dr. Marsha A. Moses.

REFERENCES

1. Shapiro, E., M.J. Becich, E. Perlman, et al., Bladder wall abnormalities in myelodysplastic bladders: A computer assisted morphometric analysis, *J. Urol.* 145:1024-1029 (1991).
2. Peters, C.A., Congenital bladder obstruction, *Problems in Urology* 8:1-15 (1994).
3. Docimo, S.G., T. Luetic, R.K. Crone, et al., Pulmonary development in the fetal lamb with severe bladder outlet obstruction and oligohydramnios: a morphometric study, *J Urol* (1989).
4. Peters, C.A., L.M. Reid, S. Docimo, et al., The role of the kidney in lung growth and maturation in the setting of obstructive uropathy and oligohydramnios, *J Urol* (1991).
5. Peters, C.A., S.G. Docimo, T. Luetic, et al., Effect of in utero vesicostomy on pulmonary hypoplasia in the fetal lamb with bladder outlet obstruction and oligohydramnios: a morphometric analysis, *J Urol* 146(4):1178-83 (1991).
6. Peters, C.A., M. Bolkier, S.B. Bauer, et al., The urodynamic consequences of posterior urethral valves, *J. Urol.* 144:122-126 (1990).
7. Susset, J.G. and C.H. Regnier, Viscoelastic properties of bladder strips: standardization of a technique, *Invest Urol* 18(8):445-50 (1981).
8. Peters, C.A. and S.B. Bauer, Evaluation and management of urinary incontinence after surgery for posterior urethral valves, *Urol. Clinics North Amer.* 17:379-387 (1990).
9. Malmquist, U., A. Arner, and B. Uvelius, Contractile and cytoskeletal proteins in smooth muscle during hyerptrophy and its reversal, *Am J Physiol* 260(5 Pt 1):C1085-1093 (1991).
10. Mattiasson, A. and B. Uvelius, Changes in contractile properties in hypertrophic rat urinary bladder, *J Urol* 128(6):1340-2 (1982).
11. Arner, A., U. Malmqvist, and B. Uvelius, Metabolism and force in hypertrophic smooth muscle from rat urinary bladder, *Am J Physiol* 258:C923-C932 (1990).
12. Buttyan, R., B.Z. Jacobs, J.G. Blaivas, et al., Early molecular response to rabbit bladder outlet obstruction, *Neurourol. Urodyn.* 11:225 (1992).
13. Workman, S.J. and B.A. Kogan, Fetal bladder histology in posterior urethral valves and the prune belly syndrome, *J Urol* 144:337 (1990).
14. Kim, K.M., B.A. Kogan, C.A. Massad, et al., Collagen and elastin in the obstructed fetal bladder, *J Urol* 146:528-531 (1991).

15. Peters, C.A., S. Vasavada, D. Dator, et al., The effect of obstruction on the developing bladder, *J. Urol.* 148:491-6 (1992).
16. Karim, O.M.A., M. Cendron, J.L. Mostwin, et al., Developmental alterations in the fetal lamb bladder subjected to partial urethral obstruction in utero, *J. Urol.* 150:1060-1063 (1993).
17. Cortivo, R., F. Pagano, G. Passerini, et al., Elastin and collagen in the normal and obstructed urinary bladder, *Br J Urol* 53(2):134-7 (1981).
18. Uvelius, B. and A. Mattiasson, Collagen content in the rat urinary bladder subjected to infravesical outflow obstruction, *J Urol* 132(3):587-90 (1984).
19. Uvelius, B., P. Lindner, and A. Mattiasson, Collagen content in the rat urinary bladder following removal of an experimental infravesical outlet obstruction, *Urol Int* 47(4):245-9 (1991).
20. Ewalt, D.H., P.S. Howard, B. Blyth, et al., Is lamina propria matrix responsible for normal bladder compliance?, *J Urol* 148:544-549 (1992).
21. Baskin, L., S. Constantinescu, J.W. Duckett Jr., et al., Type III collagen decreases in normal fetal bladder development, *Amer. Acad. Pediatr., Urology Section* Abstract No. 45(1993).
22. Coplen, D.E., E.J. Macarak, and R.M. Levin, Developmental changes in normal fetal bovine whole bladder physiology, *J. Urol.* 151:1391-1395 (1994).
23. Martin, B.J., B.S. Leppien, and W.E. Stehbens, Changes in collagen fibril morphology in experimental aneurysms and arteriovenous fistulae in sheep, *Int J Exp Pathol* 74(3):267-74 (1993).
24. Naimark, W.A., J.M. Lee, H. Limeback, et al., Correlation of structure and viscoelastic properties in the pericardia of four mammalian species, *Am J Physiol* (1992).
25. Peters, C.A., M.C. Carr, M.R. Freeman, et al. Collagen gene expression in the fetal bladder is altered by in utero bladder functional status. in *American Academy of Pediatrics, Urology Section*. 1992. San Francisco, CA.
26. Laurent, G., Dynamic state of collagen: pathways of collagen degradation in vivo and their possible role in regulation of collagen mass, *Am. J. Physiol.* 252:C1-C9 (1987).
27. Weber, K.T., C.G. Brilla, and J.S. Janicki, Myocardial fibrosis: functional significance and regulatory factors, *Cardiovasc Res* 27(3):341-8 (1993).
28. Togari, H., Y. Hashimoto, Y. Wada, et al., Increased type III/I collagen and alpha 1(I)/alpha 2(I) chain in a bronchopulmonary dysplastic lung, *Acta Paediatr Jpn* 35(2):101-7 (1993).
29. Milani, S., H. Herbst, D. Schuppan, et al., Differenital expression of matrix-metalloproteinase-1 and -2 genes in normal and fibrotic human liver, *Am. J. Pathol.* 144:528-537 (1994).
30. Dean, D.D., J. Martel-Pelletier, J.-P. Pelletier, et al., Evidence for metalloproteinase and metalloproteinase inhibitor imbalance in human osteoarthritic cartilage, *J. Clin. Invest.* 84:678-685 (1989).
31. Karim, M.A., A.G. Ferguson, B.T. Wakim, et al., In vivo collagen turnover during development of thyroxine-induced left ventricular hypertrophy, *Am. J. Physiol.* 260:C316-C326 (1991).
32. Matrisian, L.M., Metalloproteinases and their inhibitors in matrix remodeling, *Trends Genet.* 6:121-125 (1990).
33. Moses, M.A. and Y. Shing, Comparison of collagenase and its inhibitor(s) in cartilage and chondrosarcoma, *J. Cell. Biol.* 115:141 (1992).
34. Woessner, J., J.F., Matrix metalloproteinases and their inhibitors in connective tissue remodeling, *FASEB J.* 5:2145-2154 (1991).
35. Murphy, G. and A.J. Docherty, The matrix metalloproteinases and their inhibitors, *Am J Respir Cell Mol Biol* 7(2):120-5 (1992).
36. Johnson-Wint, B., A quantitative collagen film collagenase assay for large numbers of samples, *Anal. Biochem.* 104:175-181 (1980).
37. Stetler-Stevenson, W.G., H.C. Krutzsch, and L.A. Liotta, Tissue inhibitor of metalloproteinase (TIMP-2). A new member of the metalloproteinase inhibitor family, *J .Biol. Chem.* 264:17374-17378 (1989).
38. Stetler-Stevenson, W.G., P.D. Brown, M. Onisto, et al., Tissue inhibitor of metalloproteinases-2 (TIMP-2) mRNA expression in tumor cell lines and human tumor tissues, *J Biol Chem* 265(23):13933-8 (1990).

DEVELOPMENTAL BLADDER PHYSIOLOGY - DISCUSSION

Participant - I would like Dr Baskin to tell me how old the cows were that were utilized in these studies and how these specimens were procured. Are they real old cows, or medium old cows?

Larry Baskin - These animals are sent to the slaughterhouse once they have lost their economic productivity in terms of dairy farming. They are probably 4 to 5 years old. There are steers sent in that are clearly much younger.

Participant - I have a question about your definition of compliance. You started off by defining compliance as volume divided by pressure, and then you talked about Young's elastic modulus which is quite different. When you showed your data of compliance over time of development your graph did not have the units presented, so which units were used?

Larry Baskin - We used Young's elastic modulus as a proportion for compliance; in truth it is the inverse of compliance. We did the actual studies in two ways; one was by injecting air into the chamber and the other was by injecting water to achieve the desired deformation. Regardless of which method was used the results correlated nicely.

Participant - The other question I have is about the contribution of bladder wall thickness to these measurements. How did you account for its changes during development and during the course of the deformation?

Larry Baskin - That is a good question. We measured bladder wall thickness with a micrometer at resting pressures, and the measurement was highly reproducible. Obviously the bladders from first trimester animals are thinner than those from later trimesters. It is true that as you stretch the bladder, the thickness will change. However these experiments were done with very slight deformations, so that such a change in thickness is likely to be quite minor. I might add that what is nice about this system is that our findings have also been validated by Doug Coplen using a whole bladder model and by Steve Zderic who showed very similar patterns when testing was done using muscle strip methods.

William C. DeGroat - I would like to ask Larry Baskin what he thinks will happen if the bladder were not to distend with urine, and compliance were not to change, what would you expect would happen to bladder weight or bladder mass? Would you expect these parameters to change at all?

Larry Baskin - Clinically we have an excellent analogy of this in the form of bilateral ectopic ureters. This is typically seen in little girls where the ureters exit distal to the external sphincter and bladder neck, and the bladder never fills. These girls have tiny, stiff, non-compliant bladders, and suffer from severe urinary incontinence. These findings are consistent with our data which suggest that the bladder must do some work in order to gain capacity and compliance. It is very important to have a functional sphincter that will provide some outlet resistance to let the bladder do some storage "work" and allow for growth.

William C. DeGroat - So it would appear that in this case, function is really directing development? Do we have questions for Dr Baskin or Dr Kogan?

William C. DeGroat - I have a question for Barry Kogan, for I too am interested in the effects of nitric oxide. I was quite surprised to see the changes that you described in which the volume threshold was altered and also the excitability of these neural pathways was altered with the appearance of unstable bladder contractions. Do you have a possible explanation for both of those occurring simultaneously?

Barry Kogan - I will admit to having been a bit brainwashed by Dr Schmidt and Dr Tanagho at UCSF, but I really do think that sphincteric function alters bladder function in an in vivo model. I think that what we are seeing is changes in sphincteric function that are often times reflected in what we see in the detrusor. So I am not certain that these are really direct effects on the bladder itself, but rather may reflect direct changes on the sphincter itself.

William C. DeGroat - So the change in volume threshold is simply the sphincter not relaxing in this case, so that the bladder has to fill up further before it reaches its threshold?

Barry Kogan - Yes, that is what we would infer from this data. There are muscle strip studies done in vitro that don't reproduce what we have shown in vivo using the fetal lamb model with real time measurements. That is another reason why I think we may well be dealing with an indirect effect with some kind of signaling between the sphincter and the bladder body.

William C. DeGroat - There is also some evidence that nitric oxide may be present in afferent pathways in the lower urinary tract. This leads one to wonder it there might be a modulation of the lower urinary tract by alterations in these afferent pathways?

Barry Kogan - Certainly this is an area that needs to be further explored. Also there may well be effects of nitric oxide within the central nervous system itself that contribute to this observed physiology.

William C. DeGroat - Do you know if the inhibitors of nitric oxide synthetase that you used in these experiments actually get into the central nervous system?

Barry Kogan - That is a very good question, and I don't have an answer for it at this time.

Participant - Have you or anyone else looked at the percentage of collagen and elastin in children with posterior urethral valves, and seen how this changes after some kind of intervention? One might be able to prognose the future function of such bladders if one had access to this kind of information. I know that Craig Peters has shown in a clinical series that many children with valve bladders have a great deal of difficulty voiding in later years. My question is does the percentages of collagen and elastin correlate with this voiding dysfunction?

Barry Kogan - That is a terrific question and in fact we tried to look at this, Ellen Shapiro and I and a group in Toronto tried to collect the series on children who were being operated on for augmentations. I think that there is some relationship but the real problem here is that it is very, very difficult to get controls. The control bladders to use to know what the normal ratios should be are almost impossible to find. We looked at reimplant specimens, but a number of people have shown that kids with reflux may have abnormal bladders, as well. So those really don't make controls; maybe occasionally, a megaureter operation where the bladder might not be affected. But then again, we hardly ever operate on those anymore, so I think is a real weakness, and needs to be looked at.

William C. DeGroat - Dr. Kogan, before you go away, one question about the reflex distension which you saw in utero. It really is surprising that this emptying phenomenon is so efficient in utero because we think that the development of the brain and spinal pathways takes place much more slowly and develops later. You almost might expect that in these fetal bladders the behavior is more like a spinal cord injury patient in which the bladder is removed from central control and becomes inefficient. Yet your data strongly support the notion that fetal voiding is highly efficient. Do we know anything about the connections between the brain and spinal cord in the fetus in either the sheep or rat, and if so when are these connections actually made? Does the brain have any control over these pathways, or is this purely a spinal reflex pathway?

Barry Kogan - I will make two points here. One is that we only looked at the third trimester physiology, and hence our sampling are only from late in gestation. I can't answer your question directly, but there is a study that I believe we have cited in our contribution to the course syllabus looking at auditory stimulation of the fetal lamb and there was indeed evidence for a functioning central nervous system in this model.

William C. DeGroat - Is there anyone that has any data on the human fetus and whether there is evidence that these patients might have damaged spinal cords, and whether they would have normal voiding function. We know that in the human, spinal cord injury clearly produces a bladder and sphincter dyssinergia. It would be interesting to know whether that type of dyscoordination occurs if the injury occurs early on in development.

Barry Kogan - Well the only studies I have shown were the prenatal ultrasounds that where they did indeed look at fetuses with spina bifida and showed definitely abnormal patterns of bladder emptying with all varieties similar to what you might see postnatally. So there were bladders that didn't fill up at all presumably from very weak sphincteric function and bladders that had multiple, multiple contractions, and bladders that just maintained what appeared to be a type of overflow incontinence.

William C. DeGroat - Are there any other questions for Dr Kogan? He is heading back to the west coast, so it is your last chance.

William C. DeGroat - My next question is for Steve Zderic, and I would like to know if he thinks that calcium sequestration, or storage in the calciosomes, may alter compliance or affect the bladder's ability to undergo relaxation. How would you reconcile this to Barry Kogan's work with his demonstration of increased sensitivity to beta agonists?

Stephen Zderic - These are the areas that our research group will be focusing on in the next months. I think that is one of the things that is important, is to point out the model s that we are studying are very different. For example, when we work on the problem of outlet obstruction, there is an initial phase of decompensation followed by a period in which the muscle undergoes compensation (compensatory hypertrophy). Sometimes when we go to the lab to study these problems to study these problems, we must remember that our findings will be dependent upon which phase of the model we are in. For example, we talk about the bladder outlet obstruction following partial urethral occlusion, it may be that calcium sequestration systems are far more affected in this initial stun phase (decompensated) than they are later on when the bladder has regained much of its function (compensated). In other words, perhaps the model in which we are talking about which is modeled on acute induction of aortic stenosis, makes it much less applicable to long term obstruction as is seen in human pathologic conditions. Still, my feeling is that even the crude methods of partial outlet obstruction have much to teach us so long as we accept that the "obstruction" that is produced is heterogenous despite our best efforts. Some animals will be more obstructed, others less so. This makes the study of each bladder by physiologic methods essential so that one can understand what the bladders contractile capabilities are before embarking on molecular studies. You can't simply study all bladders at 7, 14, and 28 days because what you have at each point is biological variability, and that is especially true at 14 and 28 days. I have not yet carried out experiments on that specifically focus on calcium sequestration and relaxation, though these are obviously on our agenda.

William C. DeGroat - How would the bladder smooth muscle get rid of the calcium? Also, there is evidence that with deformations and natural forces, which can increase the voltage operated calcium channels. How do you propose to get so much less of the slow type calcium channels and what other channels might be involved? What additional calcium binding proteins are you looking at in your survey of calcium homeostasis?

Stephen Zderic - We have to look at the other calcium binding proteins such as calsequestrin. We are currently trying to get the antibody to calsequestrin to probe for the expression of this protein in the microsomal membrane fractions, but we have experienced technical difficulty in both rabbit and bovine systems. If the calciosomes are regulated during development, this protein's expression should also change. How does calcium get out the cell? I think that is especially interesting in light of Doug Coplen's experiments in fetal bovine bladders in which he demonstrated that tone in the first trimester bladders is much more dependent upon the presence of extracellular calcium. Compliance in the first and second trimester in fetal bovine bladders appear to have a much greater sensitivity to calcium, so I do think that calcium is getting in there, and results in the generation of tone. If this is not to happen, it must be properly sequestered. I also think that this is a mechanism that is more important in the early trimesters, and that with maturation, the fetal bladder gains the ability to control its intracellular calcium to a much greater degree. Thus the development of proper ion pumps to control cytosolic calcium appears to be an important part of the normal developmental process. I think the way we are going to have to answer many of these questions is by doing a number of studies whereby we stretch and simultaneously record and measure the intracelluar free calcium using FURA technology. I think the degree of experimental sophistication will continue to increase in order to provide the best answers to these important questions.

William C. DeGroat - I have a question about the possibility of different transmitters that might be involved in a young animal versus an older animal. The effects that you saw with aging it seemed to me were much more dramatic with the use of purinergic transmitters such as ATP than they were for acetylcholine. I think in the rat's bladder we have found that nifedipine and other L-type calcium channel blockers were much more effective at blocking the effects of purinergic contraction. So, is it possible in addition to these intracellular post-junctional effects that you described, that there also would be a change in the type of transmitter that is being released, and that purinergic transmission is more important in the young animal and less important in the older animal?

Stephen Zderic - It has been our impression that the purinergic and non-adrenergic non-cholinergic transmitters are more important in the neonatal group. We also did some work on the effects of histamine, serotonin and VIP, and that also appeared to generate greater forces in the newborn group. This may in part be related to the mechanisms by which we induce contraction because most of these agents act to increase plasma membrane influx. When one uses cholinergic stimuli, one also get a release of intracellular calcium stores in addition to extracellular influx.

Marolyn Hess - I may be asking the same question in a different way, but is there any information that for a given concentration of cytosolic calcium, there may be a difference in the sensitivity of the contractile proteins in the adult versus the neonate?

Stephen Zderic - I don't have direct evidence for this in the systems I am using, but I do believe that this is a valid concept. I alluded to this earlier in my question about myosin light chain kinase phosphorylation which is accepted by most working in this field as the predominant mechanism required for generation of contractile force in smooth muscle. But there are a certainly a number of other reports that have considered the role the alternative pathways, for the generation of contractile force. It may well be that the free cytosolic calcium ion in addition to triggering light chain phosphorylation is also exerting its effects on other pathways as well.

William C. DeGroat - Have you thought about dealing with diverted bladders and producing atrophied bladders to see if the animal reverts back to its neonatal form in terms of its physiology as well as in terms of the expression of the proteins of calcium homeostasis?

Stephen Zderic - Based on my recent readings, I think this would be most worthwhile. It really does appear that bladder smooth muscle like any muscle needs to be given the right amount of work. Too little work results in atrophy, and too much work results in hypertrophy that ultimately lead to decompensation. Mechanical forces clearly alter the expression of the proteins involved in cell calcium regulation. Recently it was shown very clearly in a model of mechanically induced atrophy that by unweighting the rats hind legs (by attaching its tail in the top of the cage to prevent the animal's hind legs from working), there is a very clear shift to from fast from SERCA to slow form SERCA in these striated muscles. This calcium sequestering ATPase appears to be an enzyme which is very much responsive to the cell's mechanical environment. It may well be that the reason that these changes are seen in the normal development is that as the animal gets bigger, the animal needs to be able to sequester the calcium within the cell in order to have it close by the filaments and avoid the wait for diffusion inwards from extracellular sources. This may also be exactly what is happening when we obstruct the animal and induce cellular hyperplasia and hypertrophy. The result is much larger cell volumes which might interfere with calcium influx. Those cells which respond by developing intracellular stores are those which undergo compensation, and regain much or all of their contractile force. Those which fail to develop these intracellular storage sites remain decompensated. This hypothesis is under active investigation in our lab at this time.

William C. DeGroat -One other point; as the fetus or newborn becomes older, and more urine is formed, why does the bladder system decide just to increase its capacity? Why don't we keep the same bladder volume, and urinate more frequently? I wonder where this decision made, and how is it controlled?

Stephen Zderic - I think this gets back to something that you alluded to in your discussion of the whole brain stem, spinal cord, and bladder preparation you described earlier. Those of us who were working with strips of detrusor and do molecular experiments on membrane fractions or nucleic acid extracts must try to remember what the bladder is a whole organ. I think much of the question that you posed is related to training, psychological imprinting and development of the external sphincter. As interesting as it is to study individual prospects of bladder function; (I look at trigone or detrusor, I study cell calcium or matrix etc), we must not forget that the bladder must function as a whole organ and this is a pretty complicated process. There must be something about the way in which the bladder neck sets its resting tone during development. As Barry Kogan alluded to, there is this interaction between the detrusor and the bladder base itself.

William C. DeGroat - I think it really is an interesting point as to how this decision is made and how the nervous system and the organ interact in some way to make them function according to the physiologic needs. I would point out that diabetes also results in a similar increase in bladder capacity.

Stephen Zderic - It would seem that it can't be purely conditioned, when it is already beginning with the fetus. Unless of course one makes the major supposition that even in utero there is cortical processing to a degree that allows for the beginnings of urinary control, a prospect that does not seem likely.

William C. DeGroat - I was pleased to see this work on experimental fetal obstruction for it is my bias that we may begin to gain insight into normal development by experimentally manipulating the process of development. Are there questions for Craig Peters?

Ed Macarak - I have two questions. In your Northern blots where you looked at the type 1 and 3 collagen types, did you separate the detrusor and lamina propia when you made these determinations?

Craig Peters - Yes, we did. Early on we did not do the separations, but now we do. We have not, however, run the lamina propria as a separate layer, but have focused on the detrusor initially. What I want to look at is not only what I want to treat.

Ed Macarak - The second question is do you have any information on the wealth of protein levels corresponding to the message?

Craig Peters - No, that is what we are working on right now, and we want to be sure we do it right. We are a little adept at Northerns than we are at Westerns looking at collagen expression, but this most clearly what we must do in our future experiments.

William C. DeGroat - It looks that by inducing the obstruction, you are pushing the developmental program ahead. It would be interesting to know whether now that the changes that are described by the Penn group of calcium handling, are also pushed ahead in the same way. It would be interesting to see whether those changes that Bill Seers mentioned that occur late in the development of the afferent and efferent system are also pushed ahead by the same type of pathology. Finally, we could just go even further which is really science fiction, to just push the development of essential pathways and cause those to mature sooner. So, I guess the question is for Steve Zderic, would you be inclined to look at some of the changes in calcium handling in these fetal and in neonatal animals just after birth, to see if they appeared much more mature following in utero obstruction?

Craig Peters - You would be requiring lots of tissue, and that is one of the big advantages of the sheep model. For each obstruction produced you spent a lot of time, effort, and money, but you get a lot of tissue for many studies if well planned out in advance.

Stephen Zderic - That's really true, and I think Craig's point is well taken. I think for the person who is just entering the world of fetal physiology and biochemistry, you do need a lot ot tissue to do the classic kinds of experiments such as ligand binding assays or enzyme assays. Once you have characterized the systems and can work the mRNA level, your tissue requirements may drop. These budgetary considerations are what drove me to visit the slaughterhouse and procure fetal and adult bovine bladders. Their size makes it possible to study one bladder using muscle strip methods, and have enough tissue left to freeze at -70C to do biochemical assays when time permits. It has been very exciting to follow these projects along in our lab over the past five years, and I think our next 5 years will be even more productive. However I would be absolutely the first to admit, I don't think we are going to find just one protein that serves as the master switch that is developmentally regulated that accounts for the complex physiologic changes described in this group of papers on normal development. One of the beautiful things about studying developmental physiology if we persist long enough in searching, we will almost certainly uncover some changes in physiological or molecular differences with normal maturation. Even a relatively "simple" concept like bladder compliance is probably not going to be explained by changes in the expression of only one protein or enzyme. But these biases aside, it really helps to have a model in which you can study both the physiology and the molecular biology. I would point out though that sometimes the antibodies are very fussy and species specific. For example I presented some Western blots done in a bovine model in which we were fortunate that these antibodies recognized the sarcolemmal calcium magnesium ATPase. These antibodies that were developed at the University of Iowa by Kevin Campbell's group, and they worked beautifully in the bovine model, yet they are more difficult to use in the rabbit. I think that is also the benefit of having people working with several different models. Research in smooth muscle signal transduction is helped greatly by the fact that so much work has been in the cardiovascular areas, hence there are many antibodies and gene probes available for many of these proteins.

William C. DeGroat - There was a presentation earlier which showed that the compliance increased in the newborn and then in the aging process, the compliance went down. You also described that the compliance was reduced in these preparations and this may well just reflect the pathology and it is not really reflecting again maturation. But, in a way, we see that as compliance goes down and the activity of the bladder goes up, and once again part of what we are seeing might be just the postnatal maturation.

Craig Peters - We have not as yet studied any of the adults.

William C. DeGroat - There is one other point, rats and cats and rodents are sort of like unusual animals. They are actually born more immature in a way than humans. I would say that the human is much more mature than many of these animals, simply because the human shows reflex voiding in response to bladder distension. To those of you who are familiar with cats and rats, you know that the newborn animal does not void on its own. The newborn requires the mother animal to lick the perineal region and the mother animal controls voiding. This is probably because it keeps the nest clean because generally the mother consumes the feces and urine. So for the first three weeks or so, if you take a neonatal rat or kitten away from its mom, it goes into urinary retention. Now, the real interesting point is, how did these animals urinate in utero? They were born without the ability to urinate on their own, they require their mom to lick the perineal region and provide this external stimulus to control voiding. Does this mean that there is some other way that they urinate in utero? Do they have a mechanism in utero that gets turned off at the time of birth and then gets turned on three or four days later? It occurred to me that it would be really interesting to try in these animals to put an obstruction around the urethra and see if the whole process of maturation of the central autonomic pathways could be pushed forward ahead in these preparations. I don't know how the sheep urinates after birth, or the cow. I presume they have voiding like humans, but I am not sure.

Marc Cendron - The sheep bladder has been demonstrated by ultrasound to have cyclical contractions which actually cause the bladder to empty and that is also shown electrophysiologically. It is something that is happening in utero.

Craig Peters - I think we are beginning to get the data to support our contention that the sheep is a little closer to the human in this regard. There is work and we have been able to reproduce it which shows that although it is an unusual voiding pattern, they do have a filling limb on their cystogram and a phasic contraction rather than a simple mass contraction. We reproduced this pattern in exactly the same way as Barry Kogan's lab. No one has yet explained what that pattern is nor has anyone recognized it as anything similar to a pathological pattern in humans.

William C. DeGroat - This is a phasic pattern so you get high frequency pressure waves. This is very much like the rat. In the rat, the urethral sphincter does not relax completely as it does in the human. It undergoes bursting activity. Some people have speculated that this the mechanism that generates isotonic contractions of the bladder against the closed sphincter, and this prevents complete bladder emptying. This may be some immature sort of pattern that you see where there is an overlap between rat and these other species.

William Steers - This is a comment on some work that our laboratory will be presenting at the society for basic urology research. We have been doing daily CMGs on rats up to four weeks. Probably what happens is that the sympathetics come in and yet the urachus closes quite late in development. As a result, the compliance drops in the first weeks, residual urine drops and once the sympathetics which are correlated with adrenergics start coming in and stimulate the hypogastric nerve; then it seems that somehow these temporally related events result in a nice relaxation in the bladder and filling at low pressures. The beauty of the rat is, you have to study it in utero. In the first week or two weeks, you are getting a glimpse of what is happening much earlier probably than in humans or the sheep, but certainly the parallels are quite similar.

Participant - I am just curious about that slide you showed of the myosin isoforms. Is it possible that this results from the recruiting of myofibroblasts rather than inducing changes in the smooth muscle itself? There is some literature describing a transition from fibroblasts to myofibroblasts to smooth muscle.

Craig Peters - I am not familiar with this. I would think that if they were recruiting from a more dedifferentiated cell type that you see a more fetal pattern. What we saw, at least, in this preparation, was a forced maturation, early maturation, rather than a more primitive fashion. Whether there are these transitions of fibroblasts into smooth muscle cells, I do not know but that would be an interesting area to look at.

Philip Ransley - I would like to limit the timing of the discussion a little bit, if I may. Until we get into molecular manipulation, we are left with practical things to do. I would very much like the speakers from today to try to answer the dilemma that the pediatric urologists face. The development affecting function or related to function leaves you with a terrible problem. Should we, be doing a vesicostomy at 23 weeks gestation and close it on day one of life, or should we be leaving the bladder intact in utero and doing a vesicostomy on day one of life? Those two management options are completely opposed, yet I am not sure from this discussion which way the evidence points.

Craig Peters - Just as an anecdotal information, I follow two patients who had successful prenatal intervention for valves at about 22 weeks, both of whom who had complete documented bladder decompression, both survived, did not have pulmonary problems and had good kidneys. One has a completely normal bladder and developed normal bladder control at about age four. The other at age three has a distinctly hypertonic typical valve bladder that is probably going to be augmented this summer. It amazes me, but it suggests that these patterns may well be variable earlier than we think, that they may be a lot more firmly fixed earlier than we would like to think, or there are other factors that have little to do with the emptying and filling that we have been talking about.

Larry Baskin - I am going to answer for Dr Kogan, because he had to leave early, and because he has an exciting answer. Perhaps we will begin to consider treating these fetal bladders with L-arginine in order to try and produce a relaxation of the external sphincter. But this is very preliminary experimental data. But I think that Barry's work in looking at Nitric Oxide and trying to lower outlet resistance using pharmacologic methods is just a start. I know that there is interest in trying to prevent in utero some of the changes seen in the bladders of patients with myelomeningoceles who have a very hyperactive sphincter. Manipulating this pathway may offer a medical alternative.

Philip Ransley - I was really thinking of the valve bladder with an anatomical obstruction. At the moment, the whole world has stepped back from the in utero intervention and the bladder has never been a great subject for discussion. I think if we are going to push, which I personally am in the UK, the obstetricians to do more and more of prenatal intervention in order to see if it can make a difference. We need perhaps to try and relate that as often as possible to the scientific studies of what's happening to these bladders at the time that intervention is done. It is not quite as simple as that because the 20 week decision is easy; the 34 week decision is not easy. I still don't have a clear picture of which way we should go.

Larry Baskin - We would all agree that intervening prenatally has done little to help the kidneys, but perhaps at 34 weeks intervention would help the bladder.

Philip Ransley - Well you actually would be helping a kidney, it's just that it would be the transplanted kidney.

William Steers - It is interesting trying to focus on the clinical correlation all of this basic science. As the person doing the urodynamic studies on the myelomeningocele patients of Stu Howards' pediatric population at the University of Virginia, I have been seeing a quite interesting phenomena, in attempting to correlate their incontinence with the urodynamic parameters. We have looked at children with thickened bladders, acontractile bladders, and bladders that don't empty, and yet these children are incontinent. When we look at these children with very rapid sequence video urodynamics, what we are seeing is type III urinary incontinence which is well described with this patient population. What is happening the nerves are intact, but the bladder does not contract. What may be happening is that when we reach a certain volume, we are seeing the urethral instability Ed McGuire described. In fact, you are seeing the outlet open, we are seeing the urethra actually relax almost in an exaggerated outlet response when the bladder isn't working very well. It is quite interesting, depending on the level of the myelomeningocele, that we have the muscle that will not work but we have half the equation there. That is there exists this exaggerated relaxation of the outlet, and the result is these children who are totally incontinent. How do you manage that? Do you intervene with the Nitric Oxide pathway and potentially increase the hypertrophic state because now we would really have outlet resistance? They would be fully continent, but wouldn't empty completely because the bladder doesn't have the motor inputs. Again, I think there is tremendous clinical implications for this work and what is happening. But I think we have to look at the whole system.

William C. DeGroat - We have talked about the physical obstruction of the urethra, but we know in adults that there is also bladder and sphincter dyssynergia in certain neurological disorders. There can be dyssynergia of the smooth muscle component so you urologists who are interested in the pediatric case and also the in utero case, are you considering the possibility of bladder/sphincter dyssynergia in utero? Is there the possibility that there can be failure of coordination between the bladder and the urethral outlet that is not related at all to obstruction, but with some neurological deficit?

Craig Peters - I think that clearly occurs. We have seen it clinically in terms of myelodysplastics who are born with already established upper tract changes and bladder changes that Ellen

Shapiro has shown that look a little like obstruction. Dr. Bauer reported on four cases, very unusual children, with massive bladder dilatations and upper tract dilatations and without any obvious neurologic diagnosis or classic form of outlet obstruction. Was this transient detrusor sphincter dyssinergia, because it certainly was not present after birth? Perhaps a transient period of incoordination and dyssynergia that produced all those changes. Maybe there was a subtle CNS lesion that eluded the current methods of diagnosis (MRI, clinical neurological examination).

William C. DeGroat - I guess, there is a temporal aspect to the development of certain central controls so that the bladder reflex can't fully develop before the sphincter-relaxation pathway. Perhaps these patients presented as a result of this sequence being thrown out of its normal course. Are there other questions? Are any of the people on the panel stimulated to try to pull all of these thoughts together to give the audience some general conclusions from this afternoon's session? Is there any one in the audience who has been able to synthesize all of this?

EXTRACELLULAR MATRIX UPDATE - OVERVIEW

Gary Striker M.D.

National Institutes of Digestive and Kidney Diseases
National Institutes of Health
Bethesda, Maryland 20892

This is a very exciting time in urology, and in biology in general. The urology departments at the Children's Hospital of Philadelphia and the University of Pennsylvania are to be congratulated for their efforts in convening this meeting. They have invited a group of speakers here who are figures in the field. I would encourage you to pick their brains and the rest of their bodies, if need be, to get the information that is unique. There is no sense reinventing all the wheels that others have picked up. One of the best examples, for instance, is that everyone knows that the bladder is full of liquid whose isotonicity changes fairly rapidly. The amount of ammonia, urea, and pH to name but a few parameters change quite drastically over time. But let us have a look around and we find that very similar types of events take place in the stomach. If we look in the stomach we see that there is the same network of proteins underneath the double layered membrane itself. So instead of gastric mucosa, there is urothelium, but there are really the same kinds of molecules present. This is an exciting time in urology because research, now at a fundamental level, has become a part of the everyday experience of the residents, fellows, and faculty in academic medical centers. This is the first conference that will focus on extracellular matrix in the bladder and I hope this is just one of the many to follow. To put things in perspective, six years ago at the NIH, the budget in urology was about 4.2 million dollars and it was mostly stone disease. No one thought that was there anything interesting in the bladder. Now the budget is around 30 million dollars and it is on a graphic upswing. It is up to you to keep it going in that direction.

Bladder diseases, as this audience knows, are pretty important. They are important in all phases of life; Alan Wein has emphasized the social cost of geriatric incontinence and John Duckett and Howard Snyder have been pointing out how important diseases of the urinary tract are prenatal and neonatal time frames. Urinary tract infections account for the largest number of admissions and office visits in the late teens and early 20s, and for those in the older age groups, incontinence and voiding disorders are the major reasons for incapacity, and the most common reason elderly Americans are placed in nursing homes. This one fact alone accounts for one of the major health care costs we have, so it is an important and clinically significant problem. So where are we going? It seems to me that the bladder is relatively neglected in overall research funding. Even within urologic research funding, a major emphasis has been placed in oncology, especially prostate cancer in recent years. There has been some interest in bladder cancer. There are some people interested in fertility and the etiology of stone disease, but the bladder has largely been left out. Kind of as an appendage somewhere in close proximity to the prostate.

But, basic research - knock out mice, knock out WT1, what happens? You will get abnormalities of the ureteric bud. Take later stage embryos, knock out WT1 again and you get abnormalities in different parts of the urinary tract. All kinds of new types of collagens are being discovered, and we know more than ever about their genetics. For example type IV collagen is now known to have six chains and they are all located head to head on different chromosomes. New types of extracellular matrix, are being discovered and you will hear about these today. Both elastin and collagen as well as the collagenases and their inhibitors, will be discussed this

Muscle, Matrix, and Bladder Function,
Edited by S. Zderic, Plenum Press, New York, 1995

morning by the people who are leaders in this field. There are many, many fields that are ripe for investigation of the bladder.

I don't think anyone understands, at this point the molecular biology of the bladder that gives the baldder its special properties. Is there a difference between the extracellular matrix, its rate synthesis, its rate of degradation, and its composition when one compares adult versus fetal or neonatal bladders? Are smooth muscle cells in the bladder wall at the dome the same as those at the base? Is the composition of the extracellular matrix the same in these two locations? We know a lot about the innervation and pharmacology of these two areas, but very little about the extracellular matrix. Remember that we know very little about the normal.

We are going to hear about wound healing this morning from Paul Ehrlich. What do we know about bladder inflammation at the molecular and cellular level? What do we know about the effects of inflammation on the epithelium, submucosa, and smooth muscle of the urinary bladder? Surprisingly little considering how common some of these conditions really are. We have that old song that ontogeny recapitulates phylogeny and we can also recall that sometimes pathology recapitulates phylogeny and ontogeny. I would encourage you as you listen to what is going on this morning to keep your eyes open. As a fellow, I came back and presented at the University of Pennsylvania an observation that we made in the kidney, on what we thought was type 1 collagen. We had been looking at pepsin digests and noticed there was a doublet of about 185 kilo daltons and we couldn't figure out what this really represented. We had titled our abstract, "post translational modification of type I collagen in etc etc" because our title was based on what was known at that point which was just type I collagen. We didn't recognize it at that time, but what we were seeing was type IV collagen.

I would encourage you all to look for what was pointed out this morning by Howard Snyder; look at the data, and look at it with skepticism. Remember that the people speaking today have not spent their professional lifetimes looking at this particular subject. No one has. I think this is an area of great opportunity and promise. So, I look forward to the rest of this meeting and look forward to this being the first of several meetings of this arca.

IS COLLAGEN REMODELING ASSOCIATED WITH BLADDER OBSTRUCTION?

H. Paul Ehrlich, Ph.D.

Division of Pediatric Surgery, Department of Surgery
Milton S. Hershey Medical Center
Pennsylvania State College of Medicine
Hershey, PA 17033-0850

INTRODUCTION

Collagen, proteoglycans and elastin combined with resident smooth muscle cells (SMC) makeup the basic structural unit of the bladder wall. Any pathologic process that alters either the content or organization of collagen within the bladder wall will modulate the ability of the bladder to store urine or efficiently empty itself (1). An obstruction that develops at the outlet of the bladder results from a narrowing of that lumen. There are numerous physiological and anatomical possibilities for producing the constricture of that lumen. In some constrictures it is proposed that connective tissue remodeling is critical in causing the narrowing of the lumenal space in the bladder outlet.

The histology of an outlet obstruction shows an increased deposition of collagen in the detrusor (2,3). The obstructed rabbit bladder outlet model shows that the elastic lumina at the serosa surface has excess collagen deposited in both the extrinsic surface as well as beneath that surface (1). A thickened connective tissue capsule on the surface of the bladder suggests an alteration of organ compliance. The appearance of increased collagen deposited beneath the surface of the detrusor is scar-like in regard to the organization of its connective tissue.

Scarring in particular regions of the trauma patient commonly leads to scar contractures and constrictures. A young child who accidently tips a vessel of hot liquid on him or her self often produces a partial thickness scald burn on the neck. Due to secondary induced dermal burn ischemia that second degree burn becomes a full thickness injury that commonly heals as a hypertrophic scar. Such a scar in the neck region usually develops a scar contracture which pull down on the chin, Figure 1. Surgical intervention is needed to restore normal motion to that scarred neck. In contrast, a partial thickness burn on the front surface of the thigh, which goes on to become a full thickness burn and develops into a hypertrophic scar, does not exhibit a scar contracture that impairs the functional motion of the upper leg. The contractibility of scar tissue is influenced by its anatomical location (4). The tissues structures adjacent to a scar will influence the size, orientation, organization and contractibility of that scar. Scars located over joints are susceptible to contractures, while scars over ridged structures such as the thigh are incapable of developing contractures.

Trauma of lumenal structures such as a lye burn of the esophagus will heal by scarring. Often a lumenal constricture occurs by the contraction of that scar tissues which obstructs the esophagus. The organization of the connective tissue matrix within the scar tissue is proposed to be responsible for causing the contraction of that tissue. This is in contrast to the concept that the volume of connective tissue making up the scar causes the constricture. The constricture results from the pulling in by the scar tissue on the flexible peripheral lumenal structure. The obstruction is not caused by the volume of the scar deposited within the lumen, but rather by the contraction of that scar tissue, caused by the reorganization of the connective tissue matrix.

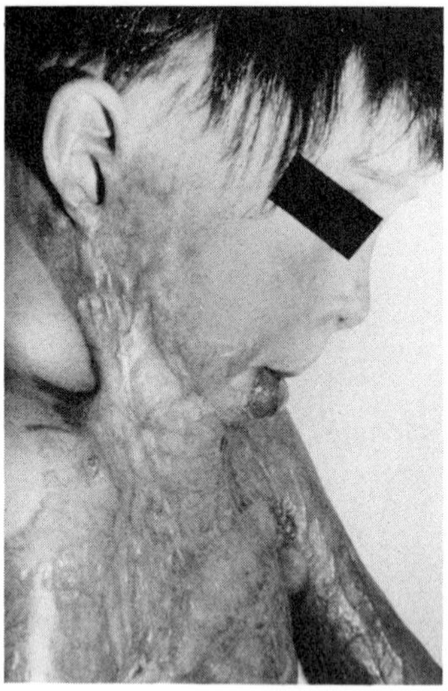

FIGURE 1. A burn has healed with a hypertrophic scar. The scar is undergoing contracture, pulling the chin down. The plastic surgeon must revise that scar to reduce the contracture and promote normal function.

HYPOTHESIS

The basis for contractile forces in scars is the generation of stress by cells residing within that scar. The mechanism for those cell generating contractile forces may be the result of either cell contraction like that seen in contracting smooth muscle cells or cell locomotion like that seen with migrating leukocytes. Secondly, the connective tissue matrix plays an important role in influencing those cellular forces?

Specialized fibroblasts have been identified in healing wounds and contracting scars (5,6). These cells have been named myofibroblasts because of their morphological features which include thick actin bundles named stress fibers (5-9). These stress fibers contain alpha-smooth muscle actin, an isoform of actin characteristic of muscle cells (10). Myofibroblasts have been described in numerous clinical and experimental situations (11-13). It is proposed that myofibroblasts produce contractile forces by a mechanism of cell contraction (8,12). The cell undergoes shortening by the contraction of its cytoplasmic stress fibers. The contraction of the cells reorientates the collagen fibrils attached to its surface (5,7,9). A problem with this hypothesis is contracted myofibroblasts have not been identified in contracting wounds or contracting scars. Contracting myofibroblasts have only been identified in an experimental situations where fibroblasts residing in a collagen lattice were splinted and then released (14). Another possibility to consider is that the contractile forces produced by fibroblasts are generated by cell locomotion. The generation of cell locomotion has been proposed to be by tractional force which was demonstrated by Harris and his coworkers (15). Fibroblasts plated on elastic silicone-film surfaces produce wrinkles beneath them. The thin film surface wrinkles occur beneath cells that retained an elongated morphology, Figure 2. There is no cellular 2contraction observed with the generation of these wrinkles by tractional forces. It is proposed that the contraction of cytoplasmic microfilaments linked to sites on thecell membrane pull on the flexible-rubber surface and produce distortions of the rubber surface in the absence of morphological distortions of the cell. Tractional forces have been proposed to be critical for the organization of collagen in early embryological development (16). Here it is proposed that tractional forces are critical for generation of lumenal constrictures.

The collagen composition of granulation tissue in healing wounds and hypertrophic scars is enriched in type III collagen (17-19). Keloids which are dermal lesions of excess scarring and normal scar do not undergo scar contractures and have relatively reduced concentrations of type III collagen. Is there a relationship between the collagen composition of the scar tissue and its susceptibility to contraction?

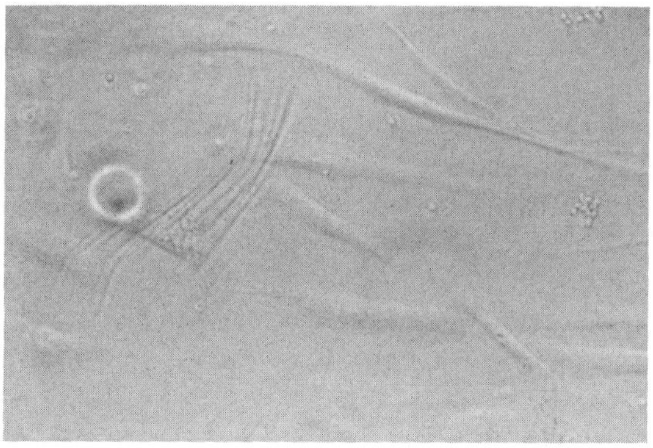

FIGURE 2. Human dermal fibroblasts were plated on to a thin film of heat polymerized silicone oil. At 6 hours after plating the fibroblasts are spread and elongated. Wrinkles appear under the cells in the absence of cell contraction.

EXPERIMENTAL MODEL

An *in vitro* model, the fibroblast populated collagen lattices (FPCL), for studying scar contracture was introduced by Bell and co-workers in 1979 (20). FPCL are manufactured by mixing fibroblasts freed from monolayer culture, serum containing culture medium and native soluble collagen solution in a Petri dish which is immediately transferred to a 37°C incubator. The collagen rapidly polymerizes, trapping cells within the fast forming matrix. With time these free floating collagen lattices undergo a steady reduction in size referred to as lattice contraction, Figure 3. Lattice contraction occurs as a consequence of the organization of collagen fibers by the resident fibroblasts (21). Collagen fibers are translocated and organized into thicker fibers which can be demonstrated by the development of collagen birefringence as lattices become smaller.

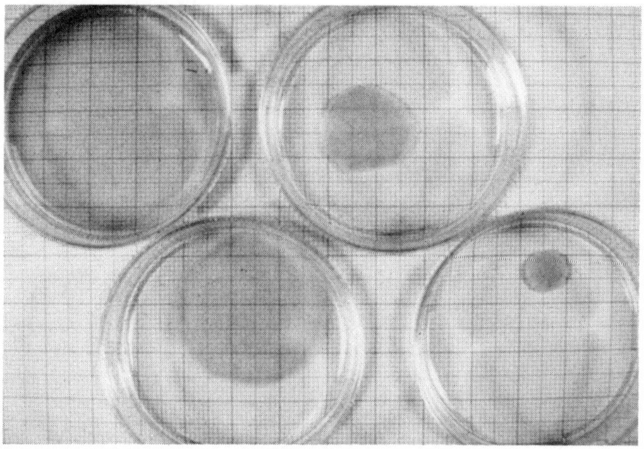

Figure 3. Two milliliter FPCL in 35 mm Petri dishes were made by mixing cultured human dermal fibroblasts with serum containing culture medium and soluble collagen. With 15,000 cells no lattice contraction occurs at 24 hours. With 50,000, 75,000 and 100,000 cells lattice contraction occurs with the greatest degree of contraction arising with the most densely populated collagen lattice.

The mechanism for the cells organizing the collagen fibrils and producing lattice contraction was examined. Agents such as cytochalasin and phalloidin which affect the dynamics of actin polymerization and microfilaments inhibit lattice contraction (20,22). Likewise colchicine and vinblastin, agents which alter the organization of microtubules also have been shown to inhibit lattice contraction (Table 1)(20,22).

Elevating the intracellular concentrations of cAMP with IL-1, cholera toxin or postaglandins inhibit lattice contraction (23). Elevated levels of cAMP inhibit the posphorylation of myosin light chains, the regulatory peptides of myosin ATPase. Reduced levels of myosin ATPase reduces the contraction of microfilaments needed for either cell contraction or cell locomotion (24). The introduction of a modulated form of the kinase which phosphorylates the myosin light chains (myosin light chain kinase, MLCK) into fibroblasts by electoporation was done to examine the effects of cAMP on lattice contraction (24). As shown in figure 4, fibroblasts with modulated MLCK contracted FPCL in the presence of cholera toxin as well as fibroblasts in the absence of cholera toxin. In untreated cells the rate of lattice contraction was enhanced by the addition of the modulated kinase.

Table 1 - FPCL were made with 100,000 human dermal fibroblasts, 10% fetal bovine serum, Delbecco's modification of Eagles medium, and 2.5 mg of rat tail tendon acid soluble collagen in a total volume of 2.0 ml in a 35 mm Petri dish. FPCL were incubated at 37°C and the diameters of the lattices were measured daily and recorded.

	INHIBITOR	NO EFFECT	STIMULATOR
SERUM			+
INTERLEUKIN 1	+		
PROSTAGLANDIN E-1	+		
CHOLERA TOXIN	+		
COLCHICINE	+		
CYTOCHALASIN B	+		
TGF_b			+
PDGF		+	
HYDROXYUREA		+	
ACTINOMYCIN D	+		
CYCLOHEXIMIDE	+		
LAURIC ACID			+

These experiments and others have shown that the cytoskeleton microfilaments were important for the reorganization of collagen fibrils and lattice contraction. Both cell contraction as well as cell locomotion are dependent upon the functioning of microfilaments (15,25). Increasing the fluidity of the cell membrane will promote changes in cellular functions which are related to greater cell locomotion (26). An enhancement of membrane fluidity would be expected to be independent of cell contraction. Free fatty acid loading of cultured fibroblasts showed enhanced lattice contraction (27). The induction of cell contraction by the addition of ATP directly to permeabilized cultured fibroblasts with and without free fatty acid supplementation in monolayer culture showed no enhancement of cell contraction (27). It is suggested that cell locomotion forces are responsible for the organization of collagenwhich is responsible for the contraction of FPCL. In another experiment the direct induction of cell contraction by the addition of ATP to permeabilized fibroblasts within collagen lattices did not produce lattice contraction (28). Such treated cells showed ruptured actin filaments within their cytoplasmic microfilaments, but no translocation of collagen fibers occurred.

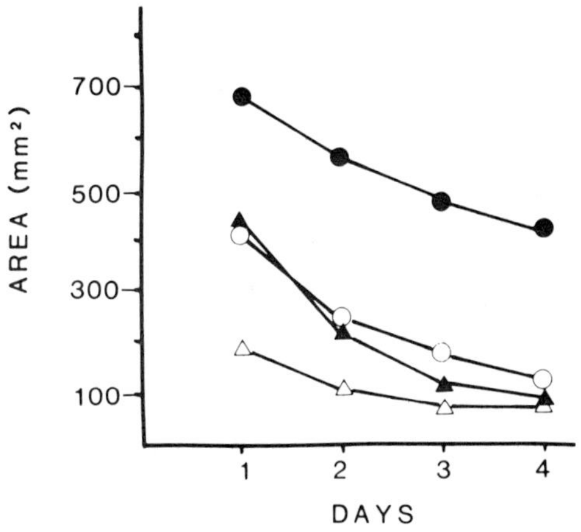

Figure 4. FPCL were made with fibroblasts either containing a modified MLCK, open symbols, or no supplement, solid symbols. FPCL were made in medium containing cholera toxin, 5 ug/ml, circles or in medium with no additions, triangles.

The mechanism for lattice contraction appears to be the contraction of actin-myosin cytoplasmic filaments and the translocation of collagen attachment sites on the membrane surface of the cell. There is no cell contraction nor is there a need for stress fibers to accomplish this organization of collagen fibrils (28). Myofibroblasts are not important for the contraction of FPCL. Myofibroblasts with alpha smooth muscle actin staining cytoplasmic stess fibers are not present in contracting FPCL.

It was found fibroblasts isolated from contracting hypertrophic scars were similar to fibroblasts derived from normal dermis at contraction FPCLs (29). However collagen extracted from contracting hypertrophic scars produced greater lattice contraction compared to

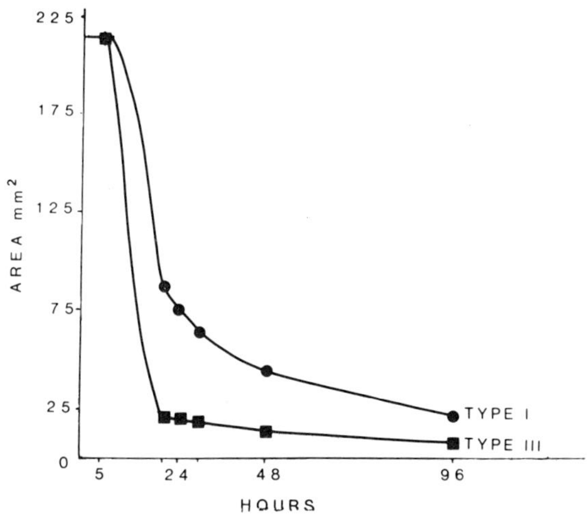

Figure 5. FPCL were made with 1.25 mg/ml of type I collagen, solid circles; 1.25 mg/ml type II collagen, solid diamonds and type 6III collagen, solid boxes.

lattices made with normal dermal extracted collagen (29). Since hypertrophic scars are enriched in type III collagen (17,19), FPCL composed of either type I or III collagens were compared for differences in lattice contraction. Type III collagen FPCL contracted faster and to a greater degree than type I collagenFPCL, Figure 5. The interpretation of those results is that fibroblasts are better at organizing type III collagen fibrils compared to organizing type I collagen fibrils (30). Tissues rich in type III collagen such as granulation tissue and young scar are more susceptible to reorganization; thus the composition of the connective tissue may be a critical for understanding the mechanism of scar contractures and lumen constrictures.

SPECULATIONS

In bladder outlet constricture it is proposed that the ongoing organization of collagen fibrils in scar tissue by either resident smooth muscle cells or fibroblasts leads to the compaction of that tissue. It is proposed that besides the volume of the newly deposited connective tissue accummulating within the lumen, the organization of the collagen into thicker more compact fibers leads to the pulling in of the bladder wall which also contributes to the narrowing of the lumen. Both the lumenal wall of the bladder outlet is remodeled and the outer diameter of the wall is reduced in circumference. The diminished luminal area is partially the result of resident cells within the scar tissue having the ability to synthesize, deposit and organize a collagen matrix enriched in Type III collagen. The resident cells compact the newly deposited collagen rich matrix, creating the force that pulls in the bladder wall. The force produced by the reorganization of the scar matrix reduces the diameter of the lumen. The volume of the scar tissue deposited plays a modest role in occluding the lumen. If cell locomotion forces were impaired or if collagen composition of that matrix was modulated, then it is postulated that bladder outlet luminal obstruction could be reduced or prevented.

REFERENCES

1.Zderic SA, Wein AJ, Monson FC, Levin RM (1991) The effects of intravesical obstruction on the bladder. Problem Urol 5:357-368.
2. Uvelius B, Persson L, Mattiasson A (1984) Smooth muscle cell hypertrophy and hyperplasia in the rat detractor after short-term intravesical outflow obstruction. J Urol 131:173.
3. Mayo ME, Hinnam Jr F (1976) Structure and function of rabbit bladder altered by chronic obstruction or cystitis. Invest Urol 14:6.
4. Brody GS, Peng TJ, Landel RF (1984) The etiology of hypertrophic scar: Another view. Plast Reconst. Surg 67:673-684.
5. Gabbiani G, Hirschel BJ, Ryan GB, Statkov PR, and Majno G (1972) Granulation tissue as a contractile organ: A study of structure and function. J Exp Med 135:719-734.
6. Hirsch BJ, Gabbiani G, Ryan GB, Majno G (1971) Fibroblasts of grannulation tissue: Immunofluorescent staining with anti-smooth muscle serum. Proc Soc Exp Biol Med 138:466-469.
7. Baur PS, Larson DL, Stacey TR (1975) The observation of myofibroblasts in hypertrophic scars. Surg Gynecol Obstet 141:22-26.
8. Darby I, Skalli O, Gabbiani G (1990) alpha smooth muscle actin is trasiently expressed by myofibroblasts during experimental wound healing. Lab Invest 63:21-29.
9. Gabbiani G, Majno G (1972) Dupuytren's contracture: Fibroblast contraction? Am J Pathol 66:131-138.
10. Rungger-Brondle E, Gabbiani G (1983) The role of cytoskeleton and cytocontractile elements in pathologial processes. Am J Pathol 110:359-392.
11. Eddy RJ, Petro JA, Tomasek JJ (1988) Evidence for the nonmuscle nature of the "myofibroblast" of granulation tissue and hypertrophic scar. An immunofluorescence study. Am J Pathol 130:252-260.
12. Majno G, Gabbiani G, Hirschel BJ, Ryan GB (1971) Contraction of granulation tissue *in vitro*: Similarity to smooth muscle. Science 173:548-550.
13. Desmouliere A, Geinoz A, Gabbiani F, Gabbiani G (1993) Transforming growth factor-1 induces alpha-smooth muscle actin expression in granulation tissue myofibroblasts and in quiescent and growing cultured fibroblasts. J Cell Biol 122:103-111.

14. Tomasek, JJ, Haaksma, CJ, Eddy, RJ, Vaughan, MB (1992) Fibroblasts contraction occurs on release of tension in attached collagen lattices: Dependency on an organized actin cytoskeleton and serum. Anat. Rec. 232:359-368.
15. Harris AK, Wild P, Stopak D (1980) Silicone rubber substrate: A new wrinkle in the study of cell locomotion. Science 280:177-179.
16. Harris AK, Stopak D, Wild P (1981) Fibroblast traction as a mechanism for collagen morphogenesis. Nature 290: 249-251.
17. Bailey AJ, Bazin S, Sims TJ, LeLous M, Nicoletis C, DeLaunay A (1975) Characterization of the collagen of human hypertrophic and normal scars. Biochem Acta 405:412-421.
18. Bailey AJ, Sims TJ, LeLous M, Bazin A (1975) Collagen polymorphism in experimental granulation tissue. Biochem Biophys Res Comm 66:1160-1165.
19. Ehrlich HP, White BS (1979) Zone precipitation chromatography: Its use in the isolation of different collagen types. Prep Biochem 9:407-425.
20. Bell E, Ivarsson B, Merrill C (1979) Production of a tissue-like structure by contraction of collagen lattices by human fibroblasts of different proliferation potential *in vitro*. Proc Natl Acad Sci (USA) 76:1274-1278.
21. Ehrlich HP (1988) Wound closure: Evidence of cooperation between fibroblasts and collagen matrix in scar contracture. Eye 2:149-157.
22. Ehrlich HP, Buttle DJ, Bernanke DH (1989) Physiological variables affecting collagen lattice contraction by humen dermal fibroblasts. Exp Mole Path 50:220-229.
23. Ehrlich HP, Griswold TR (1984) EBdr fibroblasts produce increased concentrations of cAMP within a collagen matrix. J Invest. Derm. 83:220-233.
24. Ehrlich HP, Rockwell WB, Cornwell, TL, Rajaratnan, JBM (1991) A role for myosin light chain kinase in FPCL contraction. J Cell Physol 146:1-7.
25. Ehrlich, HP, Griswold, BS, Rajaratnam, JBN. (1986) ATP-induced cell contraction with EBdr and normal dermal fibroblasts. J. Invest. Derm. 86:96-100.
26. Horowitz, AF, Hatten, ME, Burger, MM (1974) Membrane fatty acid replacement and their effect on growth and lectin-induced agglutinability. Proc. Nat. Acad. Sci. USA 71:3115-3119.
27. Rittenberg T, Ehrlich HP (1992) Free fatty acids and dialyzed serum alterations of FPCL contraction. Tiss Cell 24:243-251.
28. Ehrlich HP, Rajaratnam JBJN (1990) Cell locomotion forces versus cell contraction forces for collagen lattice contraction: An *in vitro* model of wound contraction. Tiss Cell 22:407-417.
29. Ehrlich HP (1988) The role of connective tissue matrix in wound healing. Prog Clin Biol Res 266:243-258.
30. Ehrlich HP (1988) The modulation of contraction of FPCL by types I,II, and III collagen. Tiss. Cell 20:47-50.

REGULATION OF MATRIX METALLOPROTEINASES DURING EXTRACELLULAR MATRIX TURNOVER

Marta L. Corcoran, David E. Kleiner, Jr. and William G. Stetler-Stevenson

Extracellular Matrix Pathology Section,
Laboratory of Pathology, National Cancer Institute
National Institutes of Health, Bethesda, Maryland 20892

INTRODUCTION

Extracellular matrix (ECM) turnover is a critical step in the tissue remodeling that accompanies many physiologic as well as pathologic processes. Normal remodeling events, such as wound healing and trophoblast implantation, require coordinated synthesis and removal of ECM components. These physiologic processes are strictly regulated in a spatial and temporal fashion, and are usually limited in extent. They result in a functionally intact matrix and preservation of tissue boundaries. Pathologic events may result in the extensive and uncontrolled ECM destruction that is often seen in the inflammatory collagen vascular diseases. Such processes cause the loss of normal matrix boundaries and matrix functions. Alternately, inactivation of the matrix degradative machinery may result in excessive ECM accumulation that disrupts normal tissue structure and function. Proteases responsible for matrix turnover have been identified in all four classes of proteases (seryl-, cystyl-, aspartyl- and metalloproteinases). However, the matrix metalloproteinase (MMP) family of enzymes appears to be primarily responsible for much of the coordinate or discoordinate degradation of the ECM that occurs during physiological and pathological events, respectively. Comprehending the precise points of regulation for MMP activity may facilitate therapeutic intervention in pathologic situations that characteristically favor either unregulated ECM destruction or excessive deposition of matrix components.

THE MATRIX METALLOPROTEINASE (MMP) FAMILY

The MMP enzymes are a family of zinc-atom dependent proteases, that function extracellularly at neutral pH (Birkedal-Hansen et al., 1993). Based on their substrate specificity this family can be subdivided into three groups. The interstitial collagenases include two genetically distinct proteins represented by the fibroblast (EC 3.4.24.7) and polymorphonuclear leukocyte (EC 3.4.24.34) interstitial collagenases. These enzymes degrade types I, II, III, VII, and X collagen, within the triple helical regions to generate characteristic degradation fragments. The gelatinases which include gelatinase A (EC 3.4.24.24) and gelatinase B (EC 3.4.24.35) degrade types IV, V, VII collagen, elastin, fibronectin and gelatin. They cleave type IV collagen within the triple helical region to generate characteristic 1/4 amino-terminal and 3/4 carboxy-terminal fragments. Finally, the stromelysin subgroup which includes stromelysins 1 (EC3.4.24.17), 2 (EC 3.4.24.22), and 3, matrilysin (EC 3.4.24.23) and metalloelastase which have a broader substrate specificity including laminin, fibronectin, proteoglycans, and the non-helical regions of type IV collagen.

Although the members of the MMP family possess different matrix substrate specificities, they all possess conserved functional and structural domains. These shared domains translate into similar mechanisms of proteolysis (Birkedal-Hansen et al., 1993). The sequence homologies between members of this family is greatest at the amino-terminal pro-fragment "activation locus" (PRCGVPDV) and the zinc atom binding site of the catalytic domain (VAAHEXGHXXGXXH). All of the MMPs except for matrilysin possess a carboxy-terminal domain with homology to hemopexin/vitronectin. This domain has been reported to contribute to the substrate specificity of interstitial collagenase and the stromelysins (Sanchez-Lopez et al., 1993). The gelatinases have an additional fibronectin-like gelatin binding domain which contributes to the substrate preference of these enzymes (Strongin et al., 1993). Site-directed mutagenesis of this region is known to disrupt gelatin binding and proteolytic activity. The MMP enzymes also have structural calcium atoms (Lovejoy et al., 1994). *In vitro* inhibition of MMPs is achieved by interfering with the active site zinc using metal chelating agents (EDTA or o-phenanthroline), or small peptide analogs that either mimic the "activation locus" or coordinate to the active site zinc (Van Wart and Birkedal-Hansen, 1990; Stetler-Stevenson et al., 1991). *In vivo* inhibition is accomplished by endogenous inhibitors, known as the tissue inhibitors of metalloproteinses or TIMPs. Inhibition of proteolytic activity by TIMPs is a criteria that is currently utilized to define new members of the MMP family (Nagase et al., 1992). In addition to TIMPs, serum contains a2 macroglobulin which non-specifically inhibits the action of a number of proteases including the MMPs.

The degradation of ECM ultimately depends on the balance between the local concentration of active MMP species and the levels of endogenous inhibitors. Thus, net proteolysis may be regulated at several levels. These include; (1) transcriptional and, (2) translational regulation, (3) post-translational modification and secretion of latent MMP, (4) activation of pro-MMPs and, (5) their inhibition by endogenous inhibitors, i.e. TIMPs. This review will briefly address each of these levels of regulation.

TRANSCRIPTIONAL REGULATION

Regulation of MMP expression at the transcriptional level was initially demonstrated by cytokine mediated induction of fibroblast interstitial collagenase and stromelysin mRNA expression. Cytokines such as interleukin-1 (IL-1) and tumor necrosis factor-a (TNF-a) (Brenner et al., 1989) or phorbol esters such as 12-O-tetradecanoylphorbol-13-acetate (TPA) (Angel et al., 1987) induce the expression of proto-oncogenes/ transactivating factors, c-jun and c-fos in fibroblasts. These activating proteins (AP) putatively regulate transcription via the TPA responsive element (TRE) or "AP-1 site" (TGAGTCA) located upstream of the stromelysin and collagenase genes. However, AP-1 independent mechanisms have also been reported to induce the production of fibroblast MMP mRNA (Kerr et al., 1988; Conca et al., 1991) which suggests that members of the AP-1 jun and fos family cooperate with other transactivating proteins to regulate net MMP transcriptional activity (Sassone-Corsi et al., 1989). Additionally, a fos protein induced by transforming growth factor-b (TGFb) treatment has been postulated to bind the TGFb inhibitory element (TIE) located upstream of the stromelysin and collagenase promoter that suppresses their transcription (Kerr et al., 1988; Kerr et al., 1990). The ability of the TRE to induce or suppress MMP expression is dependent on the AP-1 jun and fos transactivator protein family. Depending on the signal transduction mechanisms, the genetic characteristics of the cell, and the cellular environment, the AP-1 proteins can either antagonize or cooperate in the transactivation of MMPs.

The gelatinase A promoter is distinguished by the lack of known trans-activator sequences (AP-1, PEA-3), the presence of a unique noncanonical TATA box as well as a putative enhancer element located at nucleotides -223 to -422 relative to the translational start site (Templeton and Stetler-Stevenson, 1991). Other sites such as an SP-1 consensus sequence located 120 base pairs from the start site and an adenovirus E1A repressor element have also been reported (Frisch, 1990). A TIE element is not found in the promoter sequence of the gelatinase A gene. Consistent with this sequence information is the observation that gelatinase A transcription is not suppresed but slightly upregulated by TGFb (Brown et al., 1990; Overall et al., 1991). The unique transcriptional regulation of gelatinase A may reflect a distinct role for this enzyme in ECM turnover.

Although the precise regulatory points for transcriptional regulation of MMPs is still not fully understood, recent evidence suggests that post-transcriptional regulation may also contribute to the production of some MMPs. TPA and TGFb are reported to regulate gelatinase A production by modulating the stability of the mRNA transcripts as measured by alteration of mRNA half-life (Overall et al., 1991). Stability in collagenase and stromelysin mRNA is mediated by treatment with proinflammatory cytokines such as IL-1 or TNF (Mauviel, 1993) as well as TPA or epidermal growth factor (EGF) (Delany, 1992) . The complexity at this level of regulation is dependent on not only the transactivators induced by environmental stimulants but also the genetic characteristics of the responding cell. Thus, the specific response of a given cell type will depend on many factors that may act in concert to yield a cell-specific response.

MMP ACTIVATION

Once the mRNA is transcribed and translated, the MMP protein is secreted as a zymogen. Post-translational modifications, such as N-linked glycosylation (Wilhelm et al., 1986; Wilhelm et al., 1987) and association of MMPs with endogenous inhibitors produced by the same cells, may occur during the secretion of these enzymes. N-linked glycosylation is not essential for proteolytic activity. Progelatinase A and progelatinase B are unique because these zymogens may be secreted complexed with either TIMP-2 or TIMP-1, respectively. The function of this complex formation is not fully understood (see below).

In order for ECM proteolysis to occur, the MMP zymogen must first be converted extracellularly to an active species. A great deal is known about the biochemistry of MMP activation from *in vitro* studies of these enzymes. However, the cellular and *in vivo* mechanism for the activation of these enzymes is not as well characterized. Based on the work performed in a number of laboratories, the mechanism for mammalian MMP activation has been proposed. In the mechanism known as the "cysteine switch", an unpaired cyteine residue in the PRCGXPD sequence of the profragment maintains the latency of the zymogen by direct coordination with the active site zinc atom (Van Wart and Birkedal-Hansen, 1990). Physical or chemical interruption of this interaction is the first step in the activation cascade of these enzymes. Subsequently, there is proteolytic cleavage of the amino-terminus. Loss of this amino-terminal propeptide domain is evidence of, but not equivalent to, activation of the proenzyme. Site-specific deletion of the R or C residues in this sequence results in autoactivation of the enzyme (Sanchez-Lopez et al., 1988; Park et al., 1991) . Additionally, synthetic peptides based on PRCGXPDV sequence inhibit not only proteolytic activity (Stetler-Stevenson et al., 1991; Nomura and Suzuki, 1993) but also human tumor cell invasion (Melchiori et al., 1992) and neurite outgrowth (Muir, 1994).

The plaminogen activator (PA)/ plasmin cascade and the distinct cellular activation of latent gelatinase A are the two mechanisms currently proposed for MMP activation. Latent interstitial collagenase, stromelysin-1 and gelatinase B are activated by the plasmin cascade which can occur either at the cell surface via the urokinase PA (uPA) receptor or at a site distant from that of enzyme secretion (Figure 1) (Werb et al., 1977; He et al., 1989). The uPA receptor present on inflammatory cells has been shown to facilitate the passage of cells through basement membranes and into inflammatory sites. The activation of latent gelatinase A appears to occur via a unique cell-mediated mechanism (Figure 2).

Plasmin is a protease derived from the action of PA on plasminogen. The activation of PA can be inhibited by the endogenous PA inhibitor (PAI) (Vassalli, 1991). While the tissue type PA (tPA) is active upon secretion and can directly interact with ECM components, the uPA is secreted as a zymogen that can be activated by plasmin, kallikrein, Factor XIIa or cathepsin B. Active plasmin cleaves 84 amino acids from the latent fibroblast stromelysin (57-kDa and glycosylated 60 kDa forms) and 81 amino acids from latent interstitial collagenase (57-kDa and glycosylated 60 kDa forms) to yield the fully active stromelysin (45-kDa) and the partially active interstitial collagenase (42-kDa). Subsequently, the activated stromelysin may increase the activity of interstitial collagenase 5-8 fold by cleaving 15 amino acids from its carboxy-terminus (40-kDa) (Figure 1). Coordinate activation of MMPs has been observed following stimulation of human monocytes (Corcoran et al., 1992).

The *in vivo* mechanism of activation for latent gelatinase A is still controversial. Some reports describe plasmin/PA mediated activation of an insoluble gelatinase A preparation (Salo et al., 1982; Keskioja et al., 1992). Others have demonstrated that both purified progelatinase A and progelatinase A/TIMP-2 complex are not activated but are degraded by plasmin (Okada et al., 1990). Additional reports have demonstrated that the cell-mediated activation of exogenous or endogenous gelatinase A occurs in the presence of either general serine protease inhibitors or specific inhibitors of plasmin and uPA (Ward et al., 1991; Murphy et al., 1992;

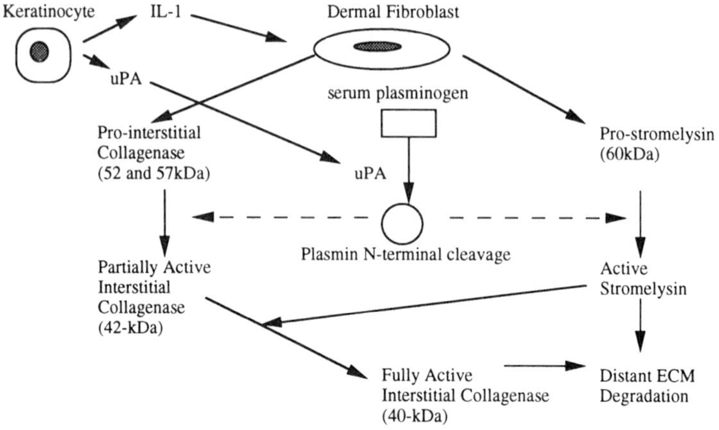

Figure 1 - Plasmin mediated MMP activation cascade. In this model, the keratinocyte derived IL-1 induces the production of pro-interstitial collagenase and pro-stromelysin from dermal fibroblasts. The serum plasminogen is converted to plasmin which subsequently cleasves the amino-terminal domain of pro-stromelysin and pro-interstitial collagenase. Active stormelysin will proteolytically degrade the matrix as well as increase the activity of interstitial collagenase. If the uPA receptor is available, this process maybe confined near the cell milieu.

Brown et al., 1993). This work also describes a cell-surface associated activator of latent gelatinase A, although the activator itself has not yet been isolated. Concanavalin A and phorbol ester treatment of tumor cells induces the expression of this cell membrane associated protein. The carboxy-terminus on latent gelatinase A appears to be required for activation since truncated enzyme which lack this domain is not activated by this mechanism (Figure 2)(Murphy et al., 1992). The cell-mediated activation of latent gelatinase A or latent gelatinase A/TIMP-2 complex can be inhibited by chelating agents as well as by addition of exogenous free TIMP-2. This mechanism is unique for gelatinase A and is not involved in the activation of gelatinase B. At present, two possible mechanisms for activation of latent gelatinase A have been proposed. It is conceivable that the activation could transpire directly via a membrane associated metalloproteinase that directly cleaves at the pro-peptide site. Alternatively, an interaction at the cell surface may induce a conformational change which result in the autoactivation of latent gelatinase A. Attempts at isolating a cell surface protease that specifically activates latent gelatinase A have to date been unsuccessful (Strongin et al., 1993).

INHIBITION OF MMPS BY TIMPS

As noted above, one of the characteristics that is now used to define new members of the matrix metalloproteinase family is inhibition of proteolysis by TIMPs. Three members of the TIMP family have been identified, cloned and sequenced; TIMP-1 (Docherty et al., 1985), TIMP-2 (DeClerck et al., 1989; Stetler-Stevenson et al., 1989), and TIMP-3 (Pavloff et al., 1992; Leco et al., 1994). TIMPs bind with high affinity in a 1:1 molar ratio to active MMPs and inhibits their proteolytic activity. The TIMPs contain 12 conserved cysteine residues that in TIMP-1 have been shown to be paired into six disulfide bridges forming six peptide loops and two knots (Williamson et al., 1990)(Figure 3). The TIMPs are approximately 40% identical at the amino acid level and presumably have similar disulfide loop patterns as that mapped for TIMP-1. Although both TIMP-1 and TIMP-2 transcripts encode a 21-kDa protein, TIMP-1 is glycosylated and migrates as a 28-kDa protein on SDS polyacrylamide gels. TIMP-3 is a unique member of this family because it is tightly bound to the extracellular matrix and is not present in the conditioned media of cells in culture (Pavloff et al., 1992; Leco et al., 1994). The high degree of homology between the amino-terminal domains of these proteins suggests that these disulfide loops are responsible for inhibiting MMP activity. This has been confirmed using truncated recombinant TIMP proteins and proteolytically derived TIMP fragments (Murphy et al., 1991; DeClerck et al., 1993). Recent mutational studies demonstrated that the region from cysteine-3 through cysteine-13 is critical for MMP inhibitory activity (O'Shea et al., 1992).

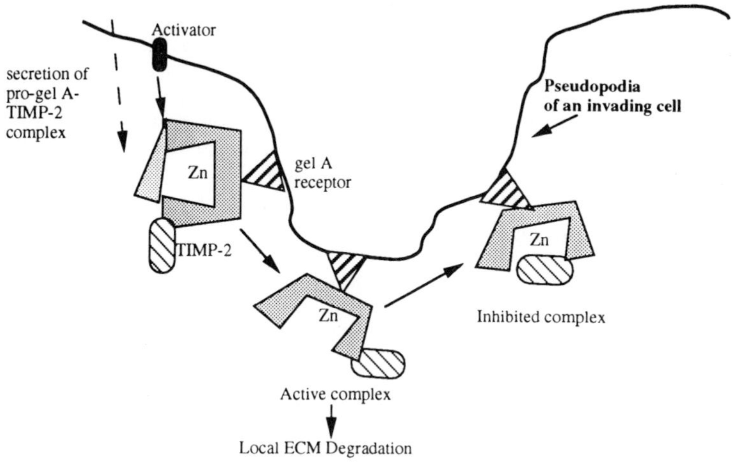

Figure 2 - Cell-Mediated activation of gelatinase:TIMP-2 complex. In this model, the latent complex possesses the amino terminal pro-piece and is carboxy-terminus associated with the carboxy-terminus of TIMP-2. The gelatinase A receptor on the pseudopod of an invading cell will bind the carboxy-terminus of the secreted glatinase A:TIMP-2 complex or free zymogen. Subsequently, the "activator" protein will activate MMP by either proteolysis of the pro-peptide or by inducing a confromational alteration in the protein that resuls in auto-proteolytic activationof the enzyme. Once activated, a second binding site becomes available for tighter binding and catalytic inhibition by TIMP-2.

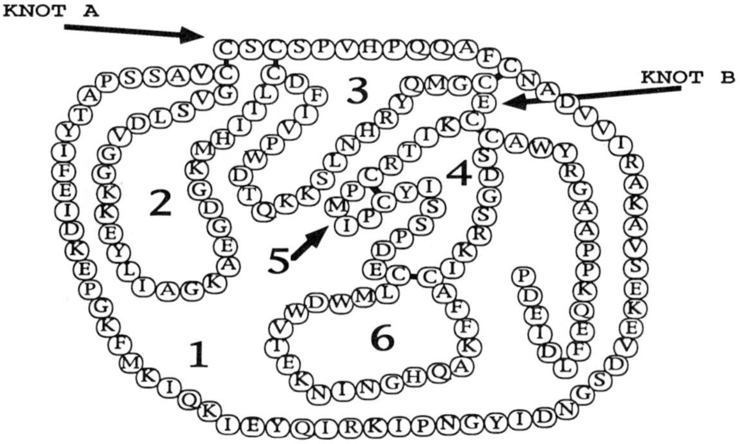

Figure 3 - Putative domain structure for TIMP-2. This figure depicts the predicted six loops formed by the characteristic disulfide bonding pattern of the TIMP family. The amino-terminal domain represents the domain responsible for binding and inhibiting the the catalytic activity of the MMPs. The carboxy-terminal protein domain represents the domain that associates with the carboxy-terminus of gelatinase A.

The transcriptional/post-transcriptional regulation for the TIMPs seem to be independent from that of MMP regulation. TIMP-1 mRNA expression is induced by a plethora of cytokines, hormones, and phorbol esters, whereas TIMP-2 seems to be constitutively expressed in many different tissues and cell lines (Leco et al., 1992; Leco et al., 1994). TIMP-3 and TIMP-1 are induced by EGF, TGFß and TPA. However, dexamethasone treatment selectively induces the expression of TIMP-3 and suppresses the EGF or TPA mediated induction of TIMP-1 (Leco et al., 1994). The differential regulation of TIMP gene expression supports the notion that individual TIMP family members may have specific physiologic roles.

In many cell lines, TIMP-2 is produced in a two fold excess over progelatinase A (unpublished observations). In this situation nearly all secreted progelatinase A is complexed to TIMP-2 either before or shortly after secretion. The interaction of TIMP-2 with latent and active gelatinase A has been examined using crosslinking agents and truncated enzyme species. Based on these studies, gelatinase A is postulated to possess two binding sites for TIMP-2 (Fridman et al., 1992; Kleiner et al., 1992). While the amino-terminal domain of TIMP-2 is known to retain inhibitory properties, our studies suggest that the carboxy-terminus of TIMP-2 directly interacts with the carboxy-terminus of gelatinase A (Fridman et al., 1992; Kleiner et al., 1992). Once the gelatinase A in the latent complex is activated, the amino-terminus of TIMP-2 interacts with a second site, presumably the active site, of gelatinase A. Additional studies from our laboratory suggest that the active gelatinase:TIMP-2 complex may be resistant to autodegradation while retaining a low level of proteolytic activity (Kleiner et al., 1993).

In addition to their function as MMP inhibitors, these proteins have other biologically significant functions. TIMP-1 (Gasson et al., 1985) and TIMP-2 (Stetler-Stevenson et al., 1992) have been documented to support the growth of erythroid precursors (CFU-E and BFU-E). Our laboratory has demonstrated that TIMP-2 plays a role in inhibiting angiogenesis, possibly by inhibiting basic fibroblast growth factor (bFGF) stimulated endothelial cell proliferation (Murphy et al., 1993). This growth suppressing activity of TIMP-2 is probably mediated by a specific TIMP-2 receptor and does not involve inhibition of MMP activity nor disruption of the b-FGF receptor. Although the important function of TIMPs as inhibitors of MMP activity is currently being realized, other potentially important biological functions of the TIMP molecules have yet to be fully investigated.

CONCLUSION

The degradation of the ECM by MMPs can be modulated at many levels including transcription and post-transcriptional regulation of the mRNA, activation of zymogen and inhibition of MMP activity by the endogenous inhibitor, TIMP. Although homology exists within these families, each protein posseses unique characteristics. In contrast to stromelysins, interstitial collagenases and gelatinase B, the activation of gelatinase A is unique because its not plasmin-dependent but appears to transpire locally at the cell surface. This local mechanism of activation tends to focus the matrix proteolysis on the immediate peri-cellular milieu. Although the zymogens of other members of this family such as stromelysin and collagenase could also be activated locally via the uPA receptor, their activation most likely occurs at sites distant from enzyme secretion. Each physiological and pathological situation has a unique tissue and cell-type specific spectrum of MMPs and TIMP expression. Recent data suggests that TIMPs may have additional biological activities that may influence physiologic and pathologic events that occur during ECM remodelling. Although each situation is distinct, an important target for therapeutic intervention in aberrant degradation or accumulation of ECM is modulation of MMP proteolytic activity. Currently available synthetic chemical inhibitors of MMP activity are the future tools for evaluating the consequences of altering MMP proteolytic activity, and therefore ECM turnover, during both physiologic and pathologic processes.

REFERENCES

Angel, P., M. Imagawa, R. Chiu, B. Stein, R. J. Imbra, H. J. Rhamsdorf, C. Jonat, P. Herrlich and M. Karin, 1987, Phorbol Ester-Inducible Genes Contain a Common Cis Element Recognized by a TPA-Modulater Trans-Acting Factor, *Cell* 49:729-739.

Birkedal-Hansen, H., W. G. I. Moore, M. K. Bodden, L. J. Windsor, B. Birkedal-Hansen, A. DeCarlo and J. A. Engler, 1993, Matrix Metalloproteinases: A Review, *Crit. Rev. Oral Biol. Med.* 4:197-250.

Brenner, D. A., M. O'Hara, P. Angel, M. Chojkier and K. Michael, 1989, Prolonged Activation of Jun and Collagenase Genes by Other Oncogenes and Phorbol Esters, *Nature* 337:661-663.

Brown, P. D., D. E. Kleiner, E. J. Unsworth and W. G. Stetler-Stevenson, 1993, Cellular activation of the 72 kDa type IV procollagenase/TIMP-2 complex, *Kidney Int.* 43:163-170.

Brown, P. D., A. T. Levy, I. M. Margulies, L. A. Liotta and W. G. Stetler-Stevenson, 1990, Independent expression and cellular processing of Mr 72,000 type IV collagenase and interstitial collagenase in human tumorigenic cell lines, *Cancer Res* 50:6184-91.

Conca, W., P. E. Auron, W. M. Aoun, N. Bennett, P. Seckinger, H. G. Welgus, S. R. Goldring, S. P. Eisenberg, J. M. Dayer, S. M. Krane and a. l. et, 1991, An interleukin 1 beta point mutant demonstrates that jun/fos expression is not sufficient for fibroblast metalloproteinase expression, *J Biol Chem* 266:16265-16268.

Corcoran, M. L., W. G. Stetler-Stevenson, P. D. Brown and L. M. Wahl, 1992, Interleukin 4 inhibition of prostaglandin E2 synthesis blocks interstitial collagenase and 92-kDa type IV collagenase/gelatinase production by human monocytes, *J Biol Chem* 267:515-9.

DeClerck, Y. A., T.-D. Yean, B. J. Ratzkin, H. S. Lu and K. E. Langley, 1989, Purification and Characterization of Two Related but Distinct Metalloproteinase Inhibitors Secreted by Bovine Aortic Endothelial Cells, *J. Biol. Chem.* 264:17445-17453.

DeClerck, Y. A., T. D. Yean, Y. Lee, J. M. Tomich and K. E. Langley, 1993, Characterization of the Functional Domain of Tissue Inhibitor of Metalloproteinases-2 (TIMP-2), *Biochem J* 289:65-69.

Delany, A. M. a. C. E. B., 1992, Post-transcriptional regulation of collagenase and stromelysin gene expression by epidermal growth factor and dexamethasone in cultured human fibroblasts, *J. Cell. Biochem* 50:400-410.

Docherty, A. J. P., A. Lyons, B. J. Smith, E. M. Wright, P. E. Stephens, T. J. R. Harris, G. Murphy and J. J. Reynolds, 1985, Sequence of Human Tissue Inhibitor of Metalloproteinases and Its Identity to Erythroid-Potentiating Activity, *Nature* 318:66-69.

Fridman, R., T. R. Fuerst, R. E. Bird, M. Hoyhtya, M. Oelkuct, S. Kraus, D. Komarek, L. A. Liotta, M. L. Berman and W. G. Stetler-Stevenson, 1992, Domain structure of human 72-kDa gelatinase/type IV collagenase. Characterization of proteolytic activity and identification of the tissue inhibitor of metalloproteinase-2 (TIMP-2) binding regions, *J Biol Chem* 267:15398-405.

Frisch, S. a. J. H. M., 1990, Positive and negative transcriptional elements of the human type IV collagenase gene, *Mol Cell Biol* 10:6524-6532.

Gasson, J. C., D. W. Golde, S. E. Kaufman, C. A. Westbrook, R. M. Hewick, R. J. Kaufman, G. G. Wong, P. A. Temple, A. C. Leary, E. L. Brown, E. C. Orr and S. C. Clark, 1985, Molecular characterization and expression of the gene encoding human erythroid-potentiating activity, *Nature* 315:768-771.

He, C., S. M. Wilhelm, A. P. Pentland, B. L. Marmer, G. A. Grant, A. Z. Eisen and G. I. Goldberg, 1989, Tissue cooperation in a proteolytic cascade activating human interstitial collagenase, *Proc. Natl. Acad. Sci.* 86:2632-2636.

Kerr, L. D., J. T. Holt and L. M. Matrisian, 1988, Growth Factors Regulate Transin Gene Expression by c-fos-Dependent and c-fos-Independent Pathways, *Science* 242:1424-1427.

Kerr, L. D., D. B. Miller and L. Matrisian, 1990, TGF-beta Inhibition of Transin/Stromelysin Gene Expression Is Mediated Through a Fos Binding Sequence, *Cell* 61:267-278.

Keskioja, J., J. Lohi, A. Tuuttila, K. Tryggvason and T. Vartio, 1992, Proteolytic Processing of the 72,000-Da Type-IV Collagenase by Urokinase Plasminogen Activator, *Exp Cell Res* 202:471-476.

Kleiner, D., Jr., A. Tuuttila, K. Tryggvason and W. G. Stetler-Stevenson, 1993, Stability analysis of latent and active 72-kDa type IV collagenase: the role of tissue inhibitor of metalloproteinases-2 (TIMP-2), *Biochemistry* 32:1583-92.

Kleiner, D., Jr., E. J. Unsworth, H. C. Krutzsch and W. G. Stetler-Stevenson, 1992, Higher-order complex formation between the 72-kilodalton type IV collagenase and tissue inhibitor of metalloproteinases-2, *Biochemistry* 31:1665-72.

Leco, K. J., L. J. Hayden, R. R. Sharma, H. Rocheleau, A. H. Greenberg and D. R. Edwards, 1992, Differential Regulation of TIMP-1 and TIMP-2 messenger RNA Expression in Normal and Ha-ras-Transformed Murine Fibroblasts, *Gene* 117:209-217.

Leco, K. J., R. Khokha, N. Pavloff, S. P. Hawkes and D. R. Edwards, 1994, Tissue Inhibitor of Metalloproteinases-3 (TIMP-3) Is an Extracellular Matrix-associated Protein with a Distinctinve Pattern of Expression in Mouse Cells and Tissues, *J. Biol. Chem.* 269:9352-9360.

Lovejoy, B., A. Cleasby, A. M. Hassell, K. Longley, M. A. Luther, D. Weigl, G. McGeehan, A. B. McElroy, D. Dewry, M. H. Lambert and S. R. Jordan, 1994, Structure of the Catalytic Domain of Fibroblst Collagenase Complexed with and Inhibitor, *Science* 263:375-377.

Mauviel, A., 1993, Cytokine Regulation of Metalloproteinase Gene Expression, *J. Cell. Biol.* 53:288-295.

Melchiori, A., A. Albini, J. M. Ray and W. G. Stetler-Stevenson, 1992, Inhibition of tumor cell invasion by a highly conserved peptide sequence from the matrix metalloproteinase enzyme prosegment, *Cancer Res* 52:2353-6.

Muir, D., 1994, Metalloproteinase-Dependent Neurite outgrowth within a synthetic extracellular matrix is induced by nerve growth factor, *Exp Cell Res* 210:243-252.

Murphy, A. N., E. J. Unsworth and W. G. Stetler-Stevenson, 1993, Tissue Inhibitor of Metalloproteinases-2 Inhibits bFGF-Induced Human Microvascular Endothelial Cell Proliferation, *J Cell Phys* 157:351-358.

Murphy, G., A. Houbrechts, M. I. Cockett, R. A. Williamson, M. O'Shea and A. J. Docherty, 1991, The N-terminal domain of tissue inhibitor of metalloproteinases retains metalloproteinase inhibitory activity, *Biochemistry* 30:8097-102.

Murphy, G., F. Willenbrock, R. V. Ward, M. I. Cockett, D. Eaton and A. J. Docherty, 1992, The C-terminal domain of 72 kDa gelatinase A is not required for catalysis, but is essential for membrane activation and modulates interactions with tissue inhibitors of metalloproteinases, *Biochem J* 283:637-641.

Nagase, H., A. J. Barrett and J. F. Woessner 1992, Nomenclature and Glossary of the Matrix Metalloproteinases, *In* Matrix Metalloproteinases and Inhibitors. Proceedings of the Matrix Metalloproteinase Conference Held at Sandestin Beach, FL, September 11-15, 1989, cd. H. Birkedal-Hansen, Z. Werb, H. G. Welgus and H. E. Van Wart, H. Birkedal-Hansen, Z. Werb, H. G. Welgus and H. E. Van Warts. Gustav Fischer Verlag, New York, 421-424.

Nomura, K. and N. Suzuki, 1993, Stereo-specific inhibition of sea urchin envelysin (hatching enzyme) by a synthetic autoinhibitor peptide with a cysteine-switch consensus sequence., *FEBS Lett.* 321:84-88.

O'Shea, M., F. Willenbrock, R. A. Williamson, M. I. Cockett, R. B. Freedman, J. J. Reynolds, A. J. Docherty and G. Murphy, 1992, Site-Directed Mutations That Alter the Inhibitory Activity of the Tissue Inhibitor of Metalloproteinases-1: Importance of the N-Terminal Region Between Cysteine 3 and Cysteine 13, *Biochemistry* 31:10146-10152.

Okada, Y., T. Morodomi, J. J. Enghild, K. Suzuki, A. Yasui, I. Nakanishi, G. Salvesen and H. Nagase, 1990, Matrix metalloproteinase 2 from human rheumatoid synovial fibroblasts. Purification and activation of the precursor and enzymic properties, *Eur. J. Biochem.* 194:721-730.

Overall, C. M., J. L. Wrana and J. Sodek, 1991, Transcriptional and post-transcriptional regulation of 72-kDa gelatinase/type IV collagenase by transforming growth factor-beta 1 in human fibroblasts. Comparisons with collagenase and tissue inhibitor of matrix metalloproteinase gene expression, *J Biol Chem* 266:14064-14071.

Park, A. J., L. M. Matrisian, A. F. Kells, R. Pearson, Z. Y. Yuan and M. Navre, 1991, Mutational Analysis of the transin (rat stromelysin) autoinhibitor region demonstrates a role for residues surrounding the "cysteine switch", *J. Biol. Chem.* 266:1584-1590.

Pavloff, N., P. W. Staskus, N. S. Kishnani and S. P. Hawkes, 1992, A New Inhibitor of Metalloproteinases from Chicken: ChIMP-3. A Third Member of the TIMP Family, *J Biol Chem* 267:17321-17326.

Salo, T., L. A. Liotta, J. Keski-Oja, T. Turpeennieme-Hujanen and K. Tryggvason, 1982, Secretion of a Basement Membrane Collagen Degrading Enzyme and Plasminogen Activator by Transformed Cells- Role in Metastasis, *Int. J. Cancer* 30:669-673.

Sanchez-Lopez, R., C. M. Alexander, O. Behrendtsen, R. Breathnach and Z. Werb, 1993, Role of zinc-binding- and hemopexin domain-encoded sequences in the substrate specificity of collagenase and stromelysin-2 as revealed by chimeric proteins, *J Biol Chem* 268:7238-47.

Sanchez-Lopez, R., R. Nicholson, M.-C. Gesnel, L. M. Matrisian and R. Breathnach, 1988, Structure-Function Relationships in the Collagenase Gene Family Member Transin, *J. Biol. Chem.* 263:11892-11899.

Sassone-Corsi, P., L. J. Ransone and I. M. . Verma, 1989, Cross-talk in Signal Transduction: TPA-inducible Factor jun/AP-1 Activates cAMP -Responsive Enhancer Elements, *Oncogene* 5:427-431.

Stetler-Stevenson, W. G., N. Bersch and D. W. Golde, 1992, Tissue inhibitor of metalloproteinase-2 (TIMP-2) has erythroid-potentiating activity, *Febs Lett* 296:231-4.

Stetler-Stevenson, W. G., H. C. Krutzsch and L. A. Liotta, 1989, Tissue Inhibitor of Metalloproteinase (TIMP-2), *J. Biol. Chem.* 264:17374-17378.

Stetler-Stevenson, W. G., J. A. Talano, M. E. Gallagher, H. C. Krutzsch and L. A. Liotta, 1991, Inhibition of human type IV collagenase by a highly conserved peptide sequence derived from its prosegment, *Am J Med Sci* 302:163-70.

Strongin, A. Y., I. E. Collier, P. A. Krasnov, L. T. Genrich, B. L. Marmer and G. I. Goldberg, 1993, Human 92 kDa Type-IV Collagenase - Functional Analysis of Fibronectin and Carboxyl-End Domains, *Kidney Int* 43:158-162.

Strongin, A. Y., B. L. Marmer, G. A. Grant and G. I. Goldberg, 1993, Plasma Membrane-Dependent Activation of the 72-kDa Type-IV Collagenase Is Prevented by Complex Formation with TIMP-2, *J Biol Chem* 268:14033-14039.

Templeton, N. S. and W. G. Stetler-Stevenson, 1991, Identification of a basal promoter for the human Mr 72,000 type IV collagenase gene and enhanced expression in a highly metastatic cell line, *Cancer Res* 51:6190-3.

Van Wart, H. and H. Birkedal-Hansen, 1990, The cysteine switch: a principle of regulation of metalloproteinase activity with potential applicability to the entire matrix metalloproteinase gene family, *Proc Natl Acad Sci U S A* 87:5578-5582.

Vassalli, J.-D., Sappino, A-P, Belin , D., 1991, The plasminogen activator/plasmin system, *J Clin Invest* 88:1067-1072.

Ward, R. V., S. J. Atkinson, P. M. Slocombe, A. J. Docherty, J. J. Reynolds and G. Murphy, 1991, Tissue inhibitor of metalloproteinases-2 inhibits the activation of 72 kDa progelatinase by fibroblast membranes, *Biochim Biophys Acta* 1079:242-6.

Werb, Z., C. L. Mainardi, C. Vater and E. D. Harris, 1977, Endogenous Activation Of Latent Collagenase By Rheumatoid Synovial Cells. Evidence for a Role of Plasminogen Activator, *New Engl. J. Med.* 296:1017-1023.

Wilhelm, S. M., I. E. Collier, A. Kronberger, A. Z. Eisen, B. L. Marmer, G. A. Grant, E. A. Bauer and G. I. Goldberg, 1987, Human skin fibroblast stromelysin: Structure, glycosylation, substrate specificity, and differential expression in normal and tumorigenic cells, *Proc. Natl. Acad. Sci. (USA)* 84:6725-6729.

Wilhelm, S. M., A. Z. Eisen, M. Teter, S. D. Clark, A. Kronenberger and G. Goldberg, 1986, Human fibroblast collagenase: Glycosylation and tissue-specific level of enzyme synthesis, *Proc. Natl. Acad. Sci. (USA)* 83:3756-3760.

Williamson, R., F. Marston, S. Angal, P. Koklitis, M. Panico, H. Morris, A. Carne, B. Smith, T. Harris and R. Freedman, 1990, Disulphide bond assignment in human tissue inhibitor of metalloproteinases (TIMP), *Biochem. J.* 268:267-274.

ELASTIC FIBERS AND THEIR ROLE IN BLADDER EXTRACELLULAR MATRIX

Joel Rosenbloom[1], Harry Koo[1], Pamela S. Howard[1], Robert Mecham[2], and Edward J. Macarak[1]

[1] Department of Anatomy & Histology
School of Dental Medicine
University of Pennsylvania
Philadelphia, PA 19104

[2] Department of Cell Biology & Physiology
School of Medicine
Washington University
St. Louis, MO 63110

INTRODUCTION

The elastic properties of many tissues of the vertebrate body are due in large part to elastic fibers in the extracellular matrix. In some tissues, such as the skin, they form a relatively small but important fraction of the dry weight (2-4%), while in other tissues, such as large arteries and certain specialized ligaments, they may comprise greater than 50%. In the electron microscope, elastic fibers are seen to be composed of two components.[1-3] An amorphous fraction lacking any apparent regular or repeating structure composes upwards of 90% of the mature fiber. The microfibrillar component consists of 10-12 nm fibrils located primarily around the periphery of the amorphous component, but also to some extent interspersed within it (Fig. 1). In addition to being distinguishable morphologically, the two components are chemically distinct. The amorphous portion of the elastic fiber is composed of a single protein, elastin, which is primarily responsible for the elastic properties. Although the exact composition of the microfibrils remains to be defined, it is likely that they are composed of several glycoproteins.[4]

COMPONENTS OF THE ELASTIC FIBER

Elastin

Because of extensive crosslinking, elastin is highly insoluble and can be purified fairly readily either by rather harsh extraction procedures, such as 0.1 M NaOH at 98°C,[5] which solubilize most other proteins but leave an insoluble elastin residue or by somewhat milder methods involving extraction with concentrated guanidine solutions and digestion with bacterial collagenase.[6] Elastin is rich in glycine (33%) and proline (10-13%), and more than 40% of the remaining amino acids are hydrophobic ones.[7,8] Crucial to the proper functioning

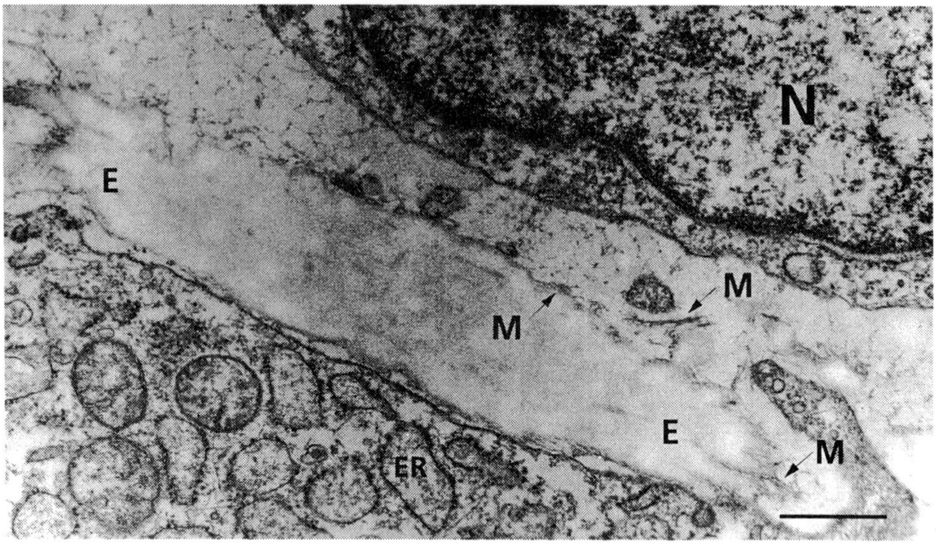

Figure. 1. Electron micrograph of 15-day chick embryo aorta. The forming elastic fiber is situated between two cells, the nucleus (N) of one of which is marked and the endoplasmic reticulum (ER) of the other. Note the microfibrils (M) on the surface of the amorphous elastin (E). Some microfibrils are interspersed within the amorphous elastin. Bar = 0.5 um. Uranyl acetate and lead citrate; x45,000.

of the elastic fiber is the high degree of crosslinking of the polypeptides. The same copper-requiring enzyme, lysyl oxidase, involved in collagen crosslinking appears to catalyze the crosslinking of elastin by oxidizing selective lysine residues to allysine.[9] The enzyme is secreted from the cell and crosslinking is probably initiated in the extracellular space rather than intracellularly. Some of the final difunctional products, such as lysinonorleucine and allysine aldol, are also found in collagen fibers, but the tetrafunctional ones, the desmosines, are found exclusively in elastin.

Tropoelastin. The isolation of a soluble form of elastin, called tropoelastin (M_r~72,000), from copper-deficient animals permitted more detailed chemical characterization.[10,11] Tryptic digestion of porcine tropoelastin produced two classes of peptides: (i) small ones rich in alanine were derived from regions destined to form the crosslinks and (ii) larger peptides rich in hydrophobic residues were derived from the regions responsible for the elastic behavior.[12,13] As shown later by sequencing of cDNA, these small crosslink peptides are spaced throughout the tropoelastin separated by the larger hydrophobic segments.

Biosynthesis of Elastin and Fiber Assembly. Under normal circumstances, elastin is synthesized in quantity only by embryonic or rapidly growing tissues, and the intracellular biosynthetic pathway appears to be the classical one followed by many secreted proteins.[14] Although debatable, it is probable that tropoelastin is incorporated without proteolytic cleavage into the insoluble fiber.[15] In a number of developing embryonic systems, a strong correlation between messenger RNA levels and the rate of elastin synthesis has been demonstrated, suggesting that the rate of synthesis is controlled by the mRNA steady-state level.[16-19] Relatively little is known concerning the detailed mechanisms involved in fiber assembly, but the notion that tropoelastin molecules are simply secreted and then diffuse onto the surface of growing fibers where they bind and become crosslinked seems inadequate to explain the variable form of elastic fibers in different tissues. Study of the earliest stages of fibrillogenesis has revealed that accumulation of microfibrils always precedes the deposition of amorphous elastin.[1,3] The distribution of the microfibrils in the plane of the developing elastic fibers gives the impression that they serve as tissue organizers for the developing elastic structures. The existence of several different structural forms of the elastic fiber also argues for a more directive role of the cell in fiber assembly. Mecham and co-workers have isolated a 67 kDa protein found on the surface of elastin-producing cells which specifically binds tropoelastin and which also exhibits beta-galactoside lectin activity.[20] Interestingly and somewhat surprisingly, the beta-galactoside, lactose, markedly decreases the binding of elastin polypeptides to the protein and dramatically inhibits elastin fiber assembly by cultured elastic cartilage chondrocytes.[21] These observations suggest that the 67 kDa protein may be a receptor involved in modulating the position of the cell surface where tropoelastin secretion is taking place and/or modulating the association of the cell surface with the growing elastic fiber, thereby affecting the architecture of the fiber.

Characterization of the Elastin Gene. As determined by sequence analysis of cDNA clones corresponding to the entire length of human,[22,23] bovine,[24,25] and rat[26] tropoelastin, there is generally good agreement at the nucleotide and encoded amino acid sequence levels among the mammalian elastins, which differ significantly, however, in some segments from those of the chicken.[27] Sequences homologous to all the tryptic peptides of porcine tropoelastin were also identified.[12] Additionally, the deduced amino acid sequences of all the cDNAs were found to have an unusual, conserved carboxy terminal segment which contains 2 cysteines and ends with 4 basic residues, and which was not previously identified in the protein sequencing. The characteristics of the carboxy terminus suggest that this portion of tropoelastin may interact strongly with other acidic matrix proteins such as those compoing the microfibrils.

The cDNA analyses demonstrated that tropoelastin consists largely of alternating hydrophobic domains responsible for the elastic properties, and lysine-rich domains which form crosslinks between molecules (Fig. 2). The lysines usually occur in pairs, supporting the suggestion that a given desmosine/isodesmosine serves to join only 2 tropoelastin molecules rather than the theoretically possible 4 molecules. The potential crosslinking sequences are not uniformly distributed and occur at closer intervals in the first 200 residues, resulting in an asymmetry to the molecule. In the bovine[25,28] and human[22,29] elastin genes, rather small exons (27-186 bp), are interspersed among large intron expanses, but the exons do not exhibit any regularity in size as is found in the fibrillar collagen genes. Significantly, the coding sequences corresponding to hydrophobic and crosslink domains of the protein are found in separate exons. With rare exceptions, these 2 types of exons alternate, so that the domain structure of the protein is a reflection of the exon organization of the gene. Comparison of the nucleotide sequences of bovine and human elastin cDNAs and genomic sequences clearly demonstrated that extensive alternative splicing of the primary transcripts occurs.[22-25] Both hydrophobic and crosslink domains are affected, so that 2 crosslink domains may be brought into apposition or the interval between crosslink domains may be increased. It is not known whether these variations have any functional significance, although clearly a tighter or looser fiber network could be produced by translation of appropriately spliced transcripts. S1 mapping experiments using elastin mRNA isolated from the developing bovine nuchal ligament have demonstrated that the splicing out of some exons occurs frequently (~50%) but in the majority of cases it is infrequent.[28] These experiments also suggested that the alternative splicing pattern in the adult may differ significantly from that occuring during development. It remains to be determined whether the splicing pattern is tissue specific. If all the alternatively spliced transcripts are translated, there will be significant primary sequence variation among individual tropoelastin molecules, and such differences could explain the finding of variant forms of tropoelastin in several species.[30-32] Currently, relatively little is known about elastin gene expression in the developing bladder (see below).

The Elastin Gene Promoter. Analyses of the 5' region flanking the translation nitiation codon revealed extremely strong conservation between the bovine and human genes (> 85% homology from -1 to -588), suggesting that this segment may have important functional roles.[28,29] No canonical TATA boxes or functional CAAT sequences were found, but the segment contains a remarkable number of potential binding sites for transcription regulatory factors, including multiple SP1 and AP2 binding sites, glucocorticoid responsive elements, and TPA and cyclic AMP (CRE) responsive elements (for review, see 33). Such elements may explain previous reports indicating that elastin gene expression can be up-

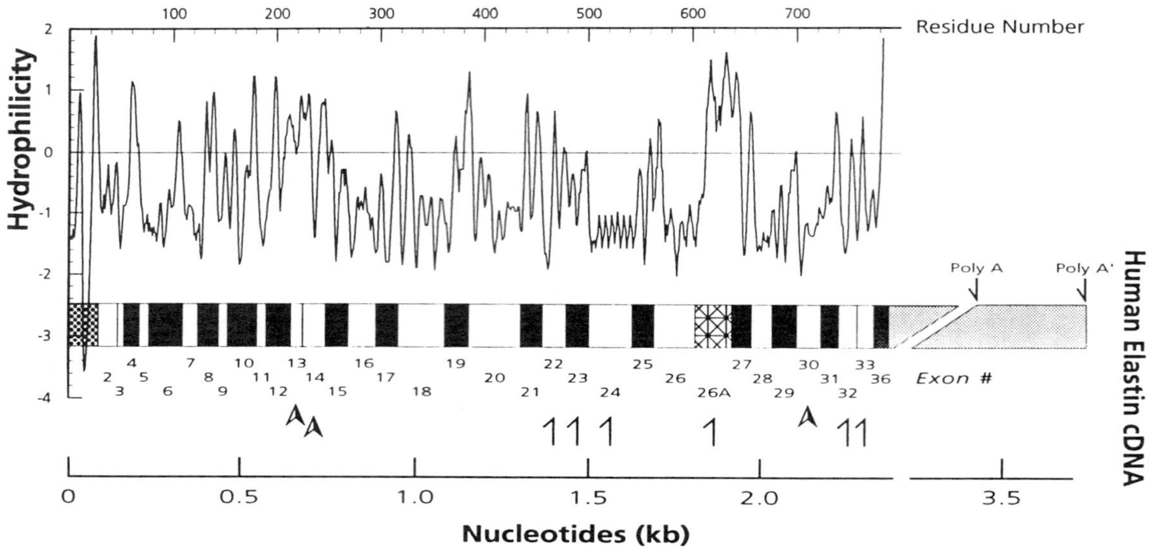

Figure 2. Diagrammatic representation of human elastin cDNA and hydropathy analysis of tropoelastin by the method of Kyte and Doolittle. The cDNA is divided into exons which are numbered. In order to preserve a numbering system consistent with homologous sequences in the bovine gene, the most 3' exon has been designated 36 in the human gene. Exons 34 and 35, found in the bovine gene, are absent in the human gene. Exons encoding hydrophobic sequences (□); exons encoding potential crosslinking sequences (■); exon 1 encoding most of the signal sequence (▩); exon encoding the carboxyl terminus (▩); 3' untranslated region (▩); exon encoding 26A (▩). A 1 symbol marks the exons known to be subject to alternative splicing in the human and bovine transcripts. A 𝐀 symbol marks exons presently known to be alternatively spliced only in the bovine transcript. In the hydropathy diagram, the hydrophilic peaks correspond for the most part to potential crosslinking domains.

regulated by glucocorticoids,[34,35] and that cAMP can abrogate the cGMP stimulation of elastin production by ligamentum nuchae fibroblasts.[36] There is also an extended sequence of alternating guanine and pyrimidine residues, a type of sequence which may be associated with Z DNA and which may be involved in transciptional regulation.[37]

Functional analysis of the human elastin promoter was carried out by constructing a panel of promoter/chloramphenicol acetyl transferase (CAT) gene chimeric constructs and assay of the panel by transient transfection in freshly isolated neonatal rat aortic smooth muscle cells which actively express elastin, as well as several other cell types.[38] The results demonstrated the presence of both up and down-regulatory elements within the 5' flanking region, and indicated that the core promoter necessary for basal expression was contained within the region -160 to -1. The positive regulatory and core promoter activities may be explained, at least in part, by the presence of multiple SP-1 and AP2 binding sites within these regions, which may act as general transcription elements. However, these experiments also suggested that all elements for tissue and developmental-specific expression of the elastin gene may not reside within the 2.26 kb promoter region tested. Comparison of the sequences of the first introns of the bovine and human elastin genes revealed regions of strong homology, suggesting the presence of a functional enhancer, but the existence of such enhancer(s) remains to be proven.

Microfibrils

On electron microscopic examination at high magnification, the microfibrils are seen to consist of apparently tubular structures, about 10-12 nm in diameter, which possess different staining properties at the ultrastructure level and susceptibilities to enzymatic digestion than the amorphous elastin. At high resolution, the microfibrils are seen to have a significant substructure, suggesting that they may be composed of more than one protein. Apparently similar microfibrils are present in many tissues, including those which contain abundant elastin as well as those in which there is no visible or immunoreactive elastin, such as the ocular ciliary zonule and the periodontal ligament. In the dermis, the microfibrillary bundles which connect the deep dermal elastic plexus with the dermal-epidermal junction region are seen to consist of microfibrils indistinguishable from those associated with elastic fibers. In their superficial distribution, no immunoreactive elastin is associated with them, but as they traverse the dermis they acquire an increasing amount of amorphous elastin.

Microfibrillar Proteins. Because of their insolubility and apparent complexity, chemical characterization of the microfibrils has progressed slowly. The problem of identifying structurally important components is compounded by the observation that several proteins found in relatively small amounts have been localized to the elastic microfibrils by immunoelectron microscopy.[39,40] While they may have some some functional role, it is very unlikely that any one of these is a major structural component of the microfibrils. Nevertheless, within the last few years significant progress has been made in solidly identifying certain proteins as structural components. The largest of these and possibly the most important is fibrillin, a 350 kDa glyprotein, reported to form an integral part of the microfibril structure.[41] Electron microscopic images of monomeric fibrillin prepared from human fibroblast cultures reveal an extended flexible molecule approximately 148 nm long and 2.2 nm wide. Multiple fibrillin molecules may align in a parallel head to tail fashion to form a major portion of the microfibrils. Molecular cloning studies have demonstrated that there are at least two distinct homologous human genes encoding fibrillin proteins, one located on chromosome 15q15-q21 and the other on chromosome 5q23-q31.[42,43] Analysis of the amino acid sequences deduced from the cloned cDNAs and genes for Fib5 and Fib15 demonstrated that the proteins contain multiple repeats of a sequence motif previously observed in epidermal growth factor, having six conserved cysteines.[44] Many of these repeats contain consensus sequences that have been associated with calcium binding through hydroxylation of asparagine or aspartic acid residues. A second motif containing eight cysteines, found in transforming growth factor ß1-binding protein[45] has also been identified in all three fibrillins. Figure 3 diagrammatically illustrates the domain structure of human fibrillin-15.[46] A very exciting finding is the demonstration of genetic linkage of the fibrillin gene on chromosome 15 to the Marfan syndrome[42,43] and the gene on chromosome 5 to congenital contractural arachnodactyly.[42] The cloning of these genes promises to facilitate greatly the determination of the function of the fibrillin proteins and their role in the pathophysiology of the elastic fiber.

A number of other proteins described as microfibril constituents appear unrelated to the fibrillins. When bovine nuchal ligament, a tissue rich in microfibrils, was extracted with reductive saline, five major bands having M_rs of 340,78,70,31, and 25 kDa were identified upon gel electrophoresis.[47] The 340 kDa component is the bovine homologue of human fibrillin 15. Immunoelectron microscopy using affinity-purified antibodies demonstrated that the 78 kDa (called microfibrillar protein 78 or MP78) and the 31 kDa protein, termed microfibril associated glycoprotein (MAGP), were components of microfibrils. The relationship of the 25 kDa and 70 species to the microfibrils remains to be established.

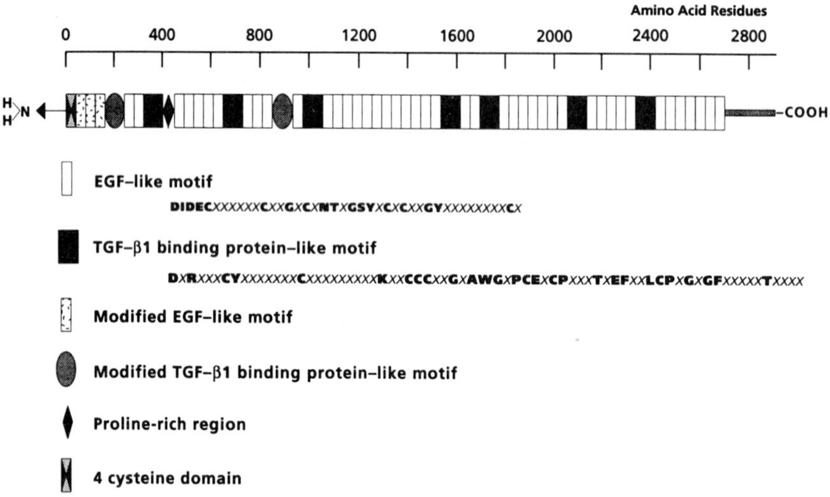

Figure 3. Diagrammatic representation of the amino acid sequence of human fibrillin-15. The different types of domains are indicated.

MAGP has been cloned and found to be distinct from fibrillin. The deduced primary structure indicates that MAGP contrains two structurally distinct regions, the amino-terminal half of the protein is rich in glutamine, proline, and acidic amino acids whereas the carboxy-terminal half contains all theirteen of the cystein residues and most of the basic amino acids.[48] The observation that MAGP can be extracted from tissues by several solvents, provided they contain a strong reducing agent, suggests that intermolecular disulfide bonding is an important feature of the associtaion of the polypeptide chains in the fibrils.

Another 52 kDa glyprotein, termed associated microfibril protein (AMP) was cloned from a chick embryo library,[49] and the encoded protein was remarkable in that it was extremely acidic with glutamic acid comprising 23% and aspartic acid 6% of the residues. Antiserum elicited to a 14 amino acid synthetic peptide sequence of the protein was shown by immunoelectron microscopy to localize specifically to ultrastructurally definable microfibrils in tissue sections from chick aortae, bovine nuchal ligament and human ocular zonules. The extremely acidic nature of the protein suggests that it may have an important function in the assembly of the very basic tropoelastin molecules.

In summary, the available evidence indicates that microfibrillar structures with apparently similar morphology are widely distributed in many tissues in the body. They appear to be complex structures, probably made up of a number of components, including glycoproteins. It is not clear whether they constitute a single type of structure with a uniform chemical composition, or whether they have different chemical compositions in different tissues. They are difficult structures to characterize on account of their small size, their intimate admixture with other matrix components, their relative insolubility, and because of the need for high resolution immunoelectron microscopy to establish that a putative component is located in the elastic fiber microfibrils.

ELASTIC FIBER COMPONENTS IN THE BLADDER

Up to the present, studies on the elastic fiber in the bladder have utilized histological analyses rather than biochemical ones. Kim et al.[50] used a point counting method to estimate the number of elastin fibers in the developing human bladder. They found that elastic fibers were difficult to find in fetuses less than 20 weeks gestation, but progressively increased in number after that age. Obstruction in the fetal bladder appeared to increase the number of fibers.[51] These measurements were of necessity semiquantitative and it was not possible to analyze for the individual components of the fibers.

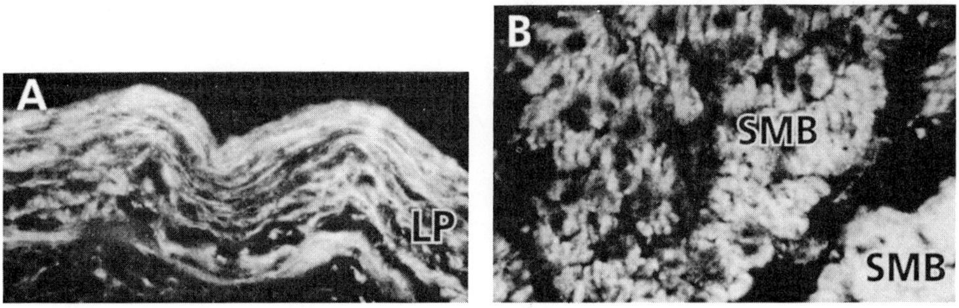

Figure. 4. A: Antielastin antibody staining in the lamina propria. The Lamina propria (LP) stains heavily with the antielastin antibody. It localizes to an area immediately beneath the non fluorescent urothelial layer at the top of the photograph. The elastin appears to be arranged in horizontal laminae parallel to the urothelial layer. **B**: Antielastin antibody staining of the detrusor smooth muscle bundles. The detrusor smooth muscle bundles heavily localize the antielastin antibody. There appears to be intracellular localization of the antibody to individual smooth muscle cells within the bundles as well as staining of the connective tissue between bundles.

We have initiated analysis of the elastic fiber in the developing bovine bladder using specific antibodies and nucleic acid probes to the component proteins. Figures 4-6 illustrate the localization of elastin, MAGP and Fib15 by immunofluorescence microscopy. These studies clearly demonstarted that in addition to elastin, MAGP and Fib15 were found in both the urothelium/lamina propria and detrusor layers. The localization of MAGP and Fib15 to the urothelium was particularly striking, and this is of some interest since their presence may be related to the highly elastic properties of the urothelium.

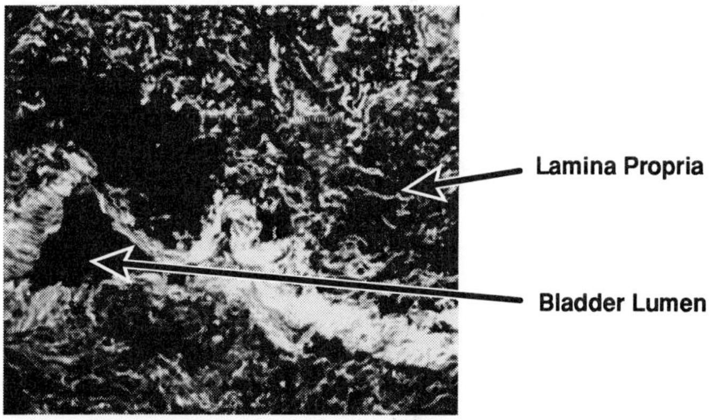

Figure. 5. Immunohistochemical staining of fetal bovine bladder using affinity purified rabbit antibody to a paptide sequence, YPDHVQYTHY, found in MAGP, and rhodamine-conjugated goat anti-rabbit IgG.

In order to begin analysis of the expression of the genes encoding these proteins, mRNA was extracted from developing bladders and the transcripts for the individual proteins quantitated by slot-blot hybridization. The results of these preliminary studies, which are summarized in Table 1, demonstrate that the genes for all three proteins are expressed during the second and third trimesters up to term. For the microfibrillar proteins MAGP and Fib15, the relative expression was somewhat higher in the urothelium/lamina propria than the detrusor layer and the maximal levels were found in the third trimester with substantially decreased expression in the calf. The expression of elastin was similar in the fetus, but relatively higher levels were found in the calf. In several other tissues, it has been well documentedthat the appearance of microfibrils precedes the deposition of elastin, and the present results are consistent with this sequence of events. The continued comparatively high level of expression of elastin in the calf may reflect growth and functional maturation of the bladder. However, more detailed studies are required to delineate firmly the temporal relationships in expression of the proteins and their functional significance.

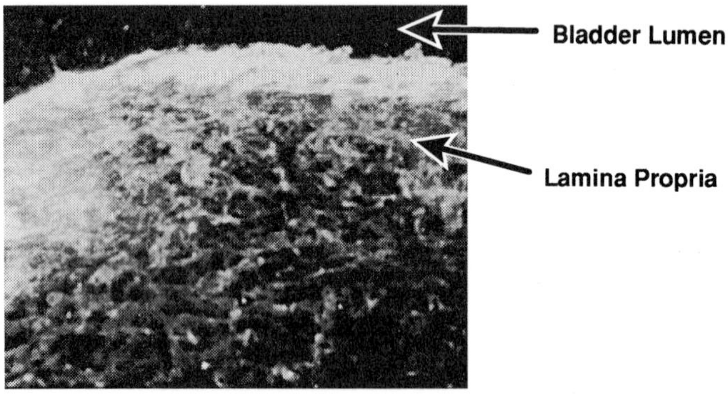

Figure. 6. Immunohistochemical staining of fetal bovine bladder using a monospecific rabbit antibody to the sequence, PVYLYPSREP, found in the human fibrillin gene on chromosome 15. the intensly staining area is the urothelium and below it is the lamina propria.

Future Directions

Detailed analysis and the role of the elastic fiber in the bladder is presently only in its initial stages and much remains to be understood. Further analyses are necessary to determine the normal complement of elastic fiber proteins and their distribution within the bladder must be mapped. In addition, definition of the stages of formation of the elastic fiber during

Table 1. Relative mRNA levels of elastic fiber proteins in developing fetal bovine bladder.[1]

	AGE			
	2nd Trimester	3rd Trimester	Term	Calf
Elastin				
Detrusor	1.27	0.78	0.67	1.00
Urothelium/LP	1.37	1.59	1.09	1.84
Fibrillin 15				
Detrusor	5.97	6.36	4.52	1.00
Urothelium/LP	8.71	10.70	6.55	1.32
MAGP				
Detrusor	3.71	3.88	2.29	1.00
Urothelium/LP	5.20	6.85	4.47	0.94

[1] Total RNA was extracted from the separated urothelium/lamina propria and detrusor layers of various age bovine bladders by the acid guanidinium thiocyanate method.[52] The relative levels of individual transcripts were determined by slot blot hybridization of serially diluted samples using specific probes which were labeled with ^{32}P by nick translation. Values are expressed in arbitrary absorbance units per mg RNA relative to the value found in the calf detrusor layer which is taken as 1.00.

development will require similar kinds of studies. The role of the elastic fiber in proper bladder function must be crucial since the bladder undergoes continuous and reversible shape change during each filling and emptying cycle. It is particularly important to determine the relationships of these proteins to the resident cells in order to understand the functional significance of such cell/matrix interactions. Since active tension is generated by smooth muscle cells within the bladder wall, there must be physical connections between the cells and the extracellular matrix to transfer the cell-generated force to the matrix and ultimately to the entire bladder wall. There is no information regarding how each of the major structural layers within the bladder associate with one another. For example, are microfibrillar proteins structural components of the basement membrane underlying the urothelium? Do the microfibrils by themselves have an anchoring function? Answeres to these and similar questions will require additional studies. This information will be crucial to understanding abnormalities which occur as a result of genetic or acquired disease.

ACKNOWLEDGMENTS

Supported by NIH Grants AR41474, AR20553, HL26499, and HL41926.

REFERENCES

1. W.H. Fahrenbach, L.B. Sandberg, and C.G. Cleary, Ultrastructural studies on early elastogenesis. Anat. Rec. 155:563 (1966).
2. T.K. Greenlee, Jr., R. Ross, and J.L. Hartman. The fine structure of elastic fibers, J. Cell Biol. 30:59 (1967).
3. H.E. Karrer, and J. Cox, Electronmicroscope study of developing chick embryo aorta. J. Ultrastruct. Res. 4:420,(1961).
4. E.G. Cleary, The microfibrillar component of the elastic fibers. Morphology and biochemistry. in "Connective Tissue Disease. Molecular Pathology of the Extracellular Matrix," U. Uitto and A.J. Perejda, eds., Marcel Dekker, New York (1987).
5. A.I. Lansing, T.B. Rosenthal, M. Alex, and W.W. Dempsey, The structure and chemical characterization of elastic fibers as revealed by elastase and by electron microscopy. J.Anat. Rec. 114: 555 (1952).
6. R. Ross, and P. Bornstein, The elastic fiber. The separation and partial characterization of its macromolecular components. J. Cell Biol. 40:366 (1969).
7. B.C. Starcher, and M.J.Galione, Purification and comparison of elastins from different animal species. Anal. Biochem 74:441 (1976).
8. H. Sage, and W.R. Gray, Studies on the evolution of elastin. Comp. Biochem. Physiol. 64B:313 (1979).
9. H.M. Kagan, Characterization and regulation of lysyl oxidase. In Regulation of Matrix Accumulation Vol I (R.P. Mecham ed.) Academic Press, Orlando: 321 (1986).
10. D.W. Smith, N. Weissman, and W.H. Carnes, Cardiovascular studies on copper-deficient swine. XII. Partial purification of a soluble protein resembling elastin. Biochem. Biophys. Res. Commun. 31:309 (1968).
11. L.B. Sandberg, N. Weissman, and D.W. Smith, The purification and partial characterization of a soluble elastin-like protein from copper-deficient porcine aorta. Biochemistry 8:2940 (1969).
12. J.A. Foster, E. Bruenger, W.R. Gray, and L.B. Sandberg, Isolation and amino acid sequences of tropoelastin peptides. J. Biol. Chem. 248:2875 (1973).
13. L.B. Sandberg, N. Weissman, and W.R. Gray, Structural features of tropoelastin related to the sites of cross-links in aortic elastin. Biochemistry 10:52 (1971).
14. V. Damiano, A.L. Tsang, G. Weinbaum, P. Christner, and J.Rosenbloom, Secretion of elastin in the embryonic chick aorta as visualized by immunoelectron microscopy. Collagen Rel. Res. 4: 153 (1984).
15. A.S. Narayanan, and R.C. Page, Demonstration of a precursor-product relationship between soluble and cross-linked elastin, and the biosynthesis of the desmosines. J. Biol. Chem. 251:1125 (1976).
16. W. Burnett, R. Eichner, and J. Rosenbloom, Correlation of functional elastin messenger ribonucleic acid levels and rate of elastin synthesis in the developing chick aorta. Biochemistry 19: 1106 (1980).
17. J.A. Foster, J.A., Rich, C.B., S. Fletcher, S.R. Karr, and A. Przybyla, Translation of chick aortic elastin messenger ribonucleic acid. Comparison to elastin synthesis in chick aorta organ culture.Biochemistry 19:857 (1980).
18. W. Burnett, A. Finnigan-Bunick, K. Yoon, and J. Rosenbloom, Analysis of elastin gene expression in the developing chick aorta using cloned elastin cDNA. J. Biol. Chem. 257:1569 (1982).
19. J.M. Davidson, S. Shibahara, C. Boyd, M.L. Mason, P. Tolstoshev, and R.G. Crystal, Elastin mRNA levels during foetal development of sheep nuchal ligament and lung. Hybridization to complementary and cDNA. Biochem. J. 220: 653 (1984).
20. R.P. Mecham, A. Hinek, R. Entwistle, D.S. Wrenn, G.L. Griffin, and R.M. Senior, Elastin binds to a multifunctional 67-kilodalton peripheral membrane protein. Biochemistry 28:3716 (1989).

21. A. Hinek, D.S. Wrenn, R.P. Mecham, and S.H. Barondes, The elastin receptor: a galactoside-binding protein. Science 239: 1539 (1988).
22. Z. Indik, H. Yeh, N. Ornstein-Goldstein, P. Sheppard, N. Anderson, J.C. Rosenbloom, L. Peltanen, and J. Rosenbloom, Alternative splicing of human elastin mRNA indicated by sequence analysis of cloned genomic and complementary DNA. Proc. Natl. Acad. Sci. 84:5680 (1987).
23. M.J. Fazio, D.R. Olsen, E.A. Kauh, B.S. Clinton, C.T. Baldwin, Z. Indik, N.O. Goldstein, H. Yeh, J. Rosenbloom, and J. Uitto, Cloning of full-length elastin cDNAs from a human skin fibroblast recombinant cDNA library: Further elucidation of alternative splicing utilizing exon-specific oligonucleotides. J. Invest. Dermatol. 91:58 (1988).
24. K. Raju, R.A. Anwar, Primary structures of bovine elastin a, b, and c deduced from the sequences of cDNA clones. J. Biol. Chem. 262:5755 (1987).
25. H. Yeh, Ornstein-Goldstein, Z. Indik, P. Sheppard, N. Anderson, N., J.C. Rosenbloom, G. Cicila, K. Yoon, and J. Rosenbloom, Sequence variation of bovine elastin messenger mRNA due to alternative splicing. Coll. Rel. Res. 7: 235 (1987).
26. R.A. Pierce, S.B. Deak, C.A. Stolle, and C.O. Boyd, Heterogeneity of rat tropoelastin mRNA revealed by cDNA cloning. Biochemistry, 29:9677 (1990).
27. G.M. Bressan, P. Argos, and Stanley, Repeating structure of chick tropoelastin revealed by complementary DNA cloning. Biochemistry, 26:1497 (1987).
28. H.Yeh, N. Anderson, N.O. Goldstein, M.O. Bashir, J.C. Rosenbloom, W. Abrams, Z. Indik, K. Yoon, W. Parks, R. Mecham, J. Rosenbloom, Structure of the bovine elastin gene and S1 nuclease analysis of alternative splicing if elastin mRNA in the bovine nuchal ligament. Biochemistry 28:2365 (1989).
29. M.M. Bashir, Z. Indik, H. Yeh, N.O. Goldstein, J.C. Rosenbloom, W. Abrams, F. Fazio, and J. Uitto., Characterization of the complete human elastin gene. Delineation of unusual features in the 5'-flanking region. J. Biol. Chem. 264, 8887 (1989).
30. C.B. Rich, and J. Foster, Isolation of tropoelastin from lathyritic chick aortae. Biochem. J., 217:581 (1984).
31. D.S. Wrenn, W.C. Parks, L.A. Whitehouse, E.C. Crouch, U. Kucich, J. Rosenbloom, J., and R.P. Mecham, Identification of multiple tropoelastin secreted by bovine cells. J. Biol. Chem. 262: 2244 (1987).
32. S.D. Chipman, B. Faris, L.M. Barone, C.A. Pratt, and C. Franzblau, Processing of soluble elastin in cultured neonatal rat smooth muscle cells. J. Biol. Chem. 260:12780 (1985).
33. E. Wingender, Transcription regulating proteins and their recognition sequences. Critical Rev. Eukaryotic Gene Exp. 1:11 (1990).
34. R. Eichner, and J. Rosenbloom, Collagen and elastin synthesis in the developing chick aorta. Arch. Biochem. Biophys. 198:414 (1979).
35. R.P. Mecham, S.L. Morris, B.D. Levy, and D.S. Wrenn, Glucocorticoids stimulat elastin production in differentiated bovine ligament fibroblasts but do not induce elastin synthesis in undifferentiated cells. J. Biol.Chem. 259:12414 (1984).
36. R.P. Mecham, B.D. Levy, S.L. Morris, J.G. Madaras, and D.S. Wrenn, Increased cyclic GMP levels lead to a stimulation of elastin production in ligament fibroblasts that is reversed by cyclic AMP. J. Biol., Chem. 260:3255 (1985).
37. A.H. Wang, G.J. Quigley, F.J. Kolpak, J.L. Crawford, J.H. VanBoom, G. Vandermarel, and A. Rich, Molecular structure of a left-handed double helical DNA fragment at atomic resolution. Nature 282:680 (1979).
38. V.M. Kahari, M.J. Fazio, Y. Q. Chen, M.M. Bashir, J. Rosenbloom, and J. Uitto, Deletion analyses of 5'-flanking region of the human elastin gene: delineation of functional promoter and regulatory cis-elements. J. Biol. Chem. 265:9485 (1990).
39. S. Inoue, C.P. Leblond, D.S. Grant, and P. Rico, The microfibrils of connective tissue: II. Immunohistochemical detection of the amyloid P component. Amer. J. Anat.176:139 (1986).

40. V.P. Werth, I.E. Ivanov, and J. Nussenzweig, Decay-accelerating factor in human skin is associated with elastic fibers. Invest. Dermatol. 91:511 (1988).
41. L.Y. Sakai, D.R. Keene and E.J. Engvall, Fibrillin, a new 350-kD glycoprotein, is a component of extracellular microfibrils. Cell Biol. 103:2499 (1986).
42. B. Lee, M. Godfrey, E. Vitale, H. Hori, M.G. Mattei, M. Sarafarazi, P. Tsipouras, F. Ramirez, and D.W. Hollister, Linkage of Marfan syndrome and a phenotypically related disorder to two different fibrillin genes. Nature 352:330 (1991).
43. D.L. Maslen, G.M. Corson, B.K. Maddox, R.W. Glanville, and L.Y. Sakai, Partial sequence of a candidate gene for the Marfan syndrome. Nature, 352:334 (1991).
44. G. Carpenter, and S. Cohen, Epidermal growth factor. J. Biol. Chem. 265:7709 (1990).
45. T. Kanzaki, TGF ß1 binding protein: a component of the large latent complex of TGFß-1 with multiple repeat sequences. Cell 61:1051 (1990).
46. L. Pereira, M. D'Alessio, F. Ramirez, J.R. Lynch, B. Sykes, T. Pangilinan, and J. Bonadio, Genomic organization of the sequence coding for fibrillin, the defective gene product in Marfan syndrome. Human Mol. Genetics 2:961 (1993).
47. M.A. Gibson, J.S. Kumaratilake, and E.G. Cleary, The protein components of the 12-nanometer microfibrils of elastic and nonelastic tissues. J. Biol. Chem. 264:4590 (1989).
48. M.A. Gibson, L.B. Sandberg, L.E. Grosso, and E.G. Cleary, Complementary DNA cloning establishes microfibril-associated glycoprotein (MAGP) to be a discrete component of the elastin-associated microfibrils. J. Biol. Chem. 266:7596 (1991).
49. S.K. Horrigan, C.B. Rich, B.W. Streeten, Z.-Y. Li, and J.A. Foster, Characterization of an associated microfibril protein through recombinant DNA techniques. J. Biol. Chem. 267: 10087 (1992).
50. K.M. Kim, B.A. Kogan, C.A. Massad, and Y.-C. Huang, Collagen and elastin in the normal fetal bladder. J. Urol. 146: 524 (1991).
51. K.M. Kim, B.A. Kogan, C.A. Massad, and Y.-C. Huang, Collagen and elastin in the obstructed fetal bladder. J. Urol. 146: (1991).
52. P. Chomczynski, and N. Sacchi, Single-step method of RNA isolation by acid guanidinium thiocyanate-phenol-chloroform extraction. Anal. Biochem. 162:156 (1987).

THE COLLAGENS AND THEIR UROLOGIC IMPLICATIONS

Edward J. Macarak, David Ewalt, Lawrence Baskin, Douglas Coplen, Harry Koo, Robert Levin, John W. Duckett, Howard Snyder, Joel Rosenbloom and Pamela S. Howard

Children's Hospital of Philadelphia
Dept. of Urology
School of Medicine
Dept. of Cell Biology
School of Dental Medicine,
University of Pennsylvania
Philadelphia, PA 19104

The bladder is designed to store urine at low pressure and expel it periodically. In order to accomplish this function, the wall of the bladder must have certain mechanical properties. The strength, elasticity and physiology of the wall is the result of the functional integration of substantial number of individual connective tissue proteins and nerve, muscle, epithelial and fibroblastic cells. Our laboratory is interested in the mechanical properties of the bladder wall and how various types of connective tissue components, and the cells which synthesize them, contribute to these properties.

Collagen is a family of proteins which make up approximately 30% of the dry weight of vertebrates. It is the major constituent of extracellular matrices. This family currently is comprised of at least 16 different collagen types and approximately 25-30 unique genes. Collagen fibers have the tensile strength of fine steel wire and it is this remarkable strength which provides tissues with their characteristic mechanical properties. Collagen is not just a structural protein. During development, there are unique tissue specific patterns of collagen expression as well as diverse molecular arrangements which mandate additional and different functional properties.

The basic building block of collagen fibers is the collagen molecule which is composed of three polypeptide chains called a chains. The existence of this basic unit was first proposed approximately 40 years ago and was called tropocollagen[1]. It is now known that collagen fibers are indeed composed of molecular units which are synthesized initially as a "Pro" form called procollagen[2]. Each of the three polypeptide chains in the most common and prototypical form of collagen (Type I) associate to form a coiled-coil configuration in the collagen molecule. The amino acids in each polypeptide chain are coiled in a left handed helix. Each of these three helical chains are in turn, like strands in rope, twisted around one another to form a right-handed tight, triple helical structure. The peptide bonds which link successive amino acids together in each a chain lie within the center of the super helix and thus are highly immune to selective attack by general proteases unless the triple helical structure which protects them is compromised. The primary structure of each polypeptide a chain generally consists of repeating triplet amino acid units consisting of Gly - Xaa - Yaa (where X and Y can be any amino acid). Glycine is required in repeating triplets because it sterically fits into the center of the helix[3]. Typically, collagens have a high imino acid content and many of the triplets contain proline in the "X" position and hydroxyproline in the "Y" position. Thus collagens characteristically contain the amino acid hydroxyproline which is shared by only a few other proteins, e.g., elastin, albeit in substantially lower quantities. Hydroxyproline

Muscle, Matrix, and Bladder Function,
Edited by S. Zderic, Plenum Press, New York, 1995

content has important biological implications since the stability of the triple helical conformation of collagen molecules is related to the relative hydroxyproline content[4].

Collagen molecules can exist as either homopolymers composed of identical alpha chains or as heteropolymers composed of dissimilar chains. For example, type II collagen which is primarily associated with cartilage[5] is a homopolymer while type I collagen is a heteropolymer composed of two a1 and one a2 chains.

Collagens Participating in 1/4 Staggered Fibrils[6]

Within the collagen family, the various types of collagen have been lumped into groups based upon the manner in which the molecules associate with one another. For example, Types, I, II, III, V and XI form fibrils with a periodic banding pattern. This pattern can be explained by the way in which molecules align with one another with a 1/4 stagger orientation. In this model, the position of each molecule is displaced 1/4 of its length from the molecules adjacent to it. In this particular form of molecular architecture, there is a space between the head and the tail of successive molecules. This space has been shown to play an important role in the process of cross-linking and mineralization.

Type I collagen is the most abundant form and is found in skin, bone, tendon, bladder, cornea and teeth. **Type II collagen** is primarily found only in embryonic and adult cartilage. **Type III collagen** is found in many of the same locations as type I including the bladder, however; it is not found in bone. It is a homotrimeric molecule composed of three identical a chains whose primary structure includes more than 333 residues of glycine as well as the presence of half cystines. These amino acids participate in the formation of intermolecular disulfide bonds. Immunohistochemistry has shown that type III collagen is the principal structural component of argyrophilic collagen fibers. Previously, these have been called reticular fibers. They are widely distributed in lymphatic tissue[7] as well as in other tissues and organs. **Type V collagen** as well as type III have been shown to be present in the same fibrils as type I collagen giving rise to the term heterotypic fibers[8]. Data from experiments on in vitro fibrillogenesis suggest that the co-assembly of different fibrillar collagen types into the same fibril can alter the biological properties of the fibril such as strength and fibril diameter[9]. **Type XI collagen** is closely related to type V; however, its distribution is thought to be largely, but not solely, limited to cartilage[10].

FACIT Collagens: Fibril-associated Collagens with Interrupted Triple Helices[11]

Collagen **Types IX, XII and XIV** are characterized by the lack of the rigid adherence to the typical collagen triplet or the Gly -Xaa - Yaa sequence. These molecules do not form quarter-staggered arrays by themselves but rather have interesting subdomains which provide for their putative biological function. One region of the molecule contains one or several triple helical domains and likely serves as a region which can interact with other triple helical domains. A second domain consists of a rigid arm that projects out and away from the triple helical interactive region. The third region does not contain any triple helical regions and probably serves to promote interactions with other matrix components[12]. Since the triple helical domains are separated by short interruptions in the triple helical structure, these collagens are called FACITS.

At the present time, it is not known whether any of these collagen types are found in the bladder wall.

SHEET-FORMING COLLAGENS

A third type of macromolecular architecture within the collagen family are "sheets." These structures serve to separate different tissue types from one another, e.g., basement membranes, or to separate individual cell types, e.g., smooth muscle basement membrane. Basement membranes or basal laminae consist of a fine meshwork of **Type IV collagen**[13] molecules which form an intricate reticulum or lace-like network. Associated with this network are other integral basement membrane proteins such as laminin, heparan sulfate proteoglycan, entactin and perhaps other components. These structures act as molecular filters and barriers to the passage of cells. Basement membranes also act to help bind epithelium to the underlying connective tissue. At the present time, there are six different Type IV a chains have been characterized. With respect to the urinary system, we know relatively little about the structure and composition of the basement membrane which separates the urothelium from the lamina propria in the bladder. Similarly, little is known about the physiological significance of the cocoon-like investment of Type IV collagen which surrounds each individual smooth muscle cell in the bladder wall. It is possible that this structure is the equivalent of the myotendonous junction found in skeletal muscle

BEADED FILAMENTS AND ANCHORING FIBRILS

Two other types of macromolecular architecture are illustrated by Types VI and VII collagen. The former forms beaded filament structures while the latter is commonly referred to as anchoring fibrils. **Type VI** collagen contains 11 Arg-Gly-Asp sequences which can bind to cell membrane receptors of the Integrin family. It appears to play an interactive role in the binding of cells to the extracellular matrix. **Type VII** collagen is found in skin at the interface between the basement membrane and the underlying matrix. It is synthesized by keratinocytes and, after some molecular processing, binds at one end to type IV collagen, the major collagenous component of basement membranes, while the other end extends into the dermis. This dermal end appears to "trap" or entangle collagen fibers and thus serve to anchor the epithelium to the collagen fibers within the dermis. Although the urothelial layer is a true stratified epithelium, it is not known how this layer is anchored to the underlying lamina propria.

We know from previous studies that the type, quantity and arrangement of collagen can alter the properties of tissues. For example, during wound healing, there is an initial change in the composition of the connective tissue. In most instances, long after healing has occurred, the composition of the wounded tissue remains different from that of a nonwounded area. This type of response which can be called a fibrosis or scarring is the result of an abnormal integration of the newly synthesized collagen into existing tissue structure.

Previous studies in the bladder have demonstrated changes in the collagen content of denervated bladders or bladders with some form of outlet obstruction [14,15,16,17,18,19]. In order to begin to understand these complex changes, it is necessary to first define the distribution of such molecules and to quantitate their individual content. The former can be accomplished by immunohistochemistry using collagen type specific antibodies while the latter can be accomplished biochemically. Once this is defined, the cells responsible for this synthesis can be identified and the molecular control mechanisms studied.

Immunohistochemistry using collagen type specific antibodies can identify the tissue location of a matrix component. This requires, however, that the primary antibodies be rigorously tested to ensure that they do not exhibit cross reactivity with other collagen types. Our laboratory has examined several examples of scarring in the genitourinary tract with distinctly different endpoints[20,21]. We have examined a group of patients with dysfunctional bladders[22]. Characteristically, these bladders have a lower than normal volumes and higher than normal bladder pressures. When these bladders were examined immuno-histochemically, we noted pronounced changes in the distribution of connective tissue as compared to controls. The principal structural alteration occurred in the detrusor muscle layer. In these neurogenic bladders, there was a pronounced pericellular infiltration by type III collagen suggesting that its accumulation may be the result of some aberrant signals which govern the synthetic behavior of cells residing within the detrusor muscle bundles.

To verify this observation quantitatively, we used a biochemical approach to disassemble the extracellular matrix and solubilize its components so they could be quantitated. We accomplished this using a reagent called cyanogen bromide which cleaves methionine residues at their carboxyl ends. Using this reagent, > 90-95% of the proteins in a sample can be solubilized. Once solubilized, the resultant peptides can be separated from one another by gel electrophoresis to permit identification. Subsequently their concentrations can be assessed by densitometry of the peptide spots. Alternatively, the concentrations of individual solubilized peptides from unique collagens can be an be quantitated by solid phase assays such as ELISA.

These studies demonstrated a major change in Type III collagen in the neurogenic bladders. In future studies, it will be important to understand the role of this collagen in the bladder wall and how its presence affects the mechanical properties of the wall. Although most forms of fibrosis and/or scarring do not involve type III collagen, there are a number of instances where changes in type III have been reported. An increase in type III has been reported in Peyronie's Disease[23], in Dupuytren's contracture[24] and in cardiac hypertrophy[25]. In animal models of chronic ultraviolet radiation, an increase in type III collagen in the dermis of the skin has also been reported.

Since collagen is a structural protein, it function in the bladder wall relates to the mechanical properties of this organ. In general, mechanical properties of the bladder can be divided into the active and passive components. The active component results from contraction of the smooth muscle cells of the wall while the passive component depends upon the composition of the connective tissue matrix. Since each of these types of properties adds to the overall functional capability of the bladder, it is useful to define the contributions of each to the overall mechanical properties of the bladder wall. To do this, we have carried out several types of experiments. Cystometrograms were performed on bovine fetal bladders which were treated in the following fashion. First, a normal cystometrogram was obtained and then a repeat analysis was performed in the presence of EGTA. By bathing the bladder with this calcium chelator, the active component or the active contractile tone provided by smooth muscle was eliminated. The cystometric curve from these bladders showed a shift to the right compared to the untreated control indicating that active smooth muscle contraction is responsible for a portion of the tone within the bladder wall. Following this, the detrusor layer was removed and a final cystometrogram performed. As shown in the figure, the curve shifts further to the right indicating that in addition to the active tone produced by smooth muscle contraction in the detrusor layer, there is a passive component which is added to the active component and which restricts the capacity of the bladder in a passive fashion. It is likely that this passive component of the detrusor layer results from the type, arrangement and quantity of collagen.

Structurally, the bladder wall consists of mucosal layer comprised of an inner urothelial layer, a lamina propria layer of dense irregular connective tissue, and a thin layer of smooth muscle, the muscularis mucosae. There are three major cell types within the mucosa: urothelium, lamina propria fibroblasts and smooth muscle cells associated with the muscularis. Beneath this layer is a rather ill-defined submucosa surrounded by large bundles of smooth muscle which make up the detrusor layer. These bundles are composed of largely smooth muscle cells; however, there are fibroblast-like cells which are also found here. Since all of these cells are capable of synthesizing connective tissue proteins, the passive mechanical properties of the bladder and the connective tissue proteins which account for them result from the integrated synthesis of a diverse group of cells. We have begun to characterize the phenotypes of these cells by isolating them and studying their connective tissue synthesis in vitro. It is of obvious importance to understand the molecular mechanisms which control connective tissue synthesis in the bladder. Since we have previously documented changes in type III collagen that are associated with profound changes in the mechanical properties of the bladder wall(Ewalt et al), the ability to study a single population of cells in vitro offers many advantages. The benefits of this type of approach can be appreciated by examining the advances which have been made in the field of atherosclerosis many of which are related to in vitro studies of the major cell types found in blood vessels: vascular endothelial and smooth muscle cells. These types of approaches should be applied to the study of compliance in the bladder and it is anticipated that new and important information will result from them.

Great progress has been made in the field of connective tissue during the past 10 years with respect to the identification and characterization of many new additional collagen genes. Unfortunately, less rapid progress has been made with respect to the biological function of these various collagens and how they interact with cells. The form of the interaction and the reciprocal information transfer which occurs as a result likely plays a role in controlling cell behavior. A better understanding of bladder function will require new information about the interaction(s) between bladder wall cells and their extracellular matrices.

REFERENCES

1. Gross, J. 1965. The behavior of collagen units as a model in morphogenesis. J. Biophys. Biochem. Cytol. Suppl. 2, 26.
2. Fessler, J. and Fessler L. 1978. Biosynthesis of procollagen. Annu. Rev. Biochem. 47:129-162.

3. Ramachandran, G.N. and Ramakkrishnan, C. Molecular Structure, In:Biochemistry of Collagen, Ramachandran, G.N. and Reddi, A.H., eds, Plenum Press, New York, p. 45, 1976.
4. Sakadibara, S., Inonye, K., Shudo, K, Kishida, Y., Kobayashi, Y., and Prockop, D.J. 1973. Synthesis of (Pro-Hyp-Gly)n of defined molecular weights. Evidence for the stabilization of collagen triple helix by hydroxyproline, Biocheim. Biophys. Acta. 303:198.
5. Miller, E.J. 1972. Structural studies on cartilage collagen employing limited cleavage and solubilization with pepsin, Biochemistry, 11:4903-4909.
6. Hodge, A.J. and Petruska, J.A. Recent studies with the electron microscope on the ordered aggregates of the tropocollagen molecule. Aspects of protein structure (Ramachandran, G.N., ed, Academic Press, New York, 289, 1963.
7. Macarak EJ, Howard PS and Lally E Production and characterization of a monoclonal antibody to human type III collagen. J. Histochem. and Cytochem. 34:1003-1011, 1986.
8. Hendrix, M.J., Hay, E.E., Von der Mark, K and Linsenmayer, T.F. 1982. Immunohistochemical localization of collagen types I and II in the developing chick cornea and tibia by electron microscopy, Invest. Opthalmol. Vis. Sci., 22:359-375.
9. Adachi, E. and Hayashi, T. 1986. In vitro formation of hybrid fibrils of type V collagen and type I collagen. Limited growth of type I into thick fibrils by type V collagen. Connect. Tissue Res. 14: 257-266.
10. Eyre D. and Wu, J. 1987. Type XI collagen, In: Structure and Function of COllagen Types (Eds. R. Mayne and R.E. Burgeson), Academic Press, New York, pp261-282, 1987.
11. Gordon, M.K. and Olsen, B.R. 1990. The contribution of collagenous proteins to tissue-specific matrix assemblies. Curr. Opin. Cell Biol. 2:833-838.
12. Van der Rest, M. and Mayne, R. Type IX collagen, In: Structure and Function of Collagen Types (eds. R. Mayne and R.E. Burgeson) Academic Press, New York, pp195-219, 1988.
13. Kefalides, NA 1973. Structure and biosynthesis of basement membranes. Int. Rev. Connect. Tissue Res.6:63-104.
14. Ghoniem GM, Regnier CH, Biancani P, Johnson L and Susset JG. 1984. Fetal bladder histology in posterior urethral valves and the prune belly syndrome. Neurol. Urodynam. 3:23-30.
15. Ghoniem GM, Bloom DA, McGuire Ej and Stewart KL 1989. Bladder compliance in meningomyelocele children. J. Urol., 141:1404-1406.
16. Gilpin SA, Gosling JA and Barnard RJ Morphologic and morphometric studies of the human obstructed, trabeculated urinary bladder. 1985. Br. J. Urol., 57:525-530.
17. Workman SJ and Kogan BA 1990. Fetal bladder histology in posterior urethral valves and the prune belly syndrome. J. Urol., 144:337-339.
18. Shapiro E, Becich MJ, Perlman E and Lepor H 1991. Bladder wall abnormalities in myelodysplastic bladders: a computer assisted morphometric analysis. J. Urol., 145:1024-1029.
19. Kim KM, Kogan BA, Massad CA and Huang Y-C 1991. Collagen and elastin in the obstructed fetal bladder. J. Urol., 146:528-531.
20. Baskin LS, Constantinescu SC, Howard PS, McAninch JW, Ewalt DH, Duckett JW, Snyder HM and Macarak EJ. Biochemical characterization and quantitation of the collagenous components of urethral stricture. J. Urol. 150:642-647,1993.
21. Ewalt, D., Constantinescu, S., Bellon, G., Borel, J.-P., Snyder, H.S., Duckett, H and Macarak, E.J. Increased collagen type III in the neurogenic non-compliant bladder. (Submitted 1994).
22. Ewalt, D.H., Howard, P.S., Blyth, B., Snyder, H.M., Duckett, J.W.., Levin, R. and Macarak, E.J. Is lamina propria matrix responsible for normal bladder compliance? J. Urol.148:544-549, 1992.
23. Somers, K.D., Sismour, E, Wright, G., Devine, C., Gilbert, D., and Horton, C. 1989. Isolation and characterization of the collagen in Peyronie's disease. J Urol.141:544-549,1992.
24. Kucharz, E.J. The Collagens: Biochemistry and Pathophysiology, Springer-Verlag, New York, 1992.
25. Weber, KT, Janicki, J, Shroff, S, Pick, R. Chen, R and Bashey, R. 1988. Collagen remodeling of the pressure-overloaded, hypertrophied non-human primate myocardium. Circ Res. 62:757-765.

EXTRACELLULAR MATRIX UPDATE - DISCUSSION

Gary Striker - What I would like to do for this section of the program is initiate a discussion about the extracellular matrix with our speakers, and ask them to challenge each other about what role these proteins play during the course of normal development, and how the extracellular matrix might change in the urinary bladder in response to outlet obstruction or other disease processes.

Ed Macarak - I have a question for Dr Ehrlich regarding the composition of the lattice and how it is influenced by the percentage of type III collagen that is combined with it. Do you have any ideas or insights about what it is about type III collagen that creates these differences in lattice contracture?

Paul Ehrlich - In the experiments that we have done, there are 2 possibilities. Fibroblasts have the capacity to move, and perhaps they have the ability to move faster if they have more receptors or binding sites for type III collagen available to them. The other possibility is that type III collagen fibers can be packed closer together due to their underlying chemistry. Both mechanisms are possible, but I would favor the notion that the type III fibers are packed closer together.

Ed Macarak - We all know now that collagen fibers are quite heterogenous, and that altering their combinations is much like altering the metals in an alloy. Thus one can produce molecular alloys with different mechanical properties by altering these types. Do you talk about doing any experiments where you could actually vary the various concentrations of these collagen types and then assess their contraction under those circumstances?

Paul Ehrlich - The only types we have looked at are the data shown today for I and III. Yes, we can look at hypertrophic scar formation in skin, where there is a variation from 20% to 30% type III collagen and in that situation we see a much greater degree of contracture in the hypertrophic scar.

Gary Striker - Do you think that these contractures represent rearrangement of the fibroblasts, or does this represent some rearrangement of type IV collagen?

Paul Ehrlich - I don't know that, because one might argue that this reflects the differences in polymerization of the types I and III collagens without even invoking yet a third type of collagen. But I agree that it would be interesting to test type IV collagen in this kind of experimental model.

Joel Rosenbloom - I would like to ask Dr Stetler-Stevenson a question. What if anything is known about the effects of mechanical forces on the metalloproteinase system, or the TIMP system?

William Stetler-Stevenson - Well we really don't know a lot about that right now, but we do known that cell shape is important in dictating or continuing the response of the metalloproteinases and the TIMP system to a number of stimuli. It has been very well demonstrated in several different laboratories that cell matrix metabolism is mediated by integrin receptors. So logically the expression of tissue collagenases and stromolysins through the alpha 3 and beta 1 receptors in fibroblasts and attachment to the vitronectin receptors in melanoma cells can also be used to modulate expression. But I am not really sure that we have much information at all about how mechanical forces result in changes in this system.

Gary Striker - Is there anyone who has tried this with smooth muscle cells so see if there can be any kind of rearrangement of matrix molecules including the fibrillogen and type IV collagens? For instance, in Paul's system, fibroblasts don't like to cover themselves up with these macromolecules, yet one might expect that smooth muscle cells would want to become enmeshed. Might they in becoming enmeshed in this matrix, also rearrange their shape and order?

Joel Rosenbloom - Well no, no one that I know is doing that yet, but certainly this would be an interesting area to look into. Particularly now that we know that we have such great increases in connective tissues in the face of obstruction, there must be some kind of signal which would be relayed to the detrusor layer where the vast majority of smooth muscle cells are found, and where the major increases in thickness of the bladder wall are found. This leads us to conclude that the smooth muscles are extremely important in matrix synthesis in the obstructed bladder, and are involved quickly and early on the obstructive process.

William Stetler-Stevenson - Joel, has anyone considered using metalloproteases as a tool for looking at the associations between various fibril associated proteins? In the sense of taking data bank tissues and digesting them various types of metalloprotease enzymes and then looking for selective removal of certain matrix elements?

William Stetler-Stevenson - Not as yet. The only approach that we have taken at all, which is not selective, is by the use of cyanogen bromide fragmentation of the matrix elements. This is a technique also described by Ed Macarak, and is employed so that we may quantitate the matrix elements. So that is a very useful approach. We don't really know much about the degradation of these macromolecules, under normal or pathological circumstances. So we really don't know much about what the sensitivity of these macromolecules are to the matrix metalloproteases. In such alterations as fibroids, there is an accumulation of the matrix because certain matrix domains appear to be protected from the metalloproteases making them less susceptible to degradation. So if this is the case in the pathogenesis of fibroids, it may well also be valid in this case too (bladder dysfunction with matrix accumulation secondary to obstruction). Perhaps this kind of assay could be developed into one with clinical utility for determining at which point (if ever) a bladder has become irreversibly fibrosed.

William Steers - I have two questions, the first of which is for Dr Ehrlich. We always seem to find the bladder behaves in the reverse of what you describe in the skin. That is if one operates the bladder under normal conditions of stretching and emptying, or if you increase the frequency of stretching emptying by denervating the bladder, or carrying out a partial cystectomy, you get very little scar formation at all. Conversely if we de-functionalize the bladder to prevent stretch, that is where we get the thickened bladder with increased collagen deposition. Can you offer a hypothesis as to why this is seen in these defunctionalized bladders? In many ways this is opposite to what you are describing in skin where you have findings of would healing with the development of contractures across mobile joint areas?

Paul Ehrlich - I don't know why that is the opposite because plastic surgeons using tissue expansion to increase the amount of skin they have available for a flap advancement use a balloon implanted below the skin which can be progressively inflated to produce a gradual increase in the volume of skin which can then be rotated down to its new location. Now in that condition, the skin is stretched a great deal, but what is interesting histologically is that on the surface of the balloon is granulation like tissue and the dermis looks totally uninteresting to the pathologist. The endo-dermal layer is actually thickened, so you have a hyperplasia of this associated layer with a normal dermis. Yet when you put it back and suture it into place, it does not contract or distort at all, in fact it becomes a normal piece of skin with normal tensile and cosmetic properties. So in the stretches and the decompressions that we are talking about, the difference between scar formation versus normal hypertrophy might be the presence of inflammation, and it is the presence of the inflammatory mediators which tip the scales in favor of a lasting fibrosis. So could it be that inflammation is the major player in determining the presence of scar development whether it be in the healing would or the bladder wall?

William Steers - My next question is for Dr Macark. As a non-pediatric urologist who deals with many women with stress urinary incontinence, I am fascinated by the near 100% stress urinary incontinence that is seen in women with the Ehlers Danlos syndrome. Recent biopsy data suggest a change in collagen types, especially type III, that is seen with aging. Can you comment on how the pathogenesis of the changes in collagen types results in hypermobility of the bladder and urethral elements to cause stress urinary incontinence? Has anyone looked at urethral collagen types as well as bladder types?

Ed Macarak - I really don't have an answer for the first part of the question. I know that no one has looked at the quantitation of the type III collagen in the urethra, and that is something that can easily be done. I think the real basic question that you are asking is how are molecules put together to give their functional attributes, and that is what we really don't know. I mentioned that people have characterized new collagen types, and new collagen genes, but the level at which we will understand how these fibers are assembled and put together is very much open to question. Some of the photomicrographs that Paul Ehrlich showed us were done using polarization microscopy, and this has been done to show a particular kind of altered structure within collagen fibers, and this is referred to as crimping. The function of this is that it permits a fiber to undergo some degree of extension within a limited range, but the range allowed by this crimp is not enough to explain the tremendous strains that many organs may be subjected to, but that is especially true when you look at the bladder wall.

Joel Rosenbloom - I'll elaborate on this point a little bit. In several instances for diseases of connective tissue, it is really a quantitative lack so that you have an absolute deficiency of the collagen type that is mutated. So depending upon the ability of other collagens to either substitute in that particular tissue or increase their functional contribution, you may or may not get a clinical deficit. For example, in osteogenesis imperfecta, you have type I collagen which is markedly altered, and yet type I collagen is found all over the body, yet the primary manifestation of this disease is in the bones. So there is an as yet unknown feature in the bone itself requires type I to be present in the correct amount and form. I do not know what the percentage of type III in bone is, and how this might contribute to this pathogenesis.

Sam Chacko - I have a question about the remodeling of the collagen matrix which would seem to be a very basic and fundamental process. You mentioned about fibroblasts or smooth muscle cells permeating or crawling through the collagen matrix, and then remodeling it. What is exactly happening here? Is the cell making collagen and adding it to that which exists or they making a collagenase that degrades the local matrix and adding more? Can you speculate on what you think might be happening in vivo?

William Stetler-Stevenson - Well, in response to injury I think there are certainly very different responses as seen in atherosclerosis. These smooth muscle cells in response to this injury must migrate following plaque formation in the aortic wall, and in order to migrate across various extracellular matrices they must produce matrix metalloprotease enzymes. But I think this an evolving concept with respect to the role of these enzymes. We must try to think of these metalloproteases not just as merely destructive but to think of them as being critical to the normal remodeling of tissues that occurs, and cell migration and motility are a part of this whole process.

Gary Striker - Paul Ehrlich, would you like to comment on this point?

Paul Ehrlich - You know it is interesting about pointing out the issues of cell locomotion versus cell migration. We are saying that cell locomotion is also involved in the organization of collagen, yet how can the cell do both things? Consider would healing as a case in point. The cells have to migrate from some place that survived the injury into the dead space where the wound is, and they must do so like a spider moves through a spider web. This process gets even more complicated if the path through the web is littered with inflammatory cells, and suddenly moving from point A to point B becomes more complex. On the other hand once its embedded in this new connective matrix that is being laid down, it can become attached to the collagen fibers and rearranged so there is a change in that same cell from the cell that was undergoing locomotion and was actually moving to a cell that becomes involved in locomotion where what moves is not the cell, but rather the extracellular matrix to produce contracture. It is fascinating how the cell can do such varied tasks with the same basic machinery. It acts initially like a spider crawling through this web, and then becomes like a fly trapped in that web that moves the spider web all around while it tries to escape.

David Ewalt - Joel, I saw the slide you had of elastin cross linked with the cell. Is there any evidence that elastin has binding to any other extracellular matrix proteins such as fibronectin or other cell surface adhesion molecules or other collagen areas?

Joel Rosenbloom - Yes, there is some reasonably compelling evidence that elastin and tropoelastin can bind to a cell surface binding protein, I hate to call it a receptor, but it has been shown to have physiologic consequences. But, particularly in our lab Bob Meecham has identified this binding protein at the cell surface and this protein may be involved in the cells ability to organize. We don't know much about than this at this time. There is weak evidence that elastin is actually cross linked to at lest one other protein, but we don';t know the nature of that. The difficulty is that in order to prove that you have to isolate and characterize the cross linking peptides in order to show that these are covalent linkages. I didn't mention this, but there a number of sequences within the microfibrillar proteins which could be substrates for lysyl oxidase which the enzyme that performs all known (currently) cross linking within the extracellular matrix. We have been able to show that we can essentially form insoluble elastin by taking purified recombinantly produced tropoelastin, and purified lysyl oxidase and combining them together. We are now starting to do these same types of experiments with the use of these microfibrillar proteins as substrates. So in summary the answer to your question is that the evidence is weak, but suggestive.

Craig Peters - I have a question for Dr Ehrlich. There is obviously a lot of interest in fetal wound healing. I wonder if you have any comments on whether your think this is a qualitatively different process or is a difference in the expression of the collagen types, or is it a difference in the breakdown of these components of the matrix?

Paul Ehrlich - We did look at some aspects of fetal wound healing. In my own personal observations of this when you look at the dermis of the fetus, it is a very different structure with very little resemblance to the mature dermis. And it would be like trying to make a wound in a jello mold, in the sense that it just comes together and doesn't require any inflammatory cells. The damaged areas are just woven together by the cells that are already there. There are some other interesting observations one of which is that there is a great deal of variance between the different species. If you compare a rabbit to a sheep, you will see that rabbit fetal wounds that are not closed with suture continue to grown in the animal and do not close themselves. Yet if you suture a rabbit wound closed, you will discover that it will heal without a scar. The only way to know there was injury here is to look for residue of suture. On the other hand in the sheep, if you create an injury and leave it open presents a beautifully healed incision. On the other hand as the sheep gets older and enters its third trimester, and you create a wound there, it will lead to a scar. So when you talk about fetal wound repairs, things get sticky because there are some very impressive differences between the species that have been studied. As far as what we can learn from it, I think there is a lot to be gained by understanding these phenomena. In this system it probably is the organization of the extracellular matrix that keeps things together, and maintains its plasticity and flexibility. This probably exerts a great influence on how these wounds are repaired with minimal scarring. If in the future we could take a tissue, and make it more plastic and more immature, and get the fusion of the existing tissue so that there is very little scar necessary in the repair process, I think you could repair a lot of things without scars.

Participant - I would like for you to speculate a little bit about the possible relationship of matrix reformation, with neuropathy as it pertains to the bladder. The basis for my question is that at least with skeletal muscle, we considered the muscle to be a slave to its nerve supply and I guess there is evidence that smooth muscle cells have the same type of relationship with their nerve supply. If the smooth muscle directs the synthesis of the collagens, is there a possibility that all this linkage can result in the neurons directing a process that leads to their own entrapment?

Gary Striker - That is a great question. Who would like to try and tackle this question?

Ed Macarak - I think that is a very distinct possibility because there are a myriad of signals which could control the synthetic behavior of these cells. The hallmark of some of these studies that have been done on connective tissue, is that in the end stage there is the appearance of fibrotic veins of connective tissue. The signal that alters the phenotype of the cell, that results in the formation of these fibrotic bands is what I think needs to be investigated. That signalling mechanism either is the result of appropriate growth factors, or mechanical input via deformations, but it is something that we need to begin to really try and understand. If we understand this mechanism, then we will be able to consider controlling the phenotype of the cell under these various conditions. If it is something mechanical, then we may be able to exercise the bladder or rest to produce the needed effect. It probably is related more to the degree of exercise and the uncontrolled contractions seen in some of these conditions may alter the cell phenotype. On the other hand if it is something as simple as the presence of nerve growth factor or some other factor, it is conceivable that one could use intravesical replacement of such factors to direct phenotypic expression. But I think the key to it all is 1) breaking down the bladder into its key cellular components and then 2) beginning to understand the control mechanisms that relate to the cell's phenotype.

Participant - I have a few observations I would like to express. If you do conventional cystometry, you can demonstrate the viscoelastic properties of the bladder wall. During maximal filling when you get to the high pressure zone and stop filling and wait, you gradually see a drop in pressure back to almost normal values. So what this suggests is that the bladder builds up pressure, but it has memory too as is evidenced by this type of pressure decay.

Ed Macarak - I cannot answer that because I think we are just at the beginning of our understanding of how the bladder wall functions in terms of the interaction between its muscular and connective tissue components. Until we understand the quantitative components of the relationship between these extracellular matrix components, and then try to relate that to cystometry and then see how this is altered by disease, we won't be able to fully answer this question.

Gary Striker - I think that is a great answer because as Ed pointed out, the bottom line is that this is work that really needs to be done.

William DeGroat - I understand from Dr Rosenbloom's comments that the elastin is permanent or semi-permanent in the bladder. So once it is produced it stays there in contrast to many of the other components of the bladder which could be modulated such as smooth muscle which could become larger or smaller depending upon the cell's needs. So if in the pediatric situation for the patient who has an obstruction, and a very large bladder, and that patient has the obstruction removed and the bladder shrinks back down. Various components of the bladder wall are now going to disappear or their presence will be reduced. Will elastin remain?

Joel Rosenbloom - I am really arguing from other systems. I don't actually know any specific studies on that kind of a situation in bladder, so we really have to look at some new experiments. In the limited studies that have been done in other tissues, active inflammation will precipitate the synthesis of elastase, but this work is really limited to studies done in PMNs. Under normal circumstances there is no enzyme that is elaborated by a smooth muscle cell that can degrade elastin.

William DeGroat - So the theory would be then that in this shrunken bladder, after relief of obstruction, there should be a much higher percentage of elastin.

Joel Rosenbloom - Right.

William DeGroat - So it would be interesting to know the relative contribution that elastin makes to the stretch and mechanical and physical properties of these systems. It would also be interesting to then ask if the bladder expands again if it will make new elastin, or can it simply use that which is already in place. If you did this several times by cycling the bladder many times and building up elastin, could you produce a rubber band like property?

Joel Rosenbloom - It is a great experimental question, but there is no answer to it at least not yet. The cells in other tissues have the capability to synthesize elastin in the adult. Ordinarily, in the few tissues that have been looked at, mostly the lung, you don't actually get a loss of elastin. Consider elastin in the pathogenesis of COPD, and you won't see a pronounced decrease in the total elastin content. But you do see a loss of the architecture of the alveoli, and puddles of elastin which have lost their properties of physiologic recoil. In atherosclerotic blood vessels you do get an increases in elastin synthesis, and that is probably more physiologic in the sense that the architecture of the vessels is much simpler. You have just a circumferential distribution of elastin which can then set up a network as opposed to the lung where a three dimensional organization is essential is formation of the alveolus. It is hard to predict what will happen in this system you have described.

Gary Striker - I think this has been a very positive and interesting session this morning, and I would like to thank the speakers and participants for their contributions.

CHANGES IN BLADDER EXTRACELLULAR MATRIX - OVERVIEW

Timothy Ratliff PhD

Urology Research Laboratories
Washington University School of Medicine
St Louis, MO 63110

For starting this session I have a couple of slides that will help me in giving a brief overview of this topic. This is one of those conceptual overviews, and not one that is rich with data. What we really ought to think about is the fact that the bladder, the way we normally think about it, tends to be a visual simplification. We think of it in terms from our textbooks of histology as a layer of glycosaminoglycans, mucosa, lamina propia, and muscle. So we always think about the cells we associate with bladder structure. However as we have seen over these past two days, there is a lot more to it than just these cell populations. So if you will just think about this analogy; think about this room here. This room has a nice aesthetically pleasing ceiling, nice wall paper, pictures, and all of those things are equivalent to those cells within the bladder wall that we normally look at. But what is behind these walls. There is steel within these walls that we don't see that extends on down to the foundation to provide the ultimate structural support and that is indeed what the extracellular matrix is. It is the supporting matrix for the cells, and I just borrowed this slide from Paul Ehrlich because it shows so nicely what we are talking about. You have the matrix, in this case collagen, that is the scaffolding or the steel structure of the bladder. Then you have the cells that are intimately associated with it in such as way that you can't always tell where the cell ends and the collagen begins.

But this is where the analogy of this building to the organs of the body stops. The matrix itself has two other parameters that are important to maintaining that facade of the cell in a way that we are used to seeing. 1) The matrix is attached to the cells by integrin receptors and 2) these receptors also signal the cell. The cell is attached to the matrix through these specific receptors and many of these receptors we know are specific for certain elements of the extracellular matrix. These are not just attachment sites, but receptors that transmit signals to the cell, and those signals tell the cell where it is and what it ought to do. Should it migrate into the wound or stop and reconfigure the collagen at a particular location? It is these signals that we don't have a clear notion of how they are initiated and how they work and how they tell the cell to go about its special needs. These signals induced by the interaction of matrix components with their receptors may also induce the production of soluble factors that can then go out and induce other cell populations to change their gene expression or attract new cell populations to the field of interest. It is this kind of process that we need to begin to try and understand. We want to know how these structures, these matrix proteins, end up signalling the cell to put our more type I collagen and especially increase type III output and consider adding type IV. In the bladder we are dealing with this pressure oriented signalling process, and we know that in the bladder we can get into situations where it is denervated and this leads to alterations in function. So how does this happen? We are not going to specifically talk about these signals, but we will set ourselves up for that with this session.

We will have three talks that deal with some clinical phenomena that point to diseases or problems that are associated with collagen. Then we will end with a couple of talks that will begin to model the bladder and its production of extracellular matrix and begin to help us understand which genes are activated and which ones are not. The bottom line is that our clinical talks will give us the associations which help us know that there is a problem in terms

Muscle, Matrix, and Bladder Function,
Edited by S. Zderic, Plenum Press, New York, 1995

of matrix composition and function. In effect this gives us a giant scientific puzzle to solve. Then in discussing the model systems, we are going to begin to put this process up to the "magnifying glass of science" as described by E.C. DeBoer. There will be presented a variety of techniques such as Western blots, Northern Blots, PCR, and cell culture. These are the modern tools we use to magnify the system so we can begin to figure out how these pieces of the puzzle all fit together. We have the basis of this now with some of the models which are being developed to gain a great deal of insight into these signaling mechanisms and activation process for the synthesis of matrix in normal and pathologic states of the bladder.

MATRIX CHANGES IN THE BLADDER ASSOCIATED WITH NORMAL AGING

Ellen Shapiro, M.D., Herbert Lepor, M.D.

Department of Urology
New York University School of Medicine
New York, NY, 10016

Bladder contractility and varying degrees of bladder distension can be observed sonographically as early as 20 weeks of gestation[1]. Postnatally, the mechanisms and neural pathways which control voiding are not fully developed. In the infant, micturition occurs spontaneously as a spinal cord reflex[2]. During the first two years of life, the afferent and efferent innervation necessary for micturition mature and the adult supraspinal voiding reflex emerges. As the cortical fibers mature, the infant develops conscious awareness of bladder activity. By about age 2, children begin to sense bladder fullness and develop daytime and subsequently nocturnal urinary continence[3]. Most children will maintain normal bladder function unless they develop a dysfunctional voiding pattern typically manifested by daytime incontinence and nocturnal enuresis. Young adults most commonly have normal bladder function but stress incontinence due to bladder neck hypermobility can be seen in the multiparous female. This type of urinary incontinence is a socially embarrassing problem for the active female. The health care expenditures for urinary incontinence are estimated to exceed 10 billion dollars per year[4]. Incontinence is a major urologic problem in the elderly and affects at least 3 million persons over the age of 65 years. The prevalence of urinary incontinence for females is 37%, and for males 19%. Women have stress incontinence as well as urge incontinence while urge incontinence predominates amongst men[5,6]. Incontinence occurs in approximately 50% of institutionalized elderly of whom 82% are women. Urodynamic evaluation demonstrates detrusor overactivity in about 60% of these elderly females[7,8,9]. In approximately 50% with detrusor overactivity, detrusor contractility is also impaired[9,10]. It is estimated that urinary incontinence is a contributing factor for the placement of as many as 50-60% of the patients in nursing homes.

What are the changes in the detrusor that contribute to urinary incontinence in our aging population? In order to address this question, we will first review the morphologic development of the detrusor. This will enablc us to appreciate alterations in the detrusor with aging. Historically, Susset et al. provided an estimate of the gender differences in collagen content in pre- and post-pubertal children and adults[11]. The mean content of collagen in these bladders was 57%. The only group that appeared to have more collagen was the females greater than 50 years of age. Gosling and Dixon have reported that clinical BPH is associated with increased infiltration of the detrusor muscle by connective tissue[12]. The findings of Susset et al is somewhat surprising since it would have been assumed that an increase in collagen infiltration would have been associated with bladder dysfunction due to bladder outlet obstruction associated with benign prostatic hyperplasia in the aging male. Gilpin et al further examined changes in detrusor smooth muscle cell morphology which accompany bladder trabeculation. They showed that smooth muscle cells undergo compensatory hypertrophy in response to outflow obstruction. In their study, connective tissue infiltration of detrusor muscle bundles was observed in bladders with the greatest amount of smooth muscle cell hypertrophy. The findings of these studies suggest that further research is necessary to understand the pathogenesis of voiding dysfunction associated with aging.

Smooth muscle first appears in the fetal bladder at about 7 weeks gestation[14]. Kogan et al. found only poorly developed thin muscle in the 19-20 week fetus[15]. They also showed that muscle thickness in the developing fetal bladder increases gradually with age. Using a grid and point count technique, the proportion of collagen to muscle in the fetus was found to be age dependent. A general downward trend in the collagen to muscle ratio during fetal development was seen. Although some fetus' had a collagen to muscle ratio of 1.1 or greater this decreased to 0.65 at the end of gestation. No gender difference were observed in their study groups.

Shapiro et al. examined normal fetal bladders and bladders of children age 1 day to 4 years using computer assisted image analysis to determine the histologic components of the bladder[16]. Although there was no statistical difference between the fetal bladders and that of the young child, there was a significant decrease in the fraction of connective tissue in the childhood bladder (0.106 $\pm$ .020) when compared to that of the fetus (0.182 $\pm$ 0.017). This difference was noted in the connective tissue:muscle ratio suggesting not only a decrease in connective tissue, but also a continued increase in detrusor smooth muscle development postnatally. There continues to be a paucity of data in the literature on the histologic components of the fetal bladder as well as the histologic findings in bladders of children, adolescents, and young adults.

Lepor et al. performed a retrospective study to further characterize the effects of aging and BPH on bladder morphometry[17]. Eighty-six bladder specimens were obtained from autopsy archives and were divided into four groups based upon age and gender. Group 1: males between the ages of 35-45 years; Group 2: males between ages 65-75 years; Group 3: females between ages of 35-45 years; and Group 4: females between the ages of 65-75 years. The age groups were selected in order to identify a group of males with and without benign prostatic hyperplasia. Using color assisted computer image analysis, the area density of smooth muscle to connective tissue was determined in these bladder specimens. The area density of smooth muscle to connective tissue in the Masson trichrome stained sections was significantly greater in the Group 1 (younger males) vs. Group 2 (older males) and in Group 3 (younger females) vs. Group 4 (older females) (Table 1). Aging was associated with a decrease in the area density of smooth muscle to connective tissue ratio in both males and females. The area density of smooth muscle to connective tissue was not significantly different in the younger males vs. females or in the older males vs. females. This morphometric study suggests that aging and not BPH is associated with a relative increase in detrusor fibrosis. Infravesical obstruction due to BPH may effect bladder function, but the mechanism appears to be unrelated to the histologic composition of the bladder. In order to unequivocally determine the effect of clinical BPH on bladder morphometry, the area densities of smooth muscle:connective tissue must be compared prospectively in older males with and without clinical BPH.

Since there are no significant differences in the histologic components of the aging bladder, and since urinary dysfunction is common in both males and females, Lepor and Machi ascertained the severity of urinary symptoms in males and females[18]. The AUA Symptom Index represents a validated intrument to quantify the severity of BPH symptoms. The AUA Symptom Index was administered to both males and females attending a health fair. The mean age of males and females completing the index was 69.3 $\pm$ 0.6 years, respectively. The mean AUA symptom scores in the males and females was 6.7 $\pm$ 0.5 and 7.5 $\pm$ 0.6, respectively. The differences between the AUA symptom scores were not statistically different between males and females in 5 consecutive advancing age groups between 55-79 years of age. The percentage of males with mild, moderate, and severe symptoms were 65%, 32%, and 4%, respectively. The percentage of females with mild, moderate, and severe symptoms was 61%, 35%, and 4%, respectively. Therefore the study suggests that the symptoms presumed to be due to BPH are more likely a manifestation of aging and not bladder outlet obstruction due to BPH.

The pathophysiology of urinary symptoms was further investigated by determining whether terazosin, an $alpha_1$ blocker, represented an effective therapy for the treatment of "prostatism-like" symptoms in women[19]. A randomized double-blind study comparing terazosin vs. placebo was undertaken in women with moderate and severe BPH-like symptoms for six weeks. The differences between the change in AUA symptom score in the placebo and terazosin treatment groups were not clinically or statistically significant. This study demonstrates that terazosin is not effective for the treatment of prostatism-like symptoms in the aging female population. Although men and women have a similar predisposition to develop urinary symptoms, this study implies that the pathophysiology and treatment of urinary symptoms is to some degree gender specific.

A recent study by Elbadawi et al. has extensively examined the electron microscopic findings of detrusor specimens from elderly men and women with urinary incontinence and voiding dysfunction[20-23]. Specimens were divided into 4 categories based upon ultrastructural features: 1) dysjunction which was associated with detrusor overactivity; 2) myohypertrophy which is associated with obstruction 3) a combination of dysjunction and myohypertrophy; and 4) the dense band pattern which was seen in patients who were symptom free. The ultrastructural changes seen with the dysjunction pattern include widened intercellular spaces, a paucity of intermediate muscle cell junctions, abundant protrusion junctions and ultraclose cell abutments. No enlarged, hypertrophic cells are seen. Protrusion junctions and abutments may be a manifestation of muscle cell differentiation associated with normal aging. These findings may also be associated with abnormal electrical coupling of muscle cells resulting in detrusor overactivity. Myohypertrophy was characterized by widely separated muscle cells with a reduction of cell junctions as well as abundant collogen between hypertrophic muscle cells. The finding of the dense band pattern was representative of normal aging of the detrusor.

From this review on the matrix changes in the bladder associated with normal aging, it is apparent that we need to further our understanding of the matrix changes that occur during the prenatal period and those that proceed postnatally and throughout one's lifetime. Understanding normal bladder development and the factors which lead to normal and abnormal alterations associated with aging, will provide the basis for developing new therapeutic strategies for the treatment of urinary dysfunction.

REFERENCES

1. Hobbins JC, Romero R, Brannum P, et al.: Antenatal diagnosis of renal anomalies with ultrasound: I. Obstructive uropathy. Am J Obstet Gynecol 148:868, 1984.

2. Goellner MH, Ziegler EE, Foman SJ: Urination during the first three years of life. Nephron 28:174, 1981.

3. Yeates WK: Bladder function in normal micturition. In: Bladder Control and Enuresis. Edited by I Kolvin, RC MacKeith, SR Meadows. London: W. Heinemann Medical Books Ltd., pp 28-36, 1973.

4. Hu, TW: Impact of urinary incontinence on health care costs. J Am Geriatr Soc 38:292-295, 1990.

5. Diokno AC, Brock BM, Brown MB, Herzog AR: Prevalence of urinary incontinence and other urological symptoms in the noninstitutionalized elderly. J Urol 136:1022-1025, 1986.

6. Diokno AC: Epidemiology of urinary incontinence in the elderly. In: Nephrology and Urology in the Aged Patient. Edited by DG Oreopoulos, MF Michelis, S. Herschorn. Boston: Kluwer Academic Publishers, pp 359-361, 1993.

7. Resnick, N.M.: Voiding dysfunction in the elderly. In: Neurourology and Urodynamics, Principles and Practice. Edited by S. V. Yalla, E.J. McGuire, A. Elbadawi and J.G. Blaivas. New York: Macmillan Publishing Co., Chapt. 19, pp 303-330, 1988.

8. Resnick, N.M. and Yalla, S.V.: Management of urinary incontinence in the elderly. New Engl. J. Med., 313:800, 1985.

9. Resnick, NM, Yalla, SV and Laurino E: The pathophysiology of urinary incontinence among institutionalized elderly persons. New Engl J Med 320:1, 1989.

10. Resnick NM, and Yalla SV: Detrusor hyperactivity with impaired contractile function. An unrecognized but common cause of incontinence in elderly patients. JAMA 257:3076, 1987.

11. Susset JG, Servot-Viguier D, Lamy F, Madernas P, and Black R: Collagen in 155 human bladders. Invest Urol 1:204, 1978.

12. Gosling JA and Dixon JS: Detrusor morphology in relation to bladder outflow obstruction and instability, In: Benign Prostatic Hypertrophy. Edited by F. Hinman, Jr., New York: Springer-Verlag, pp 666-671, 1983.

13. Gilpin SA, Gosling JA, and Barnard RJ: Morphological and morphometric studies of the human obstructed, trabeculated urinary bladder. Br J Urol 57:525-529, 1985.

14. Matsuno T, Tokunaka S, and Koyanagi T: Muscular development in the urinary tract. J Urol 132:148, 1984.

15. Kim KM, Kogan BA, Massad, CA, and Huang YI-C: Collagen and elastin in the normal fetal bladder. J Urol 146:524-527, 1991.

16. Shapiro E, Becich MJ, Perlman E, Lepor, H: Bladder wall abnormalities in myelodysplastic bladders: A computer assisted morphometric analysis. J Urol 145:1024-1029, 1991.

17. Lepor H, Sunaryadi I, Hartanto V, and Shapiro E: Quantitative morphometry of the adult human bladder. J Urol 148:414-417, 1992.

18. Lepor H, Machi G: Comparison of AUA symptom index in unselected males and females between fifty-five and seventy-nine years of age. Urology 42(1):36, 1993.

19. Lepor H, Theune C: A randomized double-blind study comparing the efficacy of terazosin in women with symptoms of prostatism. Urology, Submitted.

20. Elbadawi A, Yalla SV, and Resnick NM: Structural basis of geriatric voiding dysfunction. I. Methods of a prospective ultrastructural/urodynamic study and an overview of the findings. J Urol 150:1650-1656, 1993.

21. Elbadawi A, Yalla SV, and Resnick NM: Structural basis of geriatric voiding dysfunction. II. Aging detrusor: normal versus impaired contractility. J Urol 150:1657-1667, 1993.

22. Elbadawi A, Yalla SV, and Resnick NM: Structural basis of geriatric voiding dysfunction. III. Detrusor overactivity. J Urol 150:1668-1680, 1993.

23. Elbadawi A, Yalla SV, and Resnick NM: Structural basis of geriatric voiding dysfunction. IV. Bladder outlet obstruction. J Urol 150:1681-1695, 1993.

THE ROLE OF PROTEOGLYCANS IN BLADDER STRUCTURE AND FUNCTION

Bertram J. Levy MD. and Thomas N. Wight PhD.

Department of Urology and Pathology
University of Washington School of Medicine
Seattle, Washington

INRODUCTION

The cellular and fibrous elements of all tissues are bound together by an electrically active gel composed of polysaccharides and complex protein polysaccharides known as proteoglycans. This gel acts as the primary regulator of tissue fluid and ion transfer as well as serving as a substrate for cellular attachment and migration. In addition it functions as a stromal communication system; the "wiring" that carries the growth factor and cytokine messages to regulate growth and repair. While there exists an impressive body of knowledge on the gel compartment in other tissues (1, 2, 3, 4, 5), there has been very little interest in the extracellular matrix gel compartment of the bladder. This review will focus first on the essential principles of proteoglycan molecular structure and how that structure can determine their biological functions. The second section will then focus on the direct and indirect evidence that these compounds are present in the bladder and involved in both normal bladder function and in the pathogenesis of the trabeculation.

PROTEOGLYCAN STRUCTURE AND FUNCTION

The extracellular matrix can be divided into two compartments: fibrillar and nonfibrillar (figure 1). The fibrillar compartment will be covered elsewhere in this monograph but is basically divided into two components: the collagens and the elastic fibers. The nonfibrillar compartment (or matrix gel compartment) is largely composed of glycosaminoglycans and glycoproteins. Proteoglycans are glycoproteins with glycosaminoglycans attached through covalent linkage. Glycosaminoglycans are composed of repeating dissacharides arranged in a linear fashion. An example of a glycosaminoglycan not attached to a core protein is hyaluronan.

Proteoglycans have traditionally been divided on the basis of their dissacharide structure into four families: chondroitin sulfate, dermatan sulfate, keratan sulfate, and heparan sulfate. With the exception of keratan sulphate, the disaccharides are composed of a hexosamine and a hexuronic acid. The structural differences between these compounds are based on the type of monosaccharides and the position of the linkages. These two variables determine the interactive nature of proteoglycans and glycosaminoglycans which in many respects dictate the biomechanical properties of tissues (4).

The importance of different types of dissacharides centers around the availability of different anion groups for sulfation (the sulfate group is central to proteoglycan function). For example chondroitin has the same backbone as hyaluronan except that the N-acetylglucosamine is replaced with N-acetylgalactosamine. Sulfation occurs at the 4 and 6 position and hence the characterization of the molecule as chondroitin -4- sulfate or

Muscle, Matrix, and Bladder Function,
Edited by S. Zderic, Plenum Press, New York, 1995

chondroitin -6-sulfate. It is not uncommon however in chondroitin sulfate to have runs of 6-sulfation alternating with runs of 4-sulfation adding yet another level of diversity. While sulfation of chondroitin occurs only on the hexosamine, keratan is composed of alternating N-acetylglucosamine (rather than a hexuronic acid) and galactose. Sulfation of keratan occurs on both sugars in the 6-position. Heparan and its epimerized form, heparin, have the potential for up to four sulfation sites at the 2, 6, and 3 positions. Thus glycosaminoglycans can be characterized by the content and distribution of their charged anionic groups e.g. carboxylate -COO, the sulphate esters -O-SO3, and in the case of heparan sulfate, the sulfamino groups -N-SO3 . These anionic groups bind cations such as sodium and calcium which in turn binds water molecules within their interstice. In this way proteoglycans and hyaluronan create matrix osmotic force.

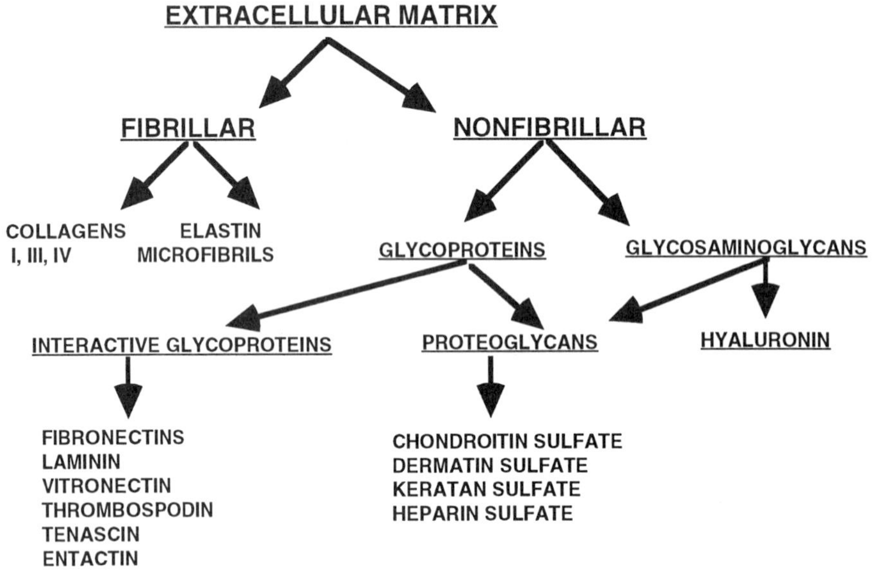

Figure 1. General organization of the extracellular matrix into fibrillar and nonfibrillar compartments.

Proteoglycans can also be characterized by the types of polysaccharide linkages. For example hyaluronan,and chondroitin are both linear molecules with alternating B-1,4 to B-1,3 linkages. Dermatan is almost identical to chondroitin except,as a result of intracellular enzymatic epimerization, the glucuronic acid is changed from a D to an L form (1). This alters the linkage from B-1,3 to A-1,3 and changes the orientation of the molecule from an equatorial to an axial configuration. The result of this change is not just a difference in the physical shape of the molecule but also the orientation of the charged groups. For example chondroitin -4-sulphate charge orientation is equitorial- axial while dermatan -4-sulfate is axial-axial (2). These diversities create sterically specific anionic barriers in the extracellular matrix that concentrate cations and macromolecules at their interface in a process called steric exclusion.

Finally the length and structure of the core glycoprotein add a third major level of diversity to proteoglycans. In fact most proteoglycans are characterized on the basis of their core proteins and end up being known by their trivial names (e.g aggrecan, versican, decorin, biglycan, syndecan, neurocan and serglycin). The amino acid sequences of a number of core proteins have been elucidated. The core proteins vary greatly in length and in number of serine-glycine sites. Serine-glycine sequences are potential binding sites for single glycosaminoglycan chains. The smallest core protein, serglycin, (- 20 kDa) contains 24 consecutive repeats of serine-glycine and is found in intracellular granules in mast cells (3). Decorin has a deduced sequence size of - 32 kDa but has only one serine -glycine binding site available for a dermatan sulfate chain (6). Biglycan is -36 kDa and has two occupied serine -glycine binding sites (6). Aggrecan is quite large at 210 kDa with -110 occupied serine-glycine sites Aggrecan is found principally in cartilage and contains three major domains. The N-terminal (G1) domain serves a a binding region for hyaluronan while the C-terminal (G3) domain is thought to bind to lectin-like cell surface ligands or other extracellular molecules. The amino acids between these two domains are referred to as the internal domain and contains a short section for keratan sulfate attachment followed by a large section for chondroitin sulfate (3). These hybrid molecules add further confusion to a classification of proteoglycans based solely on the glycosaminoglycan structure. As a result, there is a growing tendency to classify proteoglycans instead on the basis of their site of origin: intracellular proteoglycans, cell membrane proteoglycans, basement membrane proteoglycans, large interstitial proteoglycans and small interstitial proteoglycans (table 1).

Table 1. Classification of proteoglycans by site of origin

Location	trivial name	proteoglycan
INTRACELLULAR	serglycin	heparin, CSPG
CELL MEMBRANE	syndecan	HSPG, CSPG
BASEMENT MEMBRANE	perlecan	HSPG, CSPG
LARGE INTERSTITIAL	versican, aggrecan	CSPG, DSPG, KSPG
SMALL INTERSTITIAL	decorin, byglycan fibromodulin	CSPG, DSPG, KSPG

Intracellular proteoglycans are found in connective tissue mast cells and hemopoetic cells in the form of storage granules (7). Serglycin is the predominant core protein but the glycosaminoglycans sidechains vary with the site of origin. For example mucosal mast cells contain predominantly chondroitin sulphate while connective tissue mast cells contain predominantly heparan sulphate. In general the glycosaminoglycans that form these intracellular proteoglycans tend to be overly sulfated. These anionic granules concentrate cationic intracellular proteases, carboxypeptidases, and histamines and may regulate their release in host defense mechanisms and tissue repair (3).

The cell membrane proteoglycans are either an integral part of the plasma membrane (intercalated proteoglycans) or attached to the outer plasmallema through a covalent bond with phosphatydlinositol (8). The family of syndecans are intercalated proteoglycans that actually traverse both sides of the cell membrane with a 34 amino acid COOH terminus intracellular domain, a 25 amino acid residue hydrophobic transmembrane portion and a 235 amino acid extracellular portion that contains 5 serine-glycine sites which bind 3 heparan sulphate and two chondoitin sulfate/dermatan sulphate proteoglycans (3). These molecules form a bridge between the actin cytoskeleton and the extracellular matrix structures. In this way proteoglycans provide the function of cellular attachment. These attachments are continually being formed and reformed which provids an opportunity for cellular migration. For example syndecan, is expressed in the membrane of imature lymphocytes when they are in contact with bone marrow stroma but absent in the mature or circulating form (8). Proteoglycans can also bind to ligands such as fibronectin and interfere with the interaction between the ligand and the cell binding receptor in situations of cell migration. Finally there are specific receptors in the cell membrane that bind bind hyaluronan and thus attach cells to the hyaluronan rich extracellular matrix.

Cell surface proteoglycans also bind growth factors and cytokines. This interaction may prevent or facilitate the activity of growth factors (9, 10). Under the influence of stresses such as trauma, anoxia or acidosis, cells release proteolytic enzymes such as urokinase-like plasminogen activator are released. These enzymes are capable of releasing growth factors and portions of cell surface proteoglycans from the cell membrane. This model of sequestration and release is central to the concept that proteoglycans can act as regulators of growth factor activity. It also provides a basis for an intrinsic feedback mechanism for tissue growth and repair.

Within the basement membrane, the principle proteoglycans are heparan sulfate (known as perlecan), and to a lesser extent, chondroitin sulfate. They are integral components interacting with laminin and type IV collagen through their core proteins (8). These core proteins can vary in size from the large 400 kDa EHS sarcoma basement membrane to the small glomerular basement membrane and are thought to affect basement membrane permeability (3). Basement membrane heparan sulfate proteoglycan has also been shown to bind both acidic and basic FGF and provide a reservoir for these cytokines in the manner described above (11,12).

The large interstitial proteoglycans are dominated by chondroitin sulphate. These proteoglycans interact with hyaluronan to form large multimolecuar aggregates. One can envision these bottle- brush like structures as beads threaded on a hyaluronan string with their large fans of highly charge stereo specific glycosaminoglycans extending out into the extracellular space (figure 2). These glycosaminoglycan chains can then interact with extracellular macromolecules, cations ,and water while the G3 domain of the core protein remains free to interact with cell surface ligands or complement. Because the glycosaminoglycans are held in close proximity to one another on the core protein and carry negative charges, these chains tend to repel each other and provide the property of compressive reistance to the molecule. Proteoglycans such as aggrecan are found to occupy a tissue space 5 times less than there extended state in solution. Thus these compressed molecules have considerable potential energy which also contribute to their ability to resist tissue compression and deformity (3).

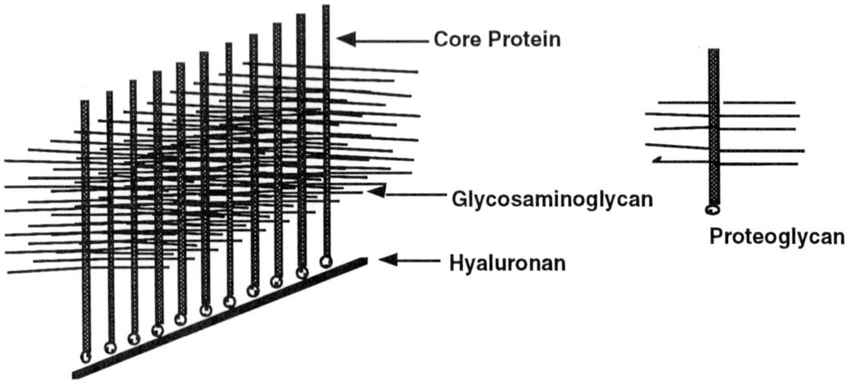

Figure 2. Large interstitial proteoglycans assembled on hyaluronan molecule.

The large interstitial proteoglycans are also responsible for the creation of the water filled compartment of the extracellular matrix. As previously mentioned, this is accomplished through their highly charged anoinic groups that bind counterions and create an osmotic imbalance that attracts water molecules (13). Though it is not as highly charged as the sulfated proteoglycans, hyaluronan entraps a thousand times its weight in water through an elaborate folding mechanism that makes its lattice dimension considerably less than its true extended length. In this way hyaluronan becomes a major determinant of the biomechanical property of tissue hydration (5).

Comper has proposed a formula for the behavior of hyaluronan in solution: the swelling pressure (or ability of the tissue to hold water) is equal to the osmotic pressure (which is the result of hyaluronan entanglement) minus the contractile pressure (which is tissue under compression) (2). Thus under tissue compression, water can be redistributed to areas of less compression. This principle may explain, as will be described later, some of the edematous tissue changes in the obstructed bladder.

The small interstitial proteoglycans are characterized by small core proteins and include decorin and byglycan which contain one or two dermatan or chondroitin sulfates (6). A third member of this group, fibromodulin, contain 4 keratan sulphate chains (3). Decorin associates near the d and e bands of collagen while fibromodulin associates near the a and c bands (14). Decorin is always associated with collagen fibrils in vivo. A single decorin molecule can ionically associate with up to four collagen fibrils and is believed to regulate fibrillogensis of collagen (15). Decorin also copurifys with TGF-B1 and is thought to provide a reservoir for this growth factor in the manner dsecribed above.

In summary the proteoglycans represent a huge library of diverse molecules with their own structural specificity which translates into specificity in their interactions with cell binding sites, collagen, elastin, and other interstitial macromolecules. These interactions in turn regulate specific tissue biomechanical properties, cell binding, and growth factor expression (table 2).

Table 2. Summary of general functions of proteoglycans.

BIOMECHANICAL PROPERTIES
water content
osmotic pressure
ionic properties
compression
viscoelasticity
compliance
resilience
CELLULAR ADHESION AND MIGRATION
GROWTH REGULATION

THE EXTRACELLULAR MATRIX COMPARTMENT IN THE BLADDER

There has been very little interest in the extracellular matrix gel compartment of the bladder and even recent exhaustive studies on bladder ultrastructural morphology refer to the extracellular matrix as "collagen, elastin, and so forth" (16). Much of the work on the bladder proteoglycans centers around comformation of their existence; their functional significance can at present only be implied from data on other tissues. These extracellular matrix proteoglycans (and glycosaminoglycans) should not be confuse with the cell surface glycosaminoglycans and glycoproteins known as the "glycocalyx " or "bladder mucin layer". While these latter compounds may be significant in cell surface defense mechanisms (17), they are not an integral part of the interstitial structure of the bladder.

As previously stated, the extracellular matrix is a three dimenssional collagen fiber network of varying rigidity (and includes elastin) imbedded in a gel-like matrix (2). It is this organization that governs to a large extent the biomechanical properties of most tissues and variations in the components of these two compartments as well as changes in their associations influence properties such as elasticity, compliance, rigidity and resilience. Thus an understanding of the organization of nonfibrillar proteoglycan components requires an understanding ot the fibrillar components as well. The following is a description of the noncellular anatomy of the bladder based on detail ultrastructural (18) and morphometric study (19) beginning first with the organization of the collagen and elastin.

The bladder can be thought of as an ornately woven collagen basket composed of different patterns at different levels and all integrated into one unit (figure 3). The pattern begins just beneath the basement membrane with a superficial 0.12 mm thick layer of closely packed flat sheets of parallel collagen fibers . This layer then abruptly terminates with tassle-like perpendicular collagen fibers which connect to a 1. 8mm thick layer of lumenally curved broader more widely spaced collagen fibers. This layer is refered to as the submucosa to distinguish it from the more superficial lamina propria. At the transition between the lamina propria and submucosa athin but definite elastin layer is interwoven within the collagen and oriented parallel to the lumen. It is unknown whether this elastin layer is analagous to the internal elastic lamina of the aorta but it is very reminiscent of the elastin band in the serosa of the rabbit bladder (20). Scanning electronmicroscopy of this serosal band in the mouse defines its structure as a flat sheet of finely woven fibers (21). In the human bladder the internal elastin band is associated with a thin layer of smooth muscle cells called the muscularis mucosa (22,23,24). Elastin microfibrils are usually present around these muscle cells and believed to be produced by them.

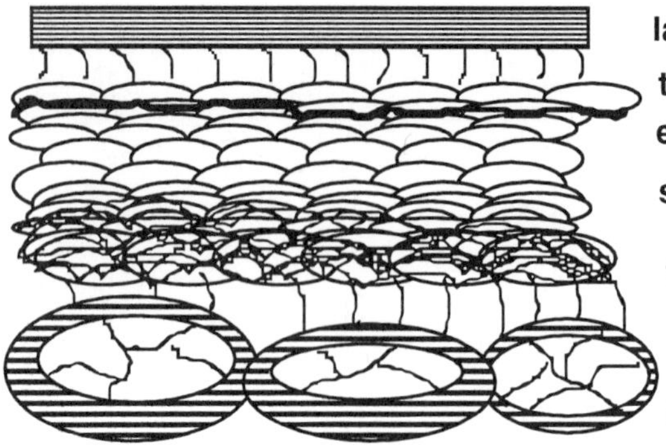

Figure 3. Model of the organization of the detrussor collagen by layer.

The fibers of the submucosa appear to condense as one moves deeper towards the muscularis propria This pattern is similar to that of the intima of arteries (25). The dense submucosal collagen fibers are then joined to the condensed fibers of the permysium. sheath with tassle like perpendicular fibers. This transition zone contains elastic fibers, collections of cell-free fat globules, and nerves. The connection between these two regions is extremely fragile and always tear during tissue preparation even with plastic embedded sections.

Finally from the perimyseal sheath fine collagen fibers project into the fasicle to form delicate microseptae between the muscle cells. These microseptae are probably a combination of type I and III collagen as opposed to the broader type I in the submucosa (26) and serves as a reminder that variations in collagen appearance throughout the bladder may be a result of either variations in collagen type or variations in collagen assembly.

Set between the collagen network of the bladder is the nonfibrillar compartment. This nonfibrillar material can be appreciated on an ultrastructural level. In the lamina propria, for example, electron- dense particulate material lies between the individual collagen fiber sheets. In the submucosa the broader collagen fibers appear ultrastructurally as bundles of collagen fasicles which are in turn composed of individual fibrils. Between these bundles are a system of channels and tributaries which are filled with fine electron-dense particulate matter. In addition, fine electron-dense material surrounds the individual collagen fibrils like halos. Pools of this electron dense material are prominent at the transition zone between the lamina propria and submucosa and around blood vessels. There are however some areas completely devoid of any material and appear to correlate with areas of focal edema seen on histologic sections. Finally within the muscularis propria, collections of fine electron dense material are present between the microseptae and the individual muscle cells.

While the identity of this fine electron dense particulate matter is unknown, there is strong evidence to suggest that this material is composed of members of the proteoglycan and hyaluronan family. Cationic dyes such as ruthenium red identify similar ultrastructural material in the lamina propria and submucosal of human colon as members of the proteoglycan family (27). Similar electron dense proteoglycan in the intima of rat carotid arteries has been demonstrated using cuprolinic blue (28). Through the use of immunogold labeled electronmicroscopy, which identifies specific proteoglycans, at least some of the electron dense amorphous clusters in the extracellular matrix of the bovine aorta intima have been shown to be chondroitin sulfate (29).

Although proteoglycans are unquestionably present in the bladder, there are suprisingly few studies confirming their presence. Rubbens et. al. (30) mechanically seperated the mucosa from the rest of the calf bladder and analyzed the chromatographically separated homogenates for glucosamine and galactosamine. A large glycoprotein band was present in both layers and probably respresents fibronectin. Heparan sulfate and hyaluronan were demonstrated in both the mucosa and the rest of the bladder wall but chondroitin 4 and 6 sulfate was present only in the wall. De Klerk (31) confirmed the presence of hyaluronan, heparan sulfate, dermatan sulfate and chondroitin sulfate.in young human bladders but could not differentially seperate their locations due to difficuties with mucosal- submucosal adherence. While these studies are extremely valuable, quantitative studies need to be repeated using adult bladders since, more than likely, the composition and distribution of proteoglycans evolve with age.

Immunochemistry studies are helpful in establishing the differential distribution of proteoglycans in the normal adult human bladder. Heparan sulfate has been immunolocalized to the epithelial surface of normal human bladder (32). Heparan sulfate has been extensively studied in the endothelium of arteries and exist in atleast three forms on cell surface syndecans (10) and two forms in the basement membrane including the large 400kDa perlecan (33, 34). Within the basement membrane, heparan sulfate functions on three levels (35): 1) as structural organizer of the other basement membrane components (type IV collagen, laminin, fibronectin and entactin); 2) as a selective filtration barrier conferred by their anionic charge; 3. selective binding of growth factors (heparin binding-EGF,acidicFGF, basicFGF,)(1,36). In the presence of capillary injury, urokinase-like plasminogen activator(plasmin) is activated and releases complexes of bFGF-HSPG from endothelial cells and basement membrane (37). These complexes are protected from further degradation by the plasmin while the uncomplexed bFGF are not (38). While similar mechanisms have not been studied in bladder, many of the elements of this process have been demonstrated. Urokinase-like plasminogen activators are present in urine, are produced by bladder urothelium (39, 40), and appear to be involved in urothelial cancer metastasis (41). Another cytokine, EGF, is produced by transitional epithelial cells, is present in urine, and stimulates growth of human bladder cancer cell lines in tissue culture (42).

The lamina propria appers to be the richest site of chondroitin sulphate proteoglycan (CSPG) in the bladder (19). There is a strikingly intense 0.02 mm band of chondroitin staining just beneath the mucosal basement membrane. On the other hand the submucosa is the richest site of decorin and hyaluronan. While decorin is evenly distributed throughout the submucosa in close association with collagen fibers, hyaluronan is especially concentrated at the transition zone between the lamina propria and submucosa. Chondroitin sulphate, hyaluronan and decorin are all localized to the perimyseal collagen sheath and along the microseptae. While this describes the general pattern of proteoglycan distribution in the adult human bladder, topographical variations are common. Decorin and hyaluronan both can be found in isolated patches within the chondroitin sulfate rich lamina propria. In addition within the submucosa, chondroitin sulfate rich areas can be seen in association with blood vessels, elastin, nerve ganglion, and around the smooth muscle cells of the muscularis mucosa. These topographical variations are believed to create unique microenvironments of functional significance. For example the small interstitial proteoglycan, decorin, is much less hydrated than the large chondroitin sulfate proteoglycan. Thus decorin concentrations in isolated patches in the lamina propria can create focal areas of mucosal noncompliance. This may explain the trabecular-like patterns seen endoscopically by clinicians on extreme filling.of normal nonobstructed bladders (personal communication) and raises the question as to whether this pattern may provide the template for the pattern of future trabeculation.

The lamina propria distribution of CSPG in the bladder follows similar subepithelial distributions in epidermis and hair follicle of skin, the intralobular subepithelial matrix of breast, and the lamina propria of intestine (43). From a functional standpoint, there is a relationship between tissues that experience compression and the presence of CSPG. Tendon provides an interesting example (44). In bovine tendons that experience both compression and tension, CSPG and small DSPG are present while in tendons which experience only tension, only the small DSPG is found. In an elegant experiment by Gillard et al. (45), rabbit tendons were surgically transposed from positions of tension only to positions of.compression and tension. Within the first two weeks CSPG and HA increased dramatically while DSPG remained the same. Tendon transfer in the opposite direction resulted in a loss of CSPG within the same time frame. It is attractive to speculate that the bladder lamina propria CSPG provides protection for the compressed mucosa during bladder filling.

The function of the intense 0.02 mm band of CSPG adjacent to the basement membrane unknown. Using a technique of complementary scanning electronmicroscopy, a system of subepithelial capillaries can be demonstrated in grooves between the underside of the urothelium and lamina propria (46). Whether the highly concentrated CSPG has a role in protecting this capillary network from compression is only speculative. Because of its high anionic charge, this band of CSPG also has the potential to provide an anionic barrier to the extravasation of urine. Ordinarily the mucosa is impermeable to urine however even in some normal subjects, permeability can be demonstrated (47). If there is a loss of the protective mucosal GAG layer, urine can readily enter the submucosa (48, 47).

The transition area between the lamina propria and submucosa is structurally unique and may have function significance. For example, the tassle like collagen fibers that join the parallel fibers of the lamina propria to the coiled fibers of the superficial submucosa. has the potential to act as a hinge during filling. Hyaluronan is most heavily concentrated at this transition point and could support this action in several ways. Hyaluronan perhaps more than any other matrix molecule is responsible for the hydration of the extracellular matrix (5). Hyaluronan, therefore, could create a loose matrix for movement of the lamina propria over the submucosa on filling and emptying. The hydrated hyaluronan also could create a swelling pressure which consists of the osmotic force trying to imbibe water minus the contractile force trying to press water out of the tissues (2). Under compression a pumping action is created that behaves as a transport mechanism for the movement of nurients, fluid, metabolites, and macromolecules through a relatively avascular stroma such as the submucosa.

The submucosa is primarily a stromal layer which is believed to be reponsible for compliance and has been shown to have a greater effect on invitro compliance than the muscularis propria. Susset and Regnier (49) using isolated human bladder strips demonstrated that the viscoelastic coefficient of the mucosal (and submucosa) strips exceeded that of the muscle strips and at maximal stretch, exerts nine times greater tensile strength. The organization of the collagen has been implicated in this function (26, 18) but hyaluronan could also play a role. Hyaluronan develops its property of viscoelasticity through its ability to entangle and form polymeric networks via possible cross-links with collagen (2). Solutions of hyaluronan display viscosity at low strain frequency and elasticity at high strain frequency. This can be translated into compliance in the filling phase and passive contraction in emptying phase (2).

The elastin band and smooth muscle cells of the muscularis mucosa are also present at the transition zone and may be interrelated. These two structures remain close to and often in direct contact with each other. Normally around these muscle cells are elastin microfibrils (18) as well as rich deposits of chondroitin sulfate and hyaluronan (19). It suggests that these smooth muscle cells are producing these matrix elements and could potentially be involved in maintainance of the elastin band. This assumption is supported by the demonstration that smooth muscle cells, in guinea pig aorta, produce elastin (50) and in monkey aorta, produce chondroitin sulfate (51). In the aging and obstructed human bladder, this microenvironment is associated with smooth muscle degeneration as evinced by large protruding apotthe vesicals and abundant pinocytotic vesicles, deposits of the cation calcium, and cellular debri (18). The presence of chondroitin sulfate in this region may play a role in these changes since these large interstitial proteoglycans appear in the early stages of tissue injury (43), are concentrated in .the interstitial matrix of early atheromatous plaque (29) and are present at the leading edge of tumor metastasis (43).

The presence of chondroitin sulfate around the elastin band may be significant from a functional standpoint. Although the function of this structure is unknown, it elastic properties could support the mucosa during filling and emptying. Elastin derives some of its elastic properties from its 70% hydrophobic nonpolar amino acids (52). It is theorized that when stretched these hydrophobic amino acids become exposed to a hydrated environment they retract back to a nonexposed state. The characteristically hydrated chondroitin sulfate around the elastin can thus enhance the elastic properties of the elastin band.

Decorin distribution in association with submucosal collagen follows similar distribution in dermis, intestine, blood vessels, cartilage and tendon (43, 3). Decorin is present in matrices rich in type I collagen (4), which is the predominant collagen type in the human bladder submucosa (26). It binds to collagen fibrils, is associated with smaller fiber size (25), and can be immunolocalized ultrastructurally to the fibril which may accounts for the halo like particles observed around bladder collagen fibrils (35). Functionally the presence of decorin is associated with tissue rigidity and as such has the potential to influence bladder compliance.

At the level of the muscularis propria, decorin, chondroitin sulfate and hyaluronan are present within the perimyseum and microseptae. Hyaluronan is generally not associated with smooth muscle (vascular, intestinal) but the bladder smooth muscle distribution is similar to the endomyseal distribution in skeletal muscle (5). The hydrated microionic environment created by the presence of chondroitin sulfate and hyaluronan gain importance when one considers that smooth muscle impulse conduction depends more on cell to cell contact than neural innervation (53). In addition, decorin may play a role in the maintainance of a stable microenvironment for the smooth muscle cells. Decorin (and bigycan) are produced by vascular smooth muscle cells (54) and copurifies with TGF-B1 in the media of CHO cell cultures (55). This association prevents the utilization of TGF-B1 by the cells. For example, infusion of decorin decreases the TGF-B1 mesangial cell proliferation of extracellular matrix in experimental glomerulonephritis (56). This interplay between decorin and TGF-B may play a regulatory role in maintaining the fairly constant extracellular width between the smooth muscle cells of the normal bladder. Conversely the release of the TGF-B1 from decorin can stimulate matrix growth.

MATRIX CHANGES IN AGING AND OBSTRUCTION

Few studies on connective tissue changes in the aging bladder have been done. Collagen concentration (per dry weight) by hydroxyproline analysis shows only a 10% increase in the aging female bladder and no change in the aging male (57) . Cortivo (58) also showed no change in collagen content between a normal and obstructed bladder inspite of a 30 per cent increase in elastin. These findings are contradictory to the histologic appearance of the aging bladder and may be explained by results in experimentally produced obstructed rat bladders. The total collagen content increased four times but since overall bladder weight increased twelve times, the relative concentration of collagen was reduced (59). Lepor (60) used morphometric analysis to measure the width of the connective tissue bands between the muscle fasicles.of the muscularis propria and demonstrated an increase in the connective tissue to muscle ratio in both aged males and females. Ultrastructural observations in the normal aged human bladder show that these widened bands are composed not of collagen but primarily of nonfibrillar matrix elements (16). Similarly in the aging submucosa the striking feature is nonfibrillar matrix deposition rather that collagen deposition (18). Instead of fine electron dense particulate matter accumulation within the channels and tributaries between the collagen fasicles, there is accumulation within the fasicles with seperation of the individual fibrils. Ultrastructural observations on the sumucosal and muscularis propria of obstructed bladders show an acceleration of this process with some areas of predominately nonfibrillar material and other areas displaying predominately collagen and elastin fiber.

We have had the opportunity to study bands of trabeculation from 8 obstructed human males (19). The collagen organization of the lamina propria is unchanged from normal, but at the transition between the lamina propria and submucosa, the perpendicular connecting fibers are no longer apparent. The internal elastic band (and muscularis mucosa) are also absent. Of interest is a similar loss of the serosal elastic lamina in the obstructed rabbit bladder (20). In the submucosa, the trabecular bands show two distinct patterns (figure 4). In the thinned convex center, there is loss of collagen while at the bases there is accumulation of collagen. At the convex center, the thinned collagen fibers appear disorganized and seperated by large oedematous spaces. At the bases the collagen forms dense plaques. These plaques extend down into the muscle bundles, merge with the perimysium of the muscle bundle, and then join the dense submucosal collagen plaques on the other side. This effectively sequesters one or two muscle bundles from the deeper muscle groups. These sequestered muscle bundles show considerable deposition of collagen between the atrophic smooth muscle cells (originally termed collagenosis by Gosling (61). On the basis of the submucosal collagen changes, we refer to the convex area can be as the lytic zone and the base as a proliferative zone (figure 4).

The differential distribution of the proteoglycans and hyaluronan shed some light on this process (19). In the lamina propria, just as the collagen pattern appears unchanged, the distribution of chondroitin sulfate is also unchanged (including the distinctive 0.02 mm. chondroitin sulfate band). In experimentally obstructed rabbit bladders; there is only a transient rise in proliferative activity of the cells in the lamina propria (as measured by invivo incorporation of H3-thymidine) and a return to no demonstrable cell division within three days (62). Further study is needed to determine the role that the lamina propria plays in response to obstruction.

In contrast to the lamina propria, the proteoglycan distribution in the submucosa of the trabeculated bladder are significantly changed from the normal submucosa. In many areas of the superficial submucosa, hyaluronan no longer abutts the transition zone. This, along with the loss of the internal elastin band and the tassle like collagen fibers, suggest that this region has lost its compliance. The heavy concentrations of hyaluronan have shifted to the deep submucosa and this creates a visoelastic domain beneath oedematous areas. In the deep submucosa, large collagen fibers, elastin fiber, and decorin accumulation creates compressive forces. Thus opposite microenvironments are juxtaposed: the noncompliant superficial submucosa and the heavy fibrillar elements of the deep submucosa opposing the hydrated viscoelastic hyaluronic rich deep submucosa. In physical chemistry terms (2): the contractile force (compression) exceeds the osmotic pressure (hyaluronan) resulting in a decrease in tissue swelling pressure and an increase in tissue fluid transfer (oedema). It is unlikely that the oedema has a direct collagenolytic effect however it may influence the release of growth factors such as bFGF (37) or proteolytic enzymes such as plasmin (1) which induce collagenolysis and wound healing. The complex interaction of these extracellular components require further study.

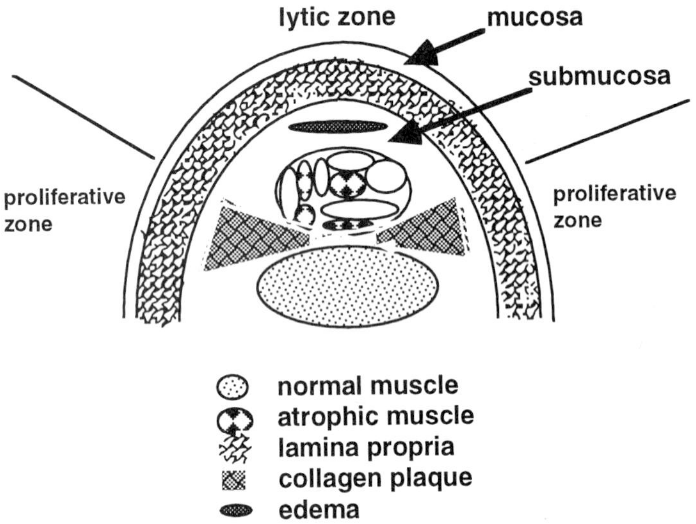

Figure 4. Schematic model of the trabecular band with division into two seperate sectors: the lytic zone and proliferative zone.

At the base of the trabeculation, the dense collagen plaques are devoid of elastin and chondroitin sulfate but rich in decorin and hyaluronan. Decorin is concentrated in other dense structures such as the interlobar septae of breast and dermis (43) as well. as in advanced atherosclerotic plaques (35). This location seems somewhat contradictory since decorin concentration is inverse to the fiber size in normal arteries (25). This discrepancy may be explained by variations in GAG chain lengths or in -4 and -6 sulfation positions in decorins with similar core proteins.

The changes within the sequestered muscle bundles are perhaps the most striking of all the trabecular changes. These bundles have traditionally been though of as degenerative. However the proliferative aspects of the matrix far exceeds the degenerative appearance of the individual muscle cells. The extracellular matrix between the atrophic smooth muscle cells is greatly expanded and stains intensely for chondroitin sulfate, hyaluronan, decorin, elastin, elastin microfibrils, type I and III collagen (16, 18, 19, 62). While still unknown in the bladder, there is considerable evidence in blood vessels that smooth muscle cells are capable of producing all of these matrix elements. It is however well documented, biochemically (59), physiologically (63), and morphologically (16), that the bladder smooth muscle cells change rapidly from contractile to proliferative cells in the presence of obstruction. What drives that transformation is not known but most likely involves growth factors since growth factors not only drive cellular proliferation but also stimulate extracellular matrix synthesis.

Except for the urothelium, where considerable study on the relationship of EGF and acid and basic FGF to bladder cancer cell line growth has been done, nerve growth factor is the only documented interstitial cytokine (64). In the obstructed rat and human bladders, nerve growth factor concentration increases significantly .Though entirely speculative, TGF-B1 is an attractive candidate for involvement in trabeculation since TGF-B1 is regularly associated with type I collagen bound decorin (55). The release of TGF-B1 from decorin occurs in the presence of an acidic environment (65). This released TGF-B1 and its latency associated peptide are then cleaved by plasmin to release the active TGF-B1. Active TGF-B1 in tissue culture, inhibits arterial smooth muscle cell proliferation and stimulates chondroitin sulfate (and collagen) expression (51). These matrix elements can then bind more growth factor and provide a reservoir for more growth factor release. This phenomenon may explain the observation that nerve growth factor in the obstructed animals does not return to normal after the obstruction is relieved (64).

If we were to entertain a hypothetical cascade of stromal events that could create the trabecular bands, we would begin with obstruction (figure 5). The obstruction causes increased pressure which exerts its influence on each topographically unique area of the bladder - those that recieve compression and those that recieve tension. In the areas rich in large interstitial proteoglycan there may be fluid shifting with oedema (based on the Comper equation), release of proteolytic ezymes and stromolysis. In the areas rich in small interstitial proteoglycans there may be release of growth factors from stromal reservoirs, such as TGF-B, which stimulate fibroblasts to produce more collagen and hyaluronan. The proliferative area may exert downward pressure while the lytic area may allow upward pressure. These forces impinge on the muscle cells which respond by either releasing proteolytic enzymes that free growth factors from proteoglycan reservoirs and/or causing muscle cell production of growth factors. These growth factors then activate the muscle cell to undergo hypertrophy. Confined physically by the collagen perimysium and the limitations of the extracellular matrix to support the rapidly increasing demands, the crowded muscle cells exist under greater pressure. These cells then actively produce matrix elements. The accumulation of new matrix thus may further isolate the smooth muscle cells, alters conduction pathways, change the biomechanical properties of the intrafasicular microseptae, and continue to stress the cells thereby perpetuating a cycle that results in eventual cellular atrophy.

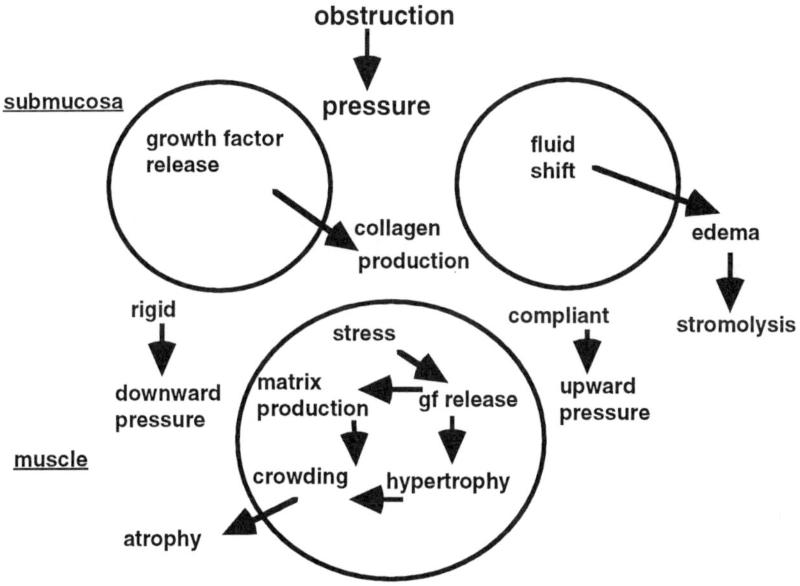

Figure 5. Proposed cascade of events in the development of trabecluation.

It would of course be too simplistic to think that only one growth factor is involved in such a process. Studies on the application of various growth factors to the rabbit ear wound healing model demonstrate individualized responses to different cytokines (66). Here PDGF-BB stimulates fibroblast expression of a provisional matrix (fibronectin and glycosaminoglycans), basic FGF stimulates neovasuclarization and collegenolysis, and TGF-B1 stimulates fibroblast expression of the mature matrix (collagen). This has led the authors to postulate a cascade of growth factor expression based on differential temporal release necessary for complete wound healing. In a similar fashion the continuim of extracellular matrix change between the smooth muscle cells from normal aged bladders to obstructed bladders (16) can be thought of as a process of wound healing starting with a provisional matrix and ending with fibrillar accumulation.

In summary there is strong direct and indirect evidence that proteoglycans and hyaluronan exist in the human bladder. They generally have a differential distribution in the different layers of the bladder wall but also show considerable topographical variation. These variations create unique microenvironments which can respond differently to the stress of compression and therefore may play an important role in the maintanance of normal bladder function and in the pathogenesis of obstruction.

REFERENCES

1. Jackson, R. L., Busch, S. J., and Cardin, A. D.: Glycosaminoglycans: molecular properties, protein interactions, and role in physiological processes. Physiol. Rev. 71: 481-539, 1991.
2. Comper, W. D., and Laurent, T. C.: Pysiological function of connective tissue polysaccharides. Physiol. Rev. 58: 255-315, 1979.
3. Wight, T. N., Heinegard, D.K., and Hascall, V. C.: Proteoglycans: Structure and Function. In Cell Biology of Extracellular Matrix. Edited by Hay, E.D. New York: Plenum Press, Chap. 2, pp. 45-78, 1991.
4. Kjellen, L., and Lindahl, U.: Proteoglycans: structures and interactions. Ann. Rev. Biochem. 60: 443-475, 1991.
5. Laurent, T. C., and Fraser, J. R. E.: Hyaluronan. FASEB. 6: 2397-2404, 1992.

6 Fisher, L. W., Termine, J. D., and Youg, M. F.: Deduced protein sequence of bone small proteoglycan I (biglycan) shows homology with proteoglycan II (decorin) and several nonconnective tissue proteins in a varitey of species. J. Biol. Chem. 264: 4571-4576, 1989.

7 Stevens, R. L., Intracellular Proteoglycans in Cells of the Immune System. In Biology of Proteoglycans. Edited by Wight, T. N., and Mecham, R. New York: Academic Press, Chap.13, pp. 367-387, 1987.

8. Wight, T. N., Kinsella, M. G., Qwarnström, E. E.: The role of proteoglycans in cell adhesion, migration and proliferation. Current Opinion Cell. Biol. 4: 793-801, 1992.

9 Ruoslahti, E., and Yamaguchi, Y.: Proteoglycans as modulators of growth activities. Cell. 64: 867 -869, 1991.

10 Vlodavsky, I., Bar-Shavit, R., Ishai-Michaeli, R., Bashkin, P., and Fuks, Z.: Extracellular sequestration and release of fibroblast growth factor: a regulatory mechanism? Trends Biochem. Sci. 16: 268-271, 1991.

11 Rifkin, D. B., and Moscatelli, D.: Recent developments in the cell biology of basic fibroblast growth factor. J. Cell Biol. 109: 4060-4066, 1989.

12 Bernfield, M., Kokenyesi, R., Kato, M., Hinkes, M. T., Spring, J., Gallo, R. L., and Lose, E. J.: Biology of the syndecans: A family of transmembrane heparan sulfate proteoglycans. Annu. Rev. Cell Biol. 8: 365- 393, 1992.

13. Arnott, S., and Mitra, A.,K.: X-ray diffraction analyses of glycosaminoglycans. In Molecular Biophysics of the Extracellular Matrix. Edited by Arnott, S., Rees, D.,A., and Morris, E.,R. New Jersey: Humana Press Chap.3, pp. 41-66.1983.

14 Scott, J. E., and Orford, C. R.: Dermatan sulphate-rich proteoglycan associates with rat tail-tendon collagen at the d Band in the gap region. Bioch. J. 197: 213-216, 1981.

15 Scott, J. E.: Proteoglycan-fibrillar collagen interactions. Biochem. J. 252: 313-323, 1988.

16 Elbadawi, A., Yalla, S. V., and Resnick, N. M.: Structural basis of geriatric voiding dysfunction: II. Aging detrussor: Normal versus impared contractility. J. Urol. 150: 1657-1667.

17 Parsons, C. L., Greenspan, C., Moore, S. W., and Mullholland, S. G.: Role of surface mucin in primary antibacterial defense bladder. Urol. 9: 48-52, 1977.

18 Levy, B. J., and Wight, T. N.: Structural changes in the aging submucosa: New morphologic critera for the evaluation of the unstable human bladder. J. Urol. 144: 1044-1055, 1990.

19 Levy, B. J., Lara, S., and Wight, T. N.: Topographical variations in the distribution of proteoglycans and hyaluronan in the submucosa of the normal and trabeculated human bladder. Submitted J. Urol.

20. Monson, F. C., Goldschmidt, M.H., Zderic, S. A.,Ruggierhi, Levin, R. M., and Wein, A. J.: Use of a previously undescribed elastic lamina of the serosa to characterize connective tissue hypertrophy of the rabbit bladder wall following partial outlet obstruction. Neurourol. urodynam. 7:385-396, 1988.
21. Ushiki,T., and Murakumo, M.: Scanning electron microscopic studies of tissue elastin components exposed by a KOH-collagenase or simple KOH digestion method. Arch. Histol. Cytol. 54: 427-436, 1991.
22. Dixon, J. S., and Gosling, J. A.: Histology and fine structure of the muscularis mucosae of the human urinary bladder. J. Anat. 136: 265-271, 1983.
23. Younes, M., Sussman, J., and True, L. D.: The usefulness of the Level of the muscularis mucosae in the staging of invasive transitional cell carcinoma of the urinary bladder. Cancer. 66: 543-548, 1990.
24. Ro, J. A., Ayala, A. G., and El-Naggar, A.: Muscularis mucosa of the urinary bladder: Importance for staging and treatment. Am. J. Surg. Pathol. 11: 668-673, 1987.
25. Merrilees, M. J., Tiang, K. M., and Scott, L.: Changes in collagen fibril diameters across artery walls including a correlation with glycosaminoglycan content. Conn. Tiss. Res. 16: 237-257, 1987.
26. Ewalt, D. H., Howard, P. S., Blyth, B., Snyder, III, H. M., Duckett, J.W., Levin, R.M., and Macarak, E.J.: Is lamina propria matrix responsible for normal bladder compliance? J. Urol. 148: 544-549, 1992.
27. Iozzo R. V., Bolender, R. P., Wight, T. N.: Proteoglycan changes in the inmercellular matrix of human colon carcinoma: An integrated biochemical and stereologic analysis. Lab. Invest. 47: 124-138, 1982.
28. Snow, A. D., Bolender, R. P., Wight, T. N., and Clowes, A. W.: Heparin modulates the composition of the extracellular matrix domain surrounding arterial smooth muscle cells. Amer. J. Path. 137: 313-330, 1990.
29. Lark, M. W.,Yeo, T. K., Mar, H., Lara, S., Hellstrom, I., Hellstrom, K. E., and Wight, T. N.: Arterial chondroitin sulfate proteoglycan: localization with monoclonal antibody. J. Histchem. Cytochem. 36: 1211-1221, 1988.
30. Rubben, H., Friedrichs, R., Stuhlsatz, H.W., Cosma, S., Lutzeyer, W.: Quantitative analysis of gycosaminoglycans in urothelium and bladder wall of calf. Urol. 22: 655-657, 1983.
31. De Klerk, D. P.: The glcosaminoglycans of human bladder cancers of varying grade and stage. J. Urol. 134: 978-981, 1985.
32. Takamura, S., Hiyake, K., and Takeuchi, J.: Immunohistochemical localization of proteoglycans in interstitial elements of the human bladder. Hinyokika Kiyo. 37: 329-334, 1991.
33. Noonan, D. M., Horigan, E. A., Ledbetter, S. R., Vogeli, M., Saski, Y., and Hassell, J. R.: Identification of cDNA clones encoding different domains of the basement membrane heparan sulfate proteoglycan. J Biol. Chem. 263: 16379-16387.
34. Murdoch, A. D., Dodge, G. R., Cohen, I., Tuan R. S., and Iozza, R. V.: Primary structure of the human heparan sulfate proteoglycan from basement membrane (HSPG2/perlecan): A chimeric molecule with multiple domains homologous to the low density lipoprotein receptor, laminin, neural cell adhesion molecules and epidermal growth factor. J. Biol. Chem. 267: 8544-8557, 1992.
35. Wight, T. N.: Cell biology of arterial proteoglycans. Arteriosclerosis. 9:1-20, 1989.
36. Yayon, A., Klagsbrun, J. D., Esko, J. D., Leder, P., and Ornitz, D. M.: Cell surface heparan-like molecules are required for bonding of basic fibroblast growth factor to its high affinity receptor. Cell. 64: 841-848, 1991.
37. Saksela, O., and Rifkin, D. B.: Release of basic fibroblast growth factor-heparan sulfate complexes from endothelial cells by plasminogen activator-mediated Proteolytic Activity. J. Cell Biol. 110: 767-775, 1990.
38. Saskela, O., Moscatelli, D., Sommer, A., and Rifkin, D. B.: Endothelial cell-derived heparan sulfate binds fibroblast growth factor and protects it from proteolytic degradation. J. Cell Biol. 107: 743-751, 1988.
39. Hasui, Y., Marutsuka, K., Nishi, S., Kitada, S., Osada, Y., and Sumiyoshi, A.: The content of urokinase-type plasminogen activator and tumor recurrence in superficial bladder cancer. J. Urol. 151: 16-20, 1994.
40. Schmitt M., Janicke F., Moniwa N., Chucholowski N., Pache L., and Graeff H.: Tumour associated urokinase-type plasminogen activator: biological and clinical significance. Biol. Chem. Hoppe. Seyler. 373, 611-622, 1992.

41. Hiti, A.L., Rideout III, W. M., Laug, W. E., and Jones, P. A.: Plasminogen activator regulation by transforming growth factor-beta in normal and neoplastic human urothelium. Cancer Commun. 2: 123-128, 1990.
42. Kuranami, M., Yamaguchi, K., Fuchigami, M., Imanishi, K., Watanabe, T., Abe, K., Asanuma, F., and Hiki, Y.: Effect of urine on clonal growth of human bladder cancer cell lines. Cancer Res. 51: 4631-4635, 1991.
43. Yeo, T. K., Brown, L., and Dvorak, H. F.: Alterations in proteoglycan sysnthesis common to healing wounds and tumors. Amer. J. Path. 138: 1437-1450, 1991.
44. Vogel, K. G., and Koob, T. J.: Structural specialization in tendons under compression. Int. Rev. Cytol. 115: 267-293, 1989.
45. Gillard, G. C., Reilly, H. C., Bell-Booth, P.G., and Flint, M. H.: The influence of mechanical forces on the glycosaminoglycan content of the rabbit flexor digitorum profundus tendon. Connect. Tissue Res. 7: 37-46, 1979.
46. Inou'e, T.: Complementary scanning electron microscopy: technical notes and applications. Arch. Histol. Cytol. 55 suppl.:45-51, 1992.
47. Chelsky, M. J., Rosen, S. I., Knight, L. C., Maurier, A.H., Hanno, P. M., and Ruggieri, M.R.: Bladder permeability in interstitial cystitis is similiar to that of normal volunteers: direct measurement by transvesical absorption of 99mtechnetium diethylentriamepentaacetic acid. J. Urol. 151:346-349, 1994.
48. Monson, F. C., Wein, A. J., McKenna, B. A.W., Whitmore, K., and Levin, R. M.: Indigocarmine as a quantitative indicator of urothelal integrity. J. Urol. 145: 842-845, 1991.
49. Susset, J. G., and Regnier, C. H.: Viscoelastic properties of bladder strips: standardization of a technique. Invest. Urol. 18: 445- ,1981.
50. Ross, R.: The Smooth muscle cell. II. Growth of smooth muscle in culture and formation of elastic fibers. J. Cell Biol. 50: 172-186, 1971.
51. Schönherr, E., Järveläinen, H. T., Sandell, L. J., and Wight, T. N.: Effects of platelet -derived growth factor and transforming growth factor-B1 on the sysnthesis of a large versican-like chondroitin sulfate proteoglycan by arterial smooth muscles. J. Biol. Chem. 266: 17640-17647, 1991.
52. Mecham, R. P., and Heuser, J. E.: The elastic fiber. In Cell Biology of Extracellular Matrix. Edited by Hay, E.D. New York: Plenum Press, Chap. 3, pp. 79-109, 1991.
53. Daniel, E. E., Daniel, V. P., Duchon, G., Garfield, R. E, Nichols, M., Malhotra, S. K., and Oki, M.: Is the nexus necessary for cell-to-cell coupling of smooth muscle? J. Mem. Biol. 28: 207-239, 1976.
54. Järveläinen, H. T., Kinsella, M. G., Wight, T. N. and Sandell, L.: Differential expression of small chondroitin/dermatin sulfate proteoglycans, PG-I/biglycan and PG-II//decorin, by vascular smooth muscle and endothelial cells in culture. J. Biol. Chem. 266: 23274-23281, 1991.
55. Yamaguchi, Y., Mann, D. M., and Ruoslahti, E.: Negative regulation of transforming growth factor-beta by the proteoglycan decorin. Nature. 346: 281-284, 1990.
56. Border, W. A., Noble, N. A., Yaamamoto, T., Harper, J. R., Yamaguchi, Y., Pierschbacher, M. D., Ruoslahti, E.: Natural inhibitor of transforming growth factor-beta protects against scarring in experimental kidney disease. Nature 360: 361-364, 1992.
57. Susset, J. G., Servot-Viguier, D., Lamy, F., Madernas, P., and Black,R.: Collagen in 155 human bladders. Inves. Urol. 16:204-206, 1978.
58. Cortivo, R., Pagano, F., Passerini, G., Abatangelo, G., and Castellani, I.: Elastin and collagen in the normal and obstructed bladder. Brit. J. Urol. 53: 134-137, 1981.
59. Uvelius, B., and Mattiasson, A.: Collagen content in the rat urinary bladder subjected to infravesical outflow obstruction. J. Urol. 132: 587-590, 1984.
60. Lepor, H., Sunaryadi, I., Hartanto, V., and Shapiro, E.: Quantitative morphometry of the adult human bladder. J. Urol. 148: 414-417, 1992.
61. Gosling, J. A., and Dixon, J. S.: Structure of trabeculated detrussor smooth muscle in cases of prostatic hypertrophy. Urol. Int. 35: 351-355, 1980.
62. Monson, F. C., McKenna, B. A., Wein, A. J., and Levin, R. M.: Effect of outlet obstruction on 3H-thymidine uptake: A biochemical and radioautographic study. J. Urol. 148: 158162, 1991.
63. Levin, R. M., High, J. and Wein, A. J.: The effect of short-term obstruction on urinary bladder function in the rabbit. J. Urol. 132: 789-791,1984.

64. Steers, W. D., Kolbeck, S., Creedon, D., and Tuttle, J.B.: Nerve-growth factor in the urinary bladder of the adult regulates neuronal form and function. J. Clin. Invest. 88: 1709-1715, 1991.
65. Harpel, J. G., Metz, C. N., Kojima, S.,and Rifkin, D. B.: Control of tranxforming growth factor-B activity: Latency vs activation. Progress in Growth Factor Res. 4, 321-335, 1992.
66. Pierce, G. F., Tarpley, J. E., Yanagihara, D., Mustoe, T.A., Fox, G. M., and Thomason, A.: Platelet-derived growth factor (BB homodimer), transforming growth factor-B1, and basic fibroblast growth factor in dermal wound healing: Neovessel and matrix formation and cessation of repair. Amer. J. Path140: 1375-1388, 1992.

CULTURED BLADDER CELLS AND THEIR RESPONSE TO MECHANICAL STRAIN

Douglas E. Coplen, Pamela S. Howard, John W. Duckett,
Howard McC. Snyder and Edward J. Macarak

Division of Urology
Children's Hospital of Philadelphia
Connective Tissue Research Institute
University of Pennsylvania
Philadelphia, PA, 19104

INTRODUCTION

The mechanism responsible for the transformation of a normal bladder into a poorly compliant storage vesicle is not well understood. However, a common feature of many poorly compliant bladders is an abnormal accumulation and distribution of matrix proteins within the bladder wall[1-4]. Histological examination of tissues from bladders with outlet obstruction is frequently associated with infiltration of connective tissue[2]. This abnormal accumulation is uniquely associated with the muscle bundles of the detrusor layer and has been shown by immunohistochemistry to be largely composed of type III collagen[4]. Biochemical studies of myelodysplastic bladders show this characteristic fibrosis to result from a quantitative increase in type III collagen[5]. Since these same bladders have altered compliance, we believe that the loss of elasticity is related to the changed collagen composition of the bladder wall. A common link in pathological conditions with poor bladder compliance is premorbid exposure of the bladder wall and its constituent cells to an abnormal physical environment, e.g., uncontrolled intermittent contractions. The impact of this abnormal environment upon bladder function is difficult to assess in whole bladder studies. It is likewise difficult to assess the relative contributions of the various tissue components within the bladder wall to the overall loss of compliance. We have previously isolated and maintained and characterized fetal bovine bladder urothelial, fibroblastic and smooth muscle cells in culture[6-8]. To begin to understand the role of mechanical forces in altering synthesis of connective tissue proteins and assess the relative responses of the individual cell types found in the bladder wall, we have applied equibiaxial deformations to the three major cell types found in the bladder wall[9]. Since the in vitro physical environment we create may partially mimic the in vivo conditions experienced by these cells, insight may be gained into the mechanisms of fibrosis in dysfunctional bladders.

METHODS

Cell culture and growth

Bladder smooth muscle cells, lamina propria fibroblasts and urothelial cells were isolated as previously described[6-8] from fetal bovine bladders obtained from a local abattoir (Moyer Packing Co., Souderton, PA). To ensure that there was no significant contamination in the cell strains they were evaluated immunohistochemically using alpha-actin and AE1 cytokeratin antibodies[10,11]. All cell strains were incubated at 37°C in an atmosphere containing 5% CO_2 and fed every 4 days with modified medium 199 (Gibco labs, Grand Island, NY) containing 50 units/ml Penicillin, 50ug/ml streptomycin and 2.5mg/ml amphotericin B. Media modifications have been previously described[6]. Smooth muscle cells were grown in the presence of 10% Newborn Calf Serum(NCS), fibroblast media was supplemented with 5% NCS, and urothelial cells were grown in modified M199 supplemented with 10% NCS, 50 units heparin/ml(Sigma, St. Louis, MO), 30ug/ml bovine pituitary extract.

Mechanical Deformation

The apparatus used for reproducible cyclic biaxial strain of cells in vitro has been previously described[9] (Figure 1). Strain chambers were prepared by clamping a sheet of Tecoflex(Thermetics, Woburn, MA) across one end of the inner cylinder using an O-ring. The system is designed to apply a precise and reproducible equal biaxial strain to the compliant membrane and the attached cells.

For each experiment the chambers were sterilized in 50% ethyl alcohol, assembled under sterile conditions and seeded with 1.3 x 10^5 cells/cm^2 in modified medium 199 supplemented with the appropriate concentration of newborn calf serum and ascorbic acid(50ug/ml). Cells were allowed to attach overnight at 37°C in an atmosphere of 5%CO_2

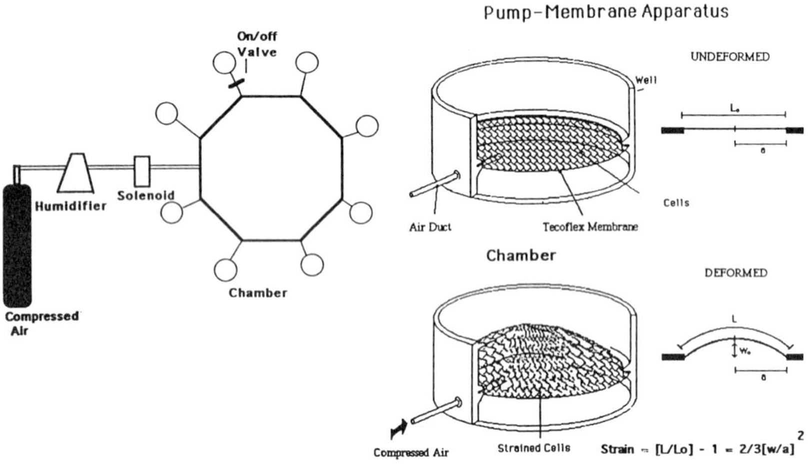

Figure 1. Diagram of apparatus that applies reproducible quantifiable biaxial deformations to cells in vitro. Compressed humidified air flows through and electrically controlled solenoid valve into the manifold inside a 37°C incubator where it is evenly distributed to up to 14 outflow ports with attached chambers. The deflection of the membrane is monitored by a dipstick.

in air. Cell attachment was confirmed by phase microscopy. Media was aspirated and replaced with 1ml of serum free endothelial cell medium (Gibco) containing 50 ug/ml ascorbic acid. Chambers containing cells to be subjected to strain were connected to the manifold. Compressed air was injected into the small chamber below the membrane and this deforms the membrane and the adherent cells in a defined manner. For this apparatus, the degree of strain and the rate of strain can be monitored. Strain equals $2/3[W_0/a]^2$, where a is the radius of the chamber and W_0 is the vertical displacement of the membrane. Identical nonstrained controls were placed in the same incubator but were not connected to the manifold. Each experiment could simultaneously use up to 14 control and 14 strained wells. Deformations in all experiments were applied cyclically at a strain rate of 4.9%. Frequency of strain varied from 5/min to 60/min.

After 7 hours, each chamber was again inspected using phase microscopy to ensure that significant cell detachment had not resulted from the deformation. Experiments were terminated by removing the media from the chambers and measuring its volume followed by trypsinizing(0.025% trypsin and 0.05% disodium ethylenediaminetetraacetic acid (EDTA)) and counting the cells using an electronic particle counter (Coulter Electronics, Hialeah Park, FL). Each chamber's media was stored at 4^oC after the addition of EDTA, N-ethylmaleimide(NEM), and phenylmethyl-sulfonylfluoride(PMSF) to final concentrations of 4mM, 10mM, and 0.9mM respectively.

Protein Quantitation

Fibronectin, collagen I, and collagen III concentration in the media of deformed and non-deformed cells was determined by competitive ELISA using the method described by Engvall[12]. Primary antibodies included an anti-human plasma fibronectin antibody raised in sheep in our laboratory, rabbit anti-bovine type I collagen (DMI, South Portland, ME) and rabbit anti-human type III collagen (DMI). Appropriate species specific alkaline phosphatase-conjugated anti-IgG(Sigma Chemical, St. Louis, MO) and Sigma 104 Phosphatase substrate were used to detect the primary antibodies. Reaction product optical density was measured at 405nm using an automated EIA reader (Bio-Tek Instruments, Winooski, VT). Standard curves were generated on a Macintosh computer using linear regression analysis (Cricket Graph, Malvern, PA). The protein concentrations of unknowns were calculated using the standard curve. Results were tabulated as protein production/cell +/- SD for each group. A two-tailed Students t test was used to evaluate statistical significance.

RESULTS

For all three cell strains, collagen I, collagen III, and fibronectin synthesis were quantitated. After initial experiments showed a significant change in protein production by strained cells over a 7 hour interval, this time course was chosen for all subsequent experiments. In the static, control state, there was no statistical difference in collagen I or collagen III production between any of the three cell types over the 7 hour period (Table 1). Lamina propria fibroblasts did produce significantly less fibronectin than either smooth muscle or urothelial cells in the unstrained state.

Table 1. Extracellular matrix production by the three cell strains in a static environment.(pg/cell/7 hours)

	Collagen I	Collagen III	Fibronectin
Bladder smooth muscle	.97 ± .33	.78 ± .28	13.6 ± 6
Lamina propria fibroblasts	.71 ± .32	.83 ± .39	3 ± 1.5*
Bladder urothelium	.51 ± .2	.51 ± .19	8.1 ± 2

* Lamina propria fibroblasts production of fibronectin was significantly less than that of either bladder smooth muscle or bladder urothelium.

Smooth muscle cells were subjected to mechanical deformation at defined frequencies for a seven hour period. The frequencies that reproduce in vivo bladder conditions are unknown so in these experiments we used a broad range(5 to 60 deformations/minute). Smooth muscle cells were acutely sensitive to deformation and showed significant increases in the synthesis of both type I and type III collagen (Table 2). There was an increase in type I collagen at all frequencies tested and this was significant when the cells were deformed at 5, 20 and 30 times per minute. Similar significant increases in type III collagen production were observed in smooth muscle cells that were strained at 10 and 20 times a minute. Smooth muscle cell production of fibronectin did not change significantly in response to mechanical strain.

We then studied bladder lamina propria fibroblasts in order to determine whether the deformation induced alterations in smooth muscle matrix synthesis were unique to that cell type. Fibroblast production of collagen I increased significantly when the cells were strained at 20 times per minute and decreased significantly when strained at 30 times per minute (Table 2). However, unlike the smooth muscle cells, fibroblast synthesis of type III collagen was unchanged at all strain frequencies. Similarly, fibroblast synthesis of fibronectin was not altered by cyclic deformation.

Table 2. Summary of alteration in production of matrix proteins for all three cell strains. Frequencies represent the number of times the cells were strained per minute.

	Collagen I	Collagen III	Fibronectin
Bladder smooth muscle	Significant increase at 5, 20 and 30x	Significant increase at 10 and 20x	No change in production at any frequency
Lamina propria fibroblasts	Significant increase at 20x and decrease at 30x	No change in production at any frequency	No change in production at any frequency
Bladder urothelium	No change in production at any frequency	No change in production at any frequency	Significant decrease at 30 and 60x

Further study showed that there was no alteration in collagen synthesis when bladder urothelial cells, the third major bladder cell type, were mechanically deformed(Table 2). These cells were the least responsive of the three cell strains. There was a significant decrease in soluble fibronectin production in urothelial cells strained at 30 and 60 times per minute.

DISCUSSION

This study represents an initial effort to gather information on the potential role of mechanical deformation in altering the expression of structurally important proteins by bladder wall cells. Our approach has been to utilize cell culture as a model system in which to characterize the behavior of bladder wall cells and determine their potential individual contributions to the overall mass of extracellular matrix in the bladder wall. Previously we have isolated and characterized three cell types[6-8] which we believe to be of paramount importance in governing the compliance of the bladder wall: 1) the lining urothelial cells, 2) the fibroblasts of the lamina propria and 3), the smooth muscle cells of the detrusor layer. The types of structural proteins synthesized by these cells were studied. All three cell types synthesized types I and III collagen which are the major structural collagens found in the bladder wall. These cells do not appear to synthesize large quantities of other collagens.

It is likely that the pathophysiology of the poorly compliant bladder will prove to be multifactorial. These bladders are often thick walled with a combination of smooth muscle hypertrophy and increased connective tissue deposition[1-4]. Significant changes in the composition of the bladder wall likely result from alteration in the synthetic patterns of one or more of these cell types. Since it would be impossible to assess the relative contributions of each of these cell types in an intact animal, we have utilized cell culture as a means to study bladder wall cells in a mechanically active environment under controlled conditions.

Previously we have utilized a system for applying equibiaxial deformations to cells to study the effects of mechanical forces on vascular tissues[13,14]. Since the bladder also sustains considerable mechanical deformation during its normal physiologic activity, we have utilized a similar system to begin to understand the potential role that forces might play in altering the matrix expression of bladder wall cells. Our rationale for this approach is based upon clinical observations in patients with outlet obstruction as well as data from animal models of outlet obstruction[15]. A common aspect of these pathologic conditions is that the bladder wall experiences uncontrolled and sudden contractions. Similarly the bladders in individuals with various forms of spinal injury or dysraphism also show comparable types of contractile responses. In animal models of outlet obstruction, the smooth muscle layer begins to hypertrophy[16]. Fibrotic changes in the bladder wall such as increased collagen content frequently occur in association with physiologically observable decreases in compliance [2,5]. These observations suggested to us that there might be a relation between uncontrolled, intermittent contractions and the fibrotic response and were our rationale for beginning a series of experiments designed to determine what effect, if any, mechanical deformation or "stretch" might have upon the secretory phenotype of bladder wall cells. Because of the association of increased collagen with the physiologic changes associated with loss of compliance, we have focused on measuring the major fibrillar collagen forms although there may be other proteins whose synthetic rates are also altered as a result of mechanical deformation.

Previously, we evaluated the significance of the intensity of strain on bladder smooth muscle cells and urothelium[7]. At lower strain intensities (less than 1.9%), we found no change in the synthetic capacities of the cells tested. Therefore we chose higher strains to begin to examine the effects of deformation on bladder wall cells. In the present study, we varied the frequency of strain and kept the strain constant at 4.9%.

It is well recognized that urothelial cells undergo considerable shape change during the normal filling and emptying cycles in the bladder. When the response of these cells was assessed, little if any change in collagen synthesis occurred as a result of mechanical deformation. Interestingly, the most significant change observed was a decrease in secreted fibronectin. Previously we have demonstrated similar changes in another epithelial cell type, the vascular endothelium[14]. When fibronectin was measured after similar lengths of mechanical deformation, secreted fibronectin was demonstrated to significantly decrease as a function of time. However, when the insoluble matrix was examined, there was an increase which corresponded to the decrease seen when the medium bathing the cells was examined. These data suggested that epithelial cells responded to the increased physical demands by facilitating the incorporation of fibronectin into the pericellular insoluble matrix. We interpreted this to represent a mechanism by which the cell could maximize its adhesive interaction with the underlying substrate to counteract the increased demands upon it as a result of the mechanical deformation. It is possible that urothelial cells, also an epithelial cell type, might respond in a similar fashion. Urothelial cells, based upon our experiments are unlikely to be responsible for altered matrix production owing to increased or altered mechanical loading as a result of unstable bladder contractions.

In contrast to the behavior of the urothelial cells was that of the smooth muscle cell. The smooth muscle cell subjected to the same types of physical demands as the urothelial cells responded much differently. Interestingly, the smooth muscle cell was acutely sensitive to deformation. Since we have no information regarding the frequency of deformation which might be experienced by bladder wall cells in vitro, we performed a series of experiments at frequencies from 5 to 60 cycles/ minute. Type I collagen synthesis was increased at frequencies of 5, 20 and 30 cycles/minute while type III was increased at 10 and 20 cycles/minute. These data raise some interesting questions.

Immunohistochemical examination of the samples from neurogenic bladders showed significant alteration in the distribution of type III collagen[4]. Previously, we demonstrated significant changes in the ratio of type III:type I collagen in tissue samples from neurogenic bladders[5]. It is possible that the mechanical environment present in these bladders initiates changes in synthesis which ultimately results in a permanent alteration in collagen type, distribution and quantity. It is notable that the smooth muscle cell, which is the putative cell responsible for initiating the histologic changes shows increased type III collagen synthesis at a frequency of 10 cycles/minute with no change in type I. If this pattern of synthesis persisted, it might result in a quantitative alteration of the matrix similar to that seen in the neurogenic bladders. While these data are provocative, we realize that there is much more information which will have to be obtained to validate this putative mechanism of fibrosis. Nevertheless, the data provide a plausible explanation for some of the events observed in vivo.

Lastly, we also subjected fibroblasts isolated from the lamina propria layer of the bladder to a similar regimen of deformation. While these cells were more responsive than urothelial cells, they were not as sensitive to deformation as smooth muscle. Fibroblasts showed significant alterations in collagen synthesis only at frequencies of 20 and 30 cycles/minute. At 20 cycles, type I synthesis increased; however, at 30 cycles/minute, there was a significant decrease.

Changes in the protein accumulation can result from an increase in production by each cell, an increase in the number of cells synthesizing protein or a decrease in the enzymatic degradation of protein. During the time course of these experiments (7 hours) there was not a change in cell number in either the strained or control wells. This suggests that the accumulated protein results from increased synthesis or decreased degradation.

To begin to understand possible regulatory mechanisms which might be responsible for these observations, we will carry out a series of experiments to compare the relative levels of matrix RNAs present in strained and control cells. Initially, total RNA will be extracted from wells at the end of the 7 hours of mechanical deformation.

Our working hypothesis is that a change in the protein production will be directly proportional to alterations in messenger RNA levels. Results would then show that relative steady-state messenger RNA levels between strained and control cells directly correlate with changes in matrix protein production. However, an absence of correlation would imply that the regulatory mechanism may not involve control at the level of transcription alone. However, additional experiments would be necessary to validate this conclusion. For example, this is an evaluation of the steady state RNA levels at the termination of the experiment. In contrast, the protein data reflects the accumulation of protein over time. Analysis of steady state RNA levels at intermediate time points during the course of the experiment would clarify if there were any transcriptional changes or accommodations throughout the duration of the experiment.

It is possible that other regulatory mechanisms are responsible for the changes in matrix protein expression. Changes in the rate of transcription, the stability of the message or a combination of these mechanisms could be operable. Further, control mechanisms at the translational level could also account for our results. Each of these alternatives has yet to be explored in our system. Another explanation is that the differences in specific matrix accumulation do not reflect any increase in matrix synthesis or gene expression, but are due to an alteration of the normal balance of enzymatic degradation and inhibition of enzymatic activity of matrix macromolecules. Studies by Lambert et al[26] showed a decrease in the steady state mRNA levels for collagenase in in vitro fibroblasts in their system. It is possible that the control of matrix levels in our experimental model are dependent on a complex set of regulatory signals involving both transcriptional and translational mechanisms of many different genes. Nevertheless, these initial studies provide a basic framework of information from which to investigate mechanical signalling and regulatory mechanisms in bladder wall cells.

Our data clearly demonstrate that some cells of the bladder can respond to mechanical deformation. The mechanism by which a cell perceives a physical force and converts it into a biological response, i.e., transduces the mechanical signal, is unknown. Other groups have studied the effects of stretch on the heart[17], lung[18], alveolar type II cells[19], smooth muscle[20], and endothelium[20]. Stretch accelerates the rate of protein synthesis in cardiac and skeletal muscle[20]. Stretching also has been demonstrated to both activate and inactivate ion channels in cells[21]. It has been proposed that modulation of ion traffic by stretch may direct cell activities. In studies performed on vascular endothelial and smooth muscle cells, the intracellular calcium release was observed to be a function of strain rate[22,23]. At very high strain rates, there was a pronounced increase in intracellular calcium as measured by the calcium binding dye, Fura II. Based upon these observations, these investigators suggest that high strain rates cause disruption of normal membrane properties and possible disruption of normal calcium channel ion traffic.

Cell-matrix interaction depends on the expression of specific cell surface receptors, the integrins[24]. These receptors interact with and influence the intracellular organization of the cytoskeleton[25]. Lambert showed variations of the _-actin mRNA in response to mechanical tension indicating that the cytoskeleton is potentially modulated by strain[26]. The direct physical association of cytoskeletal elements with transmembrane proteins, intracellular organelles, and intracellular enzymes may relay mechanical forces as a message to the interior of the cell[27].

In conclusion, we have shown that extracellular matrix production by bladder smooth muscle cells, lamina propria fibroblasts, and urothelial cells is influenced by uniform biaxial mechanical strain. This is dependent on the intensity of the strain and the frequency with which the strain is applied to the cells. Bladder smooth muscle cells show the largest and most statistically significant changes in protein production and may be the cell type responsible for the abnormal matrix accumulation evident in abnormally compliant bladders. This in vitro cellular model provides an opportunity to study the effects of mechanical forces on bladder cells at the cellular and molecular level. Since poor bladder compliance may result from the increased mechanical pressure of obstruction or the strain of uninhibited detrusor contractions this model may provide insight into the pathophysiology of these conditions.

REFERENCES

1. J.A. Gosling and J.S. Dixon, Structure of trabeculated detrusor smooth muscle in cases of prostatic hypertrophy. Urol.Int., 35:351 (1980).
2. S.A. Gilpin, J.A. Gosling, and R.J. Barnard, Morphological and morphometric studies of the human obstructed, trabeculated urinary bladder. Br.J.Urol., 57:525 (1985).
3. E. Shapiro, M.J. Becich, E. Perlman, and H. Lepor, Bladder wall abnormalities in myelodysplastic bladders: a computer assisted morphometric analysis. J.Urol. 145:1024 (1991).
4. D.H. Ewalt, P.S. Howard, B. Blyth, H.M. Snyder, J.W. Duckett, R.M. Levin, and E.J. Macarak, Is lamina propria matrix responsible for normal bladder compliance? J.Urol., 148:544 (1992).
5. D.H. Ewalt, S. Constantinescu, P.S. Howard, B. Blyth, H.M. Snyder, J.W. Duckett, and E.J. Macarak: Increased collagen type III in the neurogenic non-compliant bladder. J.Urol. 147:438A, abstract 904 (1992).
6. L.S. Baskin, P.S. Howard, J.W. Duckett, H.M. Snyder, and E.J.Macarak,: Bladder smooth muscle cells in culture: I. identification and characterization.J.Urol., 149:190 (1993).
7. L.S. Baskin, P.S. Howard, and E.J. Macarak: Effect of physical forces on bladder smooth muscle and urothelium. J.Urol., 150:601-607 (1993).
8. D.E. Coplen, P.S. Howard, J.W. Duckett, H.M. Snyder, and E.J. Macarak, Extracellular matrix synthesis by bladder lamina propria fibroblasts in vitro. in press, In Vit.Cell Biol. (1994).
9. F.K. Winston, E.J. Macarak, S.F. Gorfien, and L.E. Thibault, A system to reproduce and quantify the biomechanical environment of the cell. J. Appl. Physiol., 67:397-405 (1989).
10. O. Skalli, P. Ropraz, A. Trzeciak, G. Benzonana, D. Gillessen, and G. Gabbiani: A monoclonal antibody against a-smooth muscle actin: a new probe for smooth muscle differentiation. J.Cell Biol., 103:2787 (1986).
11. T.-T. Sun, T.I. Doran, and A. Vidrich, The use of antikeratin antibodies for the identification of cultured epithelial cells. Birth Defects, 16:183 (1980).
12. E. Engvall, Enzyme immunoassay ELISA and EMIT, In: Methods in Enzymology Immunochemical techniques, H. Van Vunakis and J. Langone, eds., Academic Press, New York (1980).
13. Gorfien, F.K. Winston, L.E. Thibault, and E.J. Macarak: Effects of biaxial deformation on pulmonary artery endothelial cells. J. Cell Physiol., 139:492-500 (.1989).
14. S.F. Gorfien, P.S. Howard, J.C. Myers, and E.J. Macarak: Cyclic biaxial strain of pulmonary artery endothelial cells causes an increase in cell layer-associated fibronectin. Am.J.Respir.Cell Mol.Biol., 3:421-429 (1990).
15. J.T. Andersen, Detrusor hyperreflexia in benign intravesical obstruction. J.Urol., 155:532 (1976).
16. R.M. Levin and W. Memberg, Functional effects of in vitro obstruction on the rabbit urinary bladder. J.Urol., 135:847 (1986).
17. X.P. Xenophontos, E.E. Gordon, and H.E. Morgan, Effect of intraventricular pressure on protein synthesis in arrested rat hearts. Am.J.Physiol., 25:C958 (1986).
18. L. Dobbs, NIH Workshop. Effects of physical forces on lung structure and metabolism, April 3, 1986, Bethesda, MD.
19. S.R. Rannels and D.E. Rannels, Alterations in type II pneumocytes cultured after partial pneumonectomy. Am.J.Physiol., 254:C684-690 (1992).

20. M. Rabinovitch, S. Beharry, T. Bothwell, and G. Jackowski: Qualitative and quantitative differences in protein synthesis comparing fetal lamb ductus arteriosus endothelium and smooth muscle with cells from adjacent vascular sites. Devel.Biol., 130:250-258 (1988).
21. J.B. Lansman, T.J. Hallam and T.J. Rink:Single stretch activated ion channels in vascular endothelial cells as mechanotransducers. Nature, 325:811-814 (1987).
22. F.K. Winston, L.E. Thibault, and E.J. Macarak, An analysis of the time-dependent changes in intracellular calcium concentration in endothelial cells in culture induced by mechanical deformation. J. Biomech.Eng., 115:160-168 (1993).
23. A.S. Lansman, L.E. Thibault, R.S. Cargill and E.J. Macarak: Strain rate dependence of intracellular calcium transients observed in mechanically deformed bovine aortic endothelial cells. submitted, 1993.
24. C.A. Buck, Cell surface receptors for extracellular molecules. Annu.Rev.Cell Biol., 3:179 (1987).
25. K. Burridge, K. Fath, T. Kelly, G. Nuckolls, and C. Turner: Focal adhesions: transmembrane junction between the extracellular matrix and the cytoskeleton. Annu.Rev.Cell.Biol., 4:487 (1988).
26. C.A. Lambert, E.P. Soudant, B.V. Nusgens, and C.M. Lapiere, Pretranslational egulation of extracellular matrix macromolecules and collagenase expression in fibroblasts by mechanical forces. Lab. Invest. 66:444-451 (1992).
27. D.E. Ingber and J. Folkman, Mechanical switching between growth and differentiation during fibroblast growth factor-stimulated angiogenesis in vitro-role of extracellular matrix. J.Cell Biol., 109:317 (1989).

ALTERATIONS IN EXTRACELLULAR MATRIX GENE EXPRESSION IN NORMAL VERSUS NON-COMPLIANT HUMAN BLADDERS

Pamela S. Howard, Ph.D.*, David H. Ewalt, M.D.+, John W. Duckett,M.D.+, Howard M. Snyder, M.D.+ and Edward J. Macarak, Ph.D.*

*Department of Anatomy and Histology
School of Dental Medicine
University of Pennsylvania
+Division of Urology,
Children's Hospital of Philadelphia
School of Medicine
University of Pennsylvania
Philadelphia, PA 19104

INTRODUCTION

Congenital or acquired obstruction of the urinary bladder can result in a stiff-walled, fibrotic organ characterized by trabeculation, low volume capacity at high pressures, and, in some cases, the occurrence of uninhibited contractions of the detrusor muscle cells. This fibrotic response is likely characterized by an alteration in the quantity and type of connective tissue (collagens and elastin) and, possibly, in the arrangement of these matrix components with one another and with the detrusor smooth muscle cells. The bladder wall becomes thickened, and the smooth muscle fascicles become infiltrated with connective tissue (2,3,4,5,6). As a result, the bladder loses compliance. This decreased compliance has been observed in up to 50% of patients suffering from spinal cord injury resulting from sacral nerve damage and 70% of myelomeningocele patients (1). Initial studies in our laboratory were carried out to characterize the connective tissue changes in non-compliant human bladders from a pediatric patient population composed of children with neurogenic, non-compliant bladders secondary to myelomeningocele. Results of these studies showed that the connective tissue infiltration within the detrusor muscle fascicles in the neurogenic bladders was specific for type III collagen (7).

The preliminary studies presented herein are an extension of the immunohistochemical and biochemical characterizations of connective tissue alterations in the pediatric neurogenic bladders. The goal of these initial molecular analyses was to determine if the connective tissue protein changes seen in the non-compliant bladder tissue samples were regulated at the level of mRNA transcription. Since our early data showed an alteration of the levels of type III collagen protein in non-compliant bladder tissue versus normal controls, our initial working hypothesis is that there is an increase in the synthesis of matrix proteins and that this expression is transcriptionally controlled. However, the increased accumulation could be the result of decreased degradation by metalloproteinases, increased expression of specific tissue inhibitors of metalloproteinases (TIMPS), or a combination of both increased matrix and inhibitor synthesis. To this end, steady state levels of mRNA for each of the extracellular matrix components (Collagens type I, III, IV, fibronectin), metalloproteinase and TIMP were analyzed for any changes in expression between non-compliant bladder tissue samples and normal controls.

Muscle, Matrix, and Bladder Function,
Edited by S. Zderic, Plenum Press, New York, 1995

METHODS

Total RNA was extracted from human bladder tissue following the method of Chomczynski and Sacchi (16). Tissue was obtained from the Urology Service of Drs. J. W. Duckett and H. M. Snyder at the Children's Hospital of Philadelphia. Full bladder thickness normal tissue to be used as controls was obtained from two patients undergoing surgery for bilateral ureteral reimplantation. Patient #20 was a 7year old female; patient #21 was a 19 month old male. Both control patients had no history of dysfunctional voiding. Non-compliant bladder tissue samples were obtained from two myelomeningocele patients who were characterized urodynamically as having low capacity, high pressure bladders. Full bladder thickness tissue was obtained at the time of vesicostomy. Patient #22 was a 10 year old male who experienced high bladder pressures (leak point: 40cm water/ 150ml), had a history of pyelonephritis and multiple urinary tract infections, but had no history of uninhibited contractions. Patient #23 was a 1 year old female who also experienced high bladder pressures (40cm water/60ml), but had no history of infection, upper tract deterioration or uninhibited contractions.

Samples were evaluated qualitatively by Northern blot and quantatively by slot blot analyses. Hybridizations were carried out under stringent conditions utilizing ^{32}P-labeled cDNA probes to collagens type I (17), type III (20), type IV (19), fibronectin (18), collagenase IV, and tissue inhibitor of metalloproteinase (TIMP-2) (13). Autoradiograms of slot blots were scanned densitometrically and the values obtained from the peak areas were plotted against RNA concentration to define the "linear" range. All values were normalized by hybridization of blots with a "housekeeping" cDNA for glyceraldehyde 3'-phosphate dehydrogenase (GAPDH) and/or a cDNA for 18S ribosomal RNA.

RESULTS

Northern blot autoradiograms of total RNA from the bladder tissue samples of control and myelomeningocele patients were qualitatively analyzed for the number and size of the different matrix mRNA transcripts expressed. Two transcripts for type III collagen mRNA were detected in all patient RNA samples corresponding to 4.8 and 5.4KB in size (Figure #1). Both transcripts appeared to be equally expressed in the control and myelomeningocele patients, indicating there is no preferential synthesis of a specific sized transcript. The difference in size of the collagen transcripts results from the usage of different polyadenylation sequences in the 3' untranslated region and not from alternative splicing, as is seen in fibronectin transcripts.

Slot blot autoradiograms were evaluated for any quantitative changes in expression of matrix RNA transcripts among patient samples (figure 2). Based upon values generated from densitometric scans of slot blot autoradiograms and linear regression analysis, relative levels of expression were determined for each patient sample. The lowest value within each group of 4 patient samples was arbitrarily expressed as 1.0,10, or 100, with the other values adjusted relative to this. Owing to the small number of patient samples, statistical significance was not determined in these preliminary studies.

Expression of Type IV collagenase and fibronectin mRNA levels was greatest of all the matrix genes analyzed in all patient samples. However variability in expression of each matrix gene analyzed occurred among the control and experimental samples as illustrated in Table 1. Analysis of steady state mRNA levels of the extracellular matrix prtoein fibronectin revealed there was no change in steady state mRNA levels between the myelomeningocele samples and the older control sample. However the younger control sample had a highly elevated fibronectin message level (figure 3).

Type IV collagen is the primary collagenous component of basement membranes. In the bladder wall, type IV collagen is present in the basement membrane underlying epithelium and also forms a "cocoon-like" investment surrounding smooth muscle cells in the muscularis mucosae and in the detrusor layer. Immunohistochemical analysis of normal and non-compliant human bladder tissue samples with a monoclonal antibody specific to type IV collagen revealed no change in the distribution of type IV with respect to detrusor smooth muscle cells (7,8.).

Quantitative analysis of the alpha 1[IV] mRNA levels in the bladder tissue samples from the two control and two myelomeningocele patients, revealed that the myelomeningocele patients expressed lower levels of type IV mRNA.

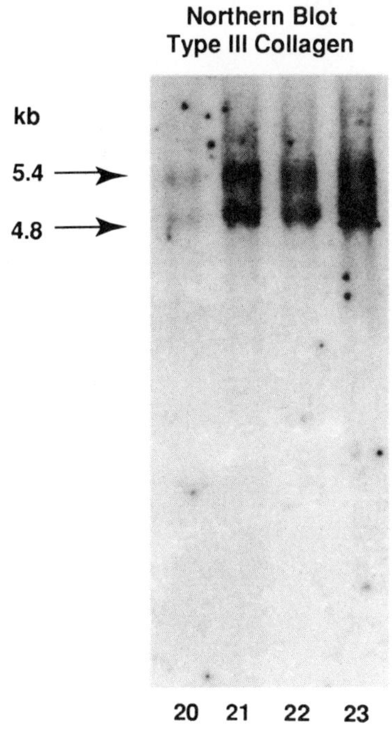

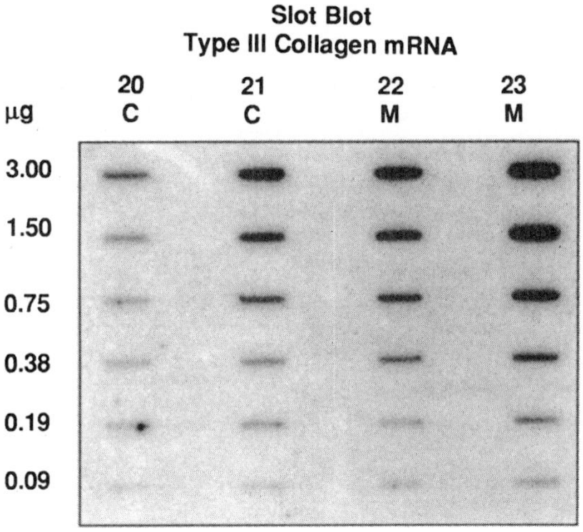

Figure 1 (left). Northern blot of total RNA extracted from bladder control and myelomeningocele tissue samples and hybridized with a radiolabeled cDNA for human type III collagen. Non-equal amounts of total RNA were loaded, resulting in variability of the hybridization signal.
Figure 2. (right). Slot blot autoradiograms of dilutions of total RNA from control (#20,21) and myelomeningocele (#22,23) bladder samples hybridized with a radiolableled cDNA probe for human type III collagen.

TABLE 1

Pt.	Dx	Age	Sex	Type I	Type III	Type IV	collag enase	TIMP -2	Fibro nectin
#20	N	7 y	F	1	1	10	192	2.5	123
#21	N	19 m	M	2.9	3.7	17.4	292	7.9	654
#22	My	10 y	M	1.8	2.6	27.2	118	1.1	132
#23	My	1 y	F	2.7	6.1	23.3	100	1	100

Table 1 shows the relative mRNA levels of collagens Type I,III, IV as well as collagenase, TIMP-2 and fibronectin in bladder samples from the four patients shown. N - normal; My - myelomeningocele.

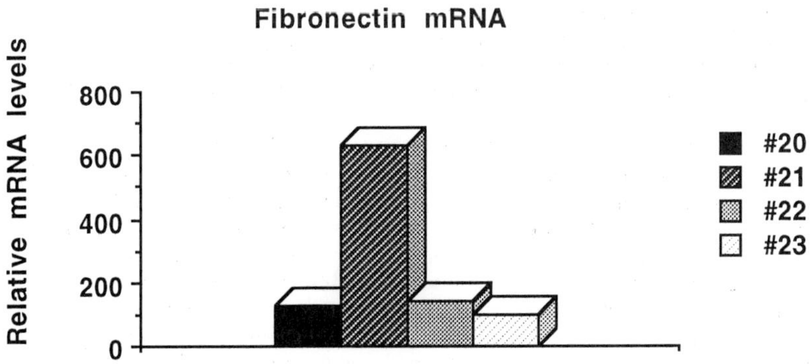

Figure 3. Relative poly-A mRNA levels of fibronectin in control (#20,21) and non-compliant (#22,23) bladder tissue from myelomeningocele patients.

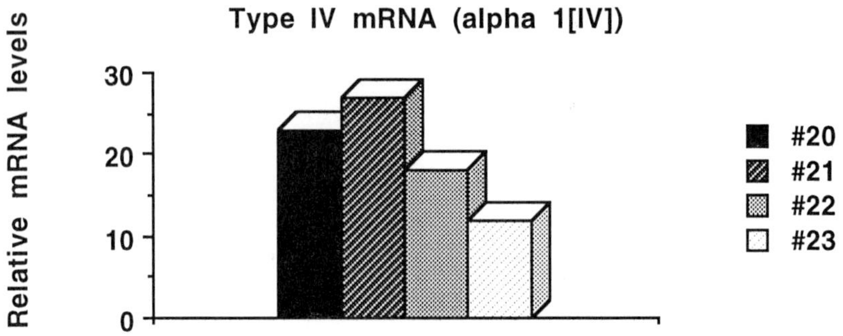

Figure 4. Relative poly-A mRNA levels of alpha 1(IV) chain of type IV collagen. Patients had greatly decreased expression of the alpha 1[IV] mRNA transcripts relative to the control patient samples.

High levels of collagenase IV mRNA were detected in all samples with the control samples elevated 1-2 fold above the myelomeningocele samples. Again, the highest level of expression occurred in the youngest control sample (figure 5). A similar pattern was also evident for the tissue inhibitor of metalloproteinawse, TIMP-2, although the absolute levels were greatly reduced in comparison to collagenase IV (figure 6). The youngest control sample had the highest levels of collagenase IV and TIMP-2 mRNA, whereas the levels decreased greatly in both myelomeningocele samples, with the lowest levels present in the youngest myelomeningocele sample.

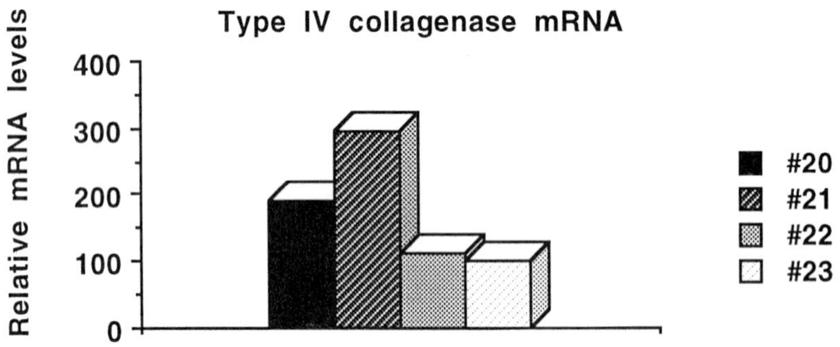

Figure 5. Relative mRNA levels of Type IV collagenase in control (#20,21) and non-compliant (#22,23) bladder tissue samples.

Comparison of expression of the alpha 1[I] collagen mRNA levels between both the control and myelomeningocele tissue samples showed variability (figure 7). However the average level was higher for myelomeningocele samples than the controls. The younger control and myelomeningocele patients expressed approximatley equal levels of alpha 1[1], whereas the older myelomeningocele sample had double the amount of mRNA than the similar aged control sample.

Expression of type III collagen mRNA among the samples showed a similar trend. Again, the average level for type III in the myelomenigocele sample was 2 fold higher than the average level of the controls (figure 8). Significantly, the youngest myelo patinets had highly elevated message levels. If levels were compared by age, again the myelmeningocele samples were elevated at least 2 fold in comparison to similar aged controls. In both myelomeningocele samples, the level of type III collagen was increased relative to the similar aged controls. In both the myelomenincogele samples, the level of type III collagen mRNA was increased relative to type I (alpha 1[I]) mRNA (figure 9), with the greatest difference occurring in the youngest patient, with approximately a 100% increase in type III mRNA relative to type I.

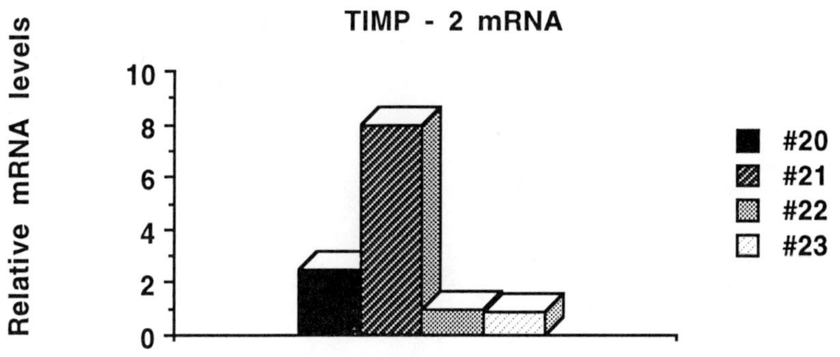

Figure 6. Tissue inhibitor of metalloproteinsase (TIMP-2) mRNA levels in control and myelomeningocele samples.

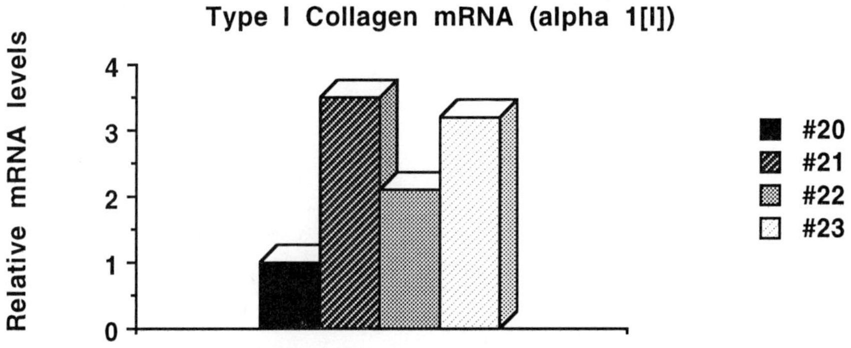

Figure 7. Relative mRNA levels of the alpha 1 [I] chain of Type I collagen in contyrol (#20,21) and non-compliant bladder (#22,23) tissue samples.

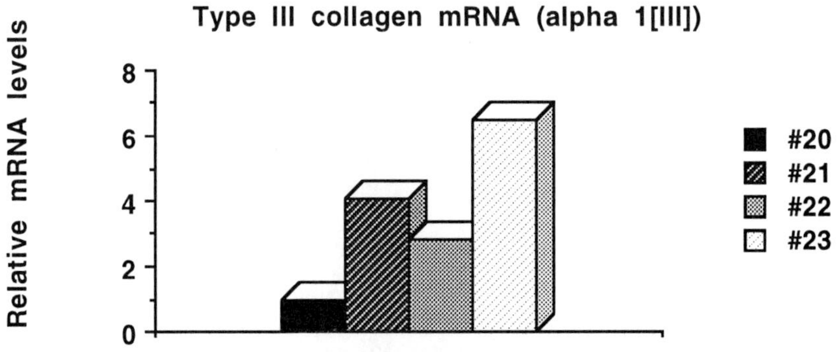

Figure 8. Relative Type III collagen mRNA levels in control and non-compliant bladder tissue.

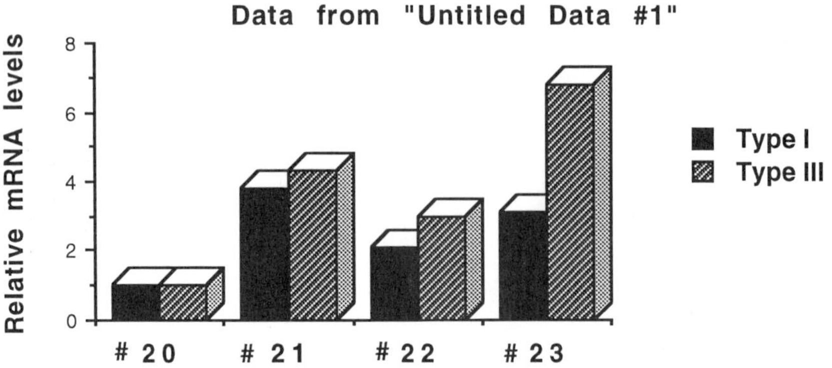

Figure 9. Comparison of relative mRNA levels for Type I and Type III collagen in each patient sample. Note that Type III mRNA levels are increased by 100% in bladder tissue from myelomeningocele patient #23.

CONCLUSIONS / DISCUSSION

These initial analyses provide some interesting data upon which to base further observations in a larger sample size.There are measurable and significant differences in collagen gene expression in bladder tissue from pediatric p[atients comparing normal and non-compliant bladders. In general, the interstitial collagen mRNAs (types I and III) are increased in non-compliant baldders, whereas type IV collagen mRNA is decreased in the non-compliant bladders relatiove to control tissue.

Specifically, Type III collagen gene expression is altered. In one myelomeningocele patient, message levels were increased by almost 100% relative to type I collagen (alpha 1[I]) mRNA. We have previously demonstrated that Type III collagen protein was modulated in neurogenic bladders. Although not enough of this tissue was available for protein analysis, the mRNA results from this patient parallels the increase in Type III collagen protein in patients with non-compliant bladders, as determined in the studies by Ewalt et al (7,8). This suggests that collagen synthesis in the bladder is transcriptionally regulated. Since there is a considerable age difference between the two myelomeningocele patient samples, future studies will use age matched controls to examine this parameter.

Decreased levels of mRNA for the alpha 1 chain of type IV collagen were measured in myelomeningocele samples relative to controls. It is possible that the reduced values for Type IV collagen mRNA in the myelo samples relfects abnormal growth of the muscle or a thinning of the existing basement membrane surrounding the smooth muscle cells. Data by Shapiro et al (3) showed either decreased number and size of muscle bundles in stillborn myelo bladder samples and possible muscle atrophy in other samples. This observed decreased could account for the changes seen here in Type IV gene expression. Interestingly, Ewalt et al (8) documented an increase (although, not significant) in 3-hydroxyproline, an isomer found in both Types IV and V collagens. Biochemically, analysis of 3-hydroxyproline levels in normal bladder tissue was 1.0mg/g (n=6, S.E. -+ 0.1) versus 1.2 mg/g (n=9, S.E. + 0.1) resulting in a 18% increase in 3-hydroxyproline in non-compliant bladder samples (8). However, there was great heterogeneiry in those samples. These findings also point out the necessity of a larger smample pool to generate statistically significant data.

Type IV collagenase and TIMP-2 mRNA levels were decreased in both myelomeningocele samples in comparison to controls. However, in general, the values were significatnly higher for collagenase in all samples than for TIMP-2. These findings are consistent with the normal balance of an enzyme and its inhibitor, in that when one level is elevated, the other is depressed. However, future studies will determine if there may be a significant change in the ratio of the enzyme and its inhibitor in non-compliant baldder tissue samples. The present study shows that there is an increase in the ratio of collagenase IV/TIMP-2 mRNAs in the non-compliant bladders relative to the controls. Fibronectin values showed a similar trend of expression to that of collagenase and TIMP-2. The youngest control sample had the highest levels of expression. Whether this is a age or developmentally regulated expression remains to be determined when more samples are able to be analyzed.

In conclusion, we have presented an approach for analyzing the possible alterations in extracellular matrix gene and prtoein expression in the non-compliant fibrotic bladders seen in patients with myelomeningocele. This preliminary study points out the importance of using well charaterized patient samples. Based on the small number of patient samples analyzed here, our future analyses will control for developmental age and gender, as well as utilizing a larger sample pool size. Urodynamic measurements of compliance levels will be determined with patients under general anesthesia to determine a true measure of loss of compliance and provide a measure for the severity of disease. These physiological measurements of compliance determined by cystometry will be correlated with physical changes as measured by quantitative biochemical and qualitattive immunohisotchemical analyses of protein and molecular analyses of matrix mRNA levels. It is hoped that these studies will provide thebasis for using changes in the ratio of matrix mRNA levels as an early diagnositc marker of pathologic changes.

ACKNOWLEDGEMENTS. This work was funded by NIH grant DK 48215.

REFERENCES

1. G.M. Ghoniem, D.A.Bloom, E.J. McGuire, and K.L.Stewart. Bladder compliance in meningomyelocele children. J. Urology 141:1404 (1989).
2. S.A. Gilpin, J. A. Gosling, and R.J. Barnard. Morphological and morphometric studies of the human obstructued, trabeculated urinary bladder. Brit. J. Urol. 57:525 (1985).
3. E.Shapiro, M.J. Becich, I.Perlman, and H.Lepor. Bladder wall abnormalities in myelodysplastic bladders: A computer assisted morphometric analysis. J. Urol 144:1024,1989
4. K.M. Kim, B.A. Kogan, C. A. Massad, and Y-C. Huang. Collagen and elastin in the obstructed fetal bladder. J. Urol. 146:528 (1991).
5. K. M. Kim, B.A. Kogan, C.A. Massad, and Y-C. Huang. Collagen and elastin in the normal fetal bladder. J. Urol. 146: 524 (1991).
6. J.G. Susset, D. Servot-Viguir, F. Lamy, P. Madernas, and R. Black.Collagen in 155 human bladders. Inv. Urol. 16:204 (1978).
7. D.H. Ewalt, P.S. Howard, B. Blyth, H.M. Snyder, J.W. Duckett, R.M. Levin, and E.J.Macarak. Is lamina propria matrix responsible for normal bladder compliance? J. Urol. 148:544 (1992).
8. D.H.Ewalt, S.C. Constantinescu, G.Bellon, J.P.Borel, H.M. Snyder, J.W. Duckett, nd E.J. Macarak. Increased Collagen Type III in the neurogenic non-compliant bladder. Manuscript submitted.
9. L.B. Baskin, P.S. Howard, H.M. Snyder, J.W.Duckett, and E.J. Macarak. Bladder smooth msucle cells in culture. I. Identification and characterization. J. Urol. 149:190 (1993).
10. D.E. Coplen, P.S. Howard, and E.J.Macarak. Characterization of a fibroblast cell from the urinary bladder wall. In Vitro Cell and Developmental Biology, in press (1994).
11. F.Ramirez, and M.DiLiberto. Complex and diversified regulatory programs control the expression of vertebrate collagen genes. FASEB J. 4:1616 (1990).
12. C.M. Alexander, and Z. Werb. Extracellular matrix degradation, *in*:: "Cell Biology of Extracellular Matrix," E. D. Hay, ed., Second Edition, Plenum Press, New York (1991).
13. W.G. Stetler-Stevenson, H.C.Krutzsch, and L.A. Liotta. Tissue inhibitor of metalloproteinase (TIMP-2). A new member of the metalloproteinase inhibitor family. J. Biol. Chem. 264:17374 (1989).
14. E.Vuorio, and B. Crombrugghe. The family of collagen genes. Ann. Rev. Biochem. 59:837,1990.15. R.O.Hynes. "Fibronectins," Springer-Verlag, New York (1990).
16. P. Chomczynski, and N. Sacchi. Single-step method of RNA isolation by acid guanidinium thiocyanate-phenol-chloroform extraction. Anal. Biochem. 162:156 (1987).

17. M. Chu, J.C. Myers, M.P. Bernard, J.F. Ding, and F. Ramirez. Cloning and characterization of five overlapping cDNA's specific for the human pro alpha 1[I] collagen chain. Nucleic Acids Research 10:5925 (1982).
18. A. Kornblihtt, K. Vibe-Pedersen, and F.E. Baralle. Isolation and characterization of cDNA clones from human and bovine fibronectins. PNAS 80:3218 (1983).
19. J.C. Myers, P.S. Howard, A.Jelen, A.S. Dion, and E.J.Macarak. Duplication of type IV collagen COOH-terminal repeats and species-specific expression of alpha 1[IV] and alpha 2 [IV] collagen genes. J. Biol. Chem. 262: 9231 (1987).
20. M. Chu, D. Weil, W. de Wet, M.Bernard, M. Sippola, and F.Rameriz. Isolation of cDNA and genomic clones encoding human pro-alpha1[III] collagen. J. Biol. Chem. 260:4357 (1985).

CHANGES IN BLADDER EXTRACELLULAR MATRIX - DISCUSSION

Timothy Ratliff - The panelists are now ready for questions from the floor or from each other. I have a question for Dr Levy. Is there a selective association for the various proteoglycans and the collagens, or different type collagens. Because what we are seeing is a shift form type I to type III collagen, do you a similar type shift in the types and distribution of the proteoglycans?

Bertram Levy - That question has not been answered. There is a preliminary study by Scott showing assembly of collagen and at various stages of assembly especially in fetal tissues, one can see a large interstitial component of proteoglycans with collagen forming around them, and as they are squeezed out, they are pushed out and other types such as *decorin* move in. But in the mature tissues the only one that has been studied thus far is *decorin* which is primarily associated with type IV collagen.

Robert Levin - Thank you Dr Levy for a wonderful talk, I have been working in the field of obstruction for many years and have not considered some of these factors in the response. The one thing is that a lot of people are showing at least in animal models, is that there is a great deal of variability in the degree of obstruction. In the animal models we see that in early phases of outlet obstruction where there is partial compensation, we don't see replacement of the smooth muscle by solid elements and we don't really see this until we get muscle that is decompensated or has reached the "end stage" of obstruction in which the changes are irreversible. Such might be the case with many of the human clinical specimens that we get. Have you ever looked in an animal model to see if there is a pattern of progression in changes in proteoglycan as you move across the spectrum of obstruction?

Bertram Levy - No and I am dying to do that study. It is a study that we would like very much to do with someone who has an animal model that is up and running to follow exactly this progression. I think that there is so much basic work that remains and I am embarrassed to say that this is the only work done looking at proteoglycan structure and function in the urinary bladder. So as you can see this is a wide open field and the work is there waiting to be done. I think that the reports that your group had on the serosal changes in the rabbit bladder after obstruction, are tantamount to what is happening in the submucosa and lamina propia in the human bladder. My own personal bias, and this may be science fiction, is that the events that happen initially are in the lamina propia and sub mucosa, and later on just as in the vascular system where you get intimal hyperplasia, you get muscle recruitment to keep the pathology going.

Robert Levin - It is very interesting to us that the early changes that we see when we do thymidine labeling are within the mucosa and sub mucosal areas and later on it does spread throughout the rest of the bladder layers. It would be wonderful to use these same techniques to look at how this progression would be altered in the face of bladders that are mildly obstructed versus those that are decompensated.

Timothy Ratliff - In that regard as proteoglycans are degraded, there could be some growth factors that might be associated with the changes that Bob Levin is seeing in terms of thymidine uptake with the earliest changes taking place in the mucosa and then migrating outwards in a radial distribution. Is there a way in which you might be able to measure the release of these degradative products that would allow for recognition of which molecule could be triggering cell hyperplasia?

Bertram Levy - Yes that might be feasible because there are now available a number of purified proteoglycans that would allow you to look at how the molecule might be degraded and what the breakdown products are. The only proteoglycan enzyme that is known to break them down is plasminogen and plasmin like urokinase.

Dale Huff - I have a question about the male gubernaculum which at 12 weeks undergoes a remarkable swelling process, and reportedly that is associated with a dramatic increase in hyaluronic acid and water. This does not occur in the female gubernaculum. Do you know of any sexually dimorphic protein that could signal this to occur in the male but not the female gubernaculum?

Bertram Levy - No I do not. I think that hyaluronidase responds to whatever cytokine is present and these cytokines do stimulate different proteoglycans and hyaluronidase, so I the cytokines is where it is at for this problem. But as to what drives the formation of one cytokine over another in this setting I do not know.

Participant - I am trying to understand the significance of the elastin layer in your model of the detrusor. It seems that tissue that lies adjacent to elastin layers might be compressed. It would be like having two rubber bands which you stretch in one direction and the materials between them are being squeezed or if you had an elastin layer separated by some tissue from the smooth muscle, that this would also be compressed as the bladder if stretched or filled. Am I reading your postulate correctly?

Bertram Levy - Well, you know the elastin is probably similar to the elastic band seen in the serosa of the rabbit bladder. Some studies have done in skin using electron microscopy in which it was shown that it is a sheet of elastin and not single fibers. So its sort of like a big hair net stretched out across the area. Now what I was intimating was that and Dr Rosenbloom can probably add a lot more to this, but that as the elastin is stretched, 70% of these hydrophobic amino acids are pulled out into a wet environment and they don't like being there. They will therefore retract back to their precontractile state. So that is what I was implying about chondroitin sulfate, that when the tissues are displaced, the physical properties change due to considerations of water packing.

Participant - What I am really interested in understanding is why certain parts of the tissue will be differentially compressed during stretch or strain while others will differentially expand. That's why I envision the tissues that lie adjacent to the elastic layer being compressed.

Participant - I have a question for Dr Ewalt. I would like to point out as we try and develop in vitro models of bladder physiology that our clinical urodynamics are strongly influenced by flow rates. This is significant because it is based on higher than normal or physiologic flow rates that the concept arose from McGuire that 40 water pressure should not be exceed if the upper urinary tract is to be preserved. Yet if the resting bladder pressure were maintained at 40 cm water pressure, you would have retrograde flow of urine. I would suspect that in reality when flow rates are physiological that the critical pressure drops to 25 to 30. Do you suppose that upper tract changes can occur with lower leak point pressures, or is there another cause to this dilatation such as a small series of bladder contractions?

David Ewalt - Well I think that the question you are asking is addressing the fundamental issue of evocative cystometry because we would all agree that if you want to show a more poorly compliant bladder, all you need to do is increase the flow rate for your CMG. And as you decrease the flow rate, the pressure curve also shows a decrease. I think it is very difficult to sort this out, and I would like to hear what Stuart Bauer decides on what rates he will fill the bladders of his spina bifida patients; is it based on projected bladder capacity or age? I know there are differences in technique, but I don't know how to explain these differences you are mentioning.

Participant - I am intrigued by the collagen that is surrounding the nerves and the blood vessels, and it seemed to be rather selective in its distribution. Is that collagen suspected in the very elastic tissues or the very inelastic tissues?

David Ewalt - The slide that I showed was of collagen type IV surrounding the nerves, and it is generally felt that this "chicken wire" collagen does not contribute much tensile strength but does line blood vessels and basement membranes. It does tend to coat the outside of nerves. But I do not know that it contributes to the tensile strength of these bladders.

Participant - Would that be a protective covering or supporting layer for the nerves in the bladder?

David Ewalt - I think that for now that would be a speculative answer only. I just don't know.

Participant - It seems to me that you have this tissue which is very heterogenous and you stretch it, and some parts of it are going to be very flexible and stretch and other parts are going to be less flexible and sort of stay in place.

David Ewalt - That is exactly what you see when you take a look at a heavily trabeculated bladder, you see some muscle bundles that are very distensible and other parts of the bladder that are not very distensible at all.

Participant - It occurred to me that the nerve fibers might be damaged if they are overstretched so you might want to have those held in place firmly by some kinds of anchors so it might be nice to know if this connective tissue is providing some of this protection.

David Ewalt - I like that explanation.

Participant - In the cardiac literature, there is some evidence that the extractability of collagens is going to be affected by these fibrotic processes. This means that the quantification process that you have employed has the potential of being somewhat selective as to which macromolecules end up being analyzed. Is this something that you have looked at or considered?

David Ewalt - Yes, we actually when we did the assay looked at some of our washed specimens, and we also looked at the debris that was obtained after the extraction of the cyanogen bromide peptide fragments. And the reason we did that was to make sure that we were not losing some of the collagens into one other precipitates during the course of the extractions. What we found, was that greater than 90% of the total collagen was within our cyanogen bromide peptide fragments. What is interesting in at least two other studies that I am familiar with from the pulmonary literature using cyanogen bromide fragmentation, and actually looked at the debris by using a second step with cyanogen bromide digestion, the type I and type III proportion did not change. In other words even if one examines the residual debris with a second digestion, the subtype distribution remains unchanged; we found this data supports our belief that the extractions do not select out one collagen type preferentially. We went so far as to model our results mathematically by considering how things might have changed if the 10% residual collagens were all type I. The answer was that even this residual 10% was all type I collagen the differences in the type I to III ratio would have still differed significantly.

William Steers - I have a question for Doug Coplen and what I am curious about is your thoughts on how mechanical stresses may affect the smooth muscle cell. How do such changes come about? Is it because during the deformation, there are openings of ion channels that are stretch sensitive which then results in secondary changes that affect nuclear expression? Is it possible that just by mechanical deformation alone you will not alter the cells ion channels, not bind to a receptor, but that strictly by changing the cell's architecture these forces will be transmitted to the nucleus to effect the kinds of changes we are seeing in this collection of papers? This would invoke the cytoskeleton as the elements that actually transmit the forces to the nucleus. Do you think that this is a possibility?

Doug Coplen - It is certainly a very intriguing possibility, but I am unaware of any data to suggest that transcription or translation are strictly mechanically linked.

William Steers - What I am thinking however is not that the mechanical forces will act at the site of transcription or translation, but that these forces will pull on the nuclear membrane. The other point is that you are not showing any changes at 7 hours, and one thing that we were amazed at when we looked at looked at the message from the growth factor is that it not is very fast and in fact peaks at 2 hours. So that you are left postulating that major ion flux changes and signal transduction happen to be relatively slow processes. The question I have is have you gone back to look at your message at other much shorter points in the time frame, because in our series of experiment by 6 hours the message was gone.

Doug Coplen - Yes, we have thought about that because it was clear that at 7 hours there was no difference. We were very intrigued by what you showed yesterday, and I suspect that something similar may be going on here. I have one question for you Bill, if that's okay with you. Yesterday during your presentation you described the cellular response to cytokines, but I wasn't sure which ones?

William Steers - These were interleukin 1 and interleukin 6. In fact the difference that is seen in response to these cytokines and growth factors is remarkably dependent upon age and stage of development. We demonstrated this in cultured smooth muscle cells from fetal, neonatal, adult, and obstructed bladder. For example, the adult and neonatal did not respond much to protein kinase C phorbol esters. The obstructed and neonatal cultured cells showed remarkable changes. So something about the stage of development and status of the tissues altered their response to the phorbol esters, and these regulatory pathways are undoubtedly pretty complex.

Craig Peters - Just a comment to Bill's question. I think there is really a good deal ot growing evidence to suggest that the forces on the cell plasma membrane can be transmitted by the cytoskeleton to the nucleus and I would site Don Coffey's and Bob Huber's work as two such examples. So I think that the work Doug Coplen has presented does fit into the concept that mechanical forces can truly alter the expression of these genes in a way that is perhaps mediated by direct transduction to the nucleus. One thing about this though is that with this system you are describing, you are really imposing passive strains on the cells, and yet in vivo, the bladder is probably generating some of its own stress and I wonder if you have thought about how the imposition of active forces might alter the results you have shown us. As a clinical example consider the child with posterior urethral valves whose bladder is doing pressure work, and compare him to a child with massive reflux whose bladder is doing volume based work. Both of these children have big distended bladders, but the processes are very different and the bladders behave very differently.

Doug Coplen - There is also the third possibility such as the child with myelodysplasia who has both a physiologic outflow obstruction and the intermittent pressure surges of inhibited contractions. But I think your point is right on about the limits of this model. We are the first to admit that this model is not meant to reproduce what is going on in the bladder. The reason we have chosen to study these cells is to see what message is being produced in response to mechanical deformation that can lead to the biochemical alterations in matrix that we have seen in the clinical specimens that David Ewalt so nicely described. But I think your point is well taken that these cells should be studied not just by passive deformation, but by subjecting them to intermittent pressurization and other such mechanical stimuli too.

William DeGroat - I realize that you are not trying to duplicate what happens in situ, and in regard to that why run these systems at a very high rate of stretch? If you were interested in blood vessels or the cardiac system, you might want to use these kinds of frequencies, but for the urinary bladder these seem high. The bladder is usually going to be subjected to sustained stretch. Why not use a constant sustained stretch or use a pharmacologic agent to induce a sustained rise in tension?

Doug Coplen - I really don't know what would happen if we studied the cells under these conditions, but I think these are important points. We can surely design a system that will slowly stretch the cells over a 2 or 4 hour period, and then rapidly allow them to return to their baseline or natural state. But we have not studied the system under these conditions as of yet. We chose these frequencies because it seemed in vascular smooth muscle that there changes at these frequencies, so based on prior data it seemed a logical place to start. Now it is up to us to fine tune this system and look at how other manipulations of these mechanical variables will alter the results.

William DeGroat - Were any of these responses graded in terms of the frequency of activation?

Doug Coplen - Yes with regard to the amount of protein produced. There does indeed seem to be a threshold, in other words the results were not linear and increasing the frequency five fold did not increase collagen output five fold.

William DeGroat - If there was a linear relationship between frequency and protein what would be the interpretation of that finding? Is this a pathological process? Is this actually irritating the cells by running the stretches at the higher frequencies or do you think it is physiological?

Doug Coplen - I think those a crucial question, and I am afraid that based on this data I presented today we don't have the answers yet.

William Steers - I would like to point out that we are stretching living cells in these experiments, and that these cells do have viscoelastic properties. Another parameter that must be considered in this type of work is rate of strain. Biological materials, the cytosol, the cytoskeleton, and the membranes all have these viscoelastic properties, and if displaced by strain, can compensate by rearrangement. This is however a time dependent phenomena, and if the strain rate is too fast then what one can get is cellular damage secondary to ion fluxes. From experiments that have been done not on bladder cells, but on neurons, and vascular smooth muscle, this has been the case. Once you reach a certain critical strain rate, then the ability of the cell and the ion channels to maintain normal form and function is compromised.

Timothy Ratliff - Pam I appreciated your presentation and I have one question for you. It is a technical question, and has to do with standardization of the experiments and how to best express changes in mRNA expression. How exactly did you go about this?

Pam Howard - We did normalize our results and did so using the standard accepted practice of comparing the mRNA signals to the housekeeping gene glyceraldehyde 3 phosphate dehydrogenase.

Timothy Ratliff - And I have one final question for you Pam and that is did you use total RNA or message RNA?

Pam Howard - This was total RNA that we used.

Participant - Has anyone looked at inflammatory cells in these clinical bladder specimens, and in particular has anyone looked at mast cells? In interstitial cystitis it is believed that tension develops in the muscle secondary to histamine release from mast cell degranulation. I was wondering if in these specimens there was any evidence to suggest such a mechanism?

Pam Howard - I know that in the studies we have done where we looked at the immunohistochemical staining for the collagen subtypes, we found some evidence for inflammatory cells in the lamina propia and directly below the urothelium. However this was quite variable among the various specimens, whereas the changes in matrix deposition were more consistent.

Participant - I would like to go back to the discussion about stretch induced changes in the model described by Doug Coplen. My bias is that there are ion channels being activated by the stretch that result in these changes. If that is so then perhaps these channels can be activated in a pharmacologic manner, or they may be blocked with a variety of agents such as calcium channel blockers. Have you tried to activate the cell using KCL depolarization, bethanechol stimulation, or stretched the cells in the presence of calcium channel blockers? These types of experiments would help sort out the issues raised by Bill Steers about whether we are dealing with ion channel activation, and second messengers or a direct effect of the cytoskeletal distortion on the nuclear membrane?

Doug Coplen - Ed has been working on this vascular system since 1987 or 1988. I am not sure he has any preliminary data on ion channels in those cells, but we certainly don't in bladder smooth muscle. I agree with your concern about extrapolating data from vascular smooth muscle to bladder. In our group at CHOP Steve Zderic has shown over the past 5 years that bladder smooth muscle is very different from vascular smooth muscle, so we all agree on that point.

Participant - No question, we all agree that bladder smooth muscle is quite unique. The real question is how do they differ in terms of the forces be they mechanical or pharmacological that stimulate the synthesis of collagen.

Timothy Ratliff - Are there any other questions or comments? I would like to thank the panel for their presentations and contribution to our lively discussion. Steve Zderic, do have any final thoughts?

Stephen Zderic - I would close with one brief comment noting my interest in Bill Steers' observation about stress incontinence in women with Ehlers' Danlos syndrome. I think that all of us who are at major referral centers have access to our colleagues in medical genetics clinics where there are large populations of patients with nature's exceptions to the rules. If one looks through the classic textbook The Metabolic Basis of Inherited Disease, one sees a classic photo of a young male with a huge bladder diverticulum who also had a diagnosis of Ehlers' Danlos syndrome. How do these diseases of the connective tissues affect urinary tract function in the absence of obstruction? Will some of them be diagnosed in a prenatal ultrasound screening? Often by turning to nature's exceptions to the rule much can be learned about normal function as has been seen over and over again in the study of metabolic diseases. Even if the function is normal, it we study these tissues and see changes, we must try and ask why the physiologic performance changes in the face of obstruction. As Joel Rosenbloom pointed out, in osteogenesis imperfecta all type I collagen is affected, yet it is only in the bone where the disease is manifested. Why should this be so? These are just some of the fascinating questions that urologic investigators will try and answer over the ensuing years. I would like to close by thanking our speakers, moderators, and audience for their enthusiastic participation over the past two days.

REGENERATIVE BLADDER AUGMENTATION: A REVIEW OF THE INITIAL PRECLINICAL STUDIES WITH PORCINE SMALL INTESTINAL SUBMUCOSA

Bradley P. Kropp M.D.[1], Stephen Badylak M.D., Ph.D., D.V.M.[2], Karl B. Thor Ph.D.[3]

[1]Department of Pediatric Urology, James Whitcomb Riley Hospital for Children, Indiana University School of Medicine, Indianapolis, Indiana
[2]Hillenbrand Biomedical Engineering Center. Purdue University, West Lafayette, Indiana
[3]Lilly Research Laboratories, A Division of Eli Lilly and Company, Indianapolis, Indiana

INTRODUCTION

The gastrointestinal tract has been the mainstay for bladder reconstruction in part due to the poor results with alloplastic materials.[1] However, bowel is not the perfect substitute and is associated with infection, intestinal obstruction, mucus production, electrolyte abnormalities and carcinogenic conversion. Therefore, a substance that enlarges the bladder, does not violate the gastrointestinal tract, is non-immunogenic, and promotes bladder regeneration would be a useful alternative in genitourinary reconstruction.

The concept of bladder regeneration is not a new phenomena and has been well documented in the literature over the past century.[2] The major obstacle has been finding a biomaterial that will act as a suitable scaffold for this natural process to occur. Synthetic non-biodegradable biomaterials such as silicone rubber, polytetrafluoroethylene, and polypropylene have been unsuccessful because of host foreign body reactions.[3-7] As a consequence of failures with non-biodegradable materials, synthetic biodegradable materials, that would allow the host bladder time for regeneration but then dissolve prior to the onset of any foreign body reaction, have been applied experimentally and showed marked improvement over non-biodegradable materials.[8] Furthermore, collagen-rich biodegradable materials such as placenta, amnion and pericardium have been used with even more encouraging experimental results.[9-12] However, none of these materials have surfaced for clinical use.

Another collagen-base biomaterial which has recently emerged is small intestinal submucosal (SIS) grafts.[13] This is a xenogenic membrane harvested from pig small intestine in which the tunica mucosa is removed from the inner surface, and the serosa and tunica muscularis are removed from the outer surface. This produces a thin, translucent graft (0.1 mm to 0.2 mm wall thickness) composed of the stratum compactum and muscularis mucosa of the tunica mucosa with the attached submucosa. This collagen-rich membrane has been previously shown to function well as an arterial or venous graft with rapid replacement by native tissues.[13-15] It has also been shown to have excellent host compatibility and remodeling when submucosal bladder injections of minced SIS were performed in pigs.[16] Also SIS has never been shown to cause host immunogenic responses in over 600 cross-species transplants or in the direct challenge test.[17] The present review describes preliminary

Muscle, Matrix, and Bladder Function,
Edited by S. Zderic, Plenum Press, New York, 1995

findings related to compatibility and functional regeneration of host bladder using porcine SIS as a bladder wall substitute in rat and canine models.

GENERAL METHODS

Preparation of SIS

Small intestinal submucosa (SIS) was prepared as previously outlined in detail by Badylak et al.[13] Briefly, sections of porcine small intestine are harvested within ten minutes of euthanasia and immediately placed in 0.9% saline solution. Sections are then cut into 10 to 20 cm lengths and the mesentery removed. The small intestine is then everted (inside out) and the tunica mucosa mechanically removed. This results in a clean separation of the superficial tunica mucosa from the deeper stratum compactum. The small intestinal segment is then everted again (the stratum compactum faces the lumen as in the original orientation) and the serosa and tunica muscularis are removed from its outer surface. This leaves a whitish membrane with intrinsic lymphatic and vascular tags evident. It is then rinsed in saline and disinfected with peracetic acid and stored in sterile water until used as a graft material. Storage time for the graft material ranged from 2 weeks to 3 months. It should be noted that preparation of SIS is a mechanical process (similar to preparation of sausage casing) and involves no biochemical or enzymatic digestion.

Surgical Augmentation Technique

The bladder augmentation technique used for the rats and dogs were similar and are outlined below. Male Sprague-Dawley rats (250-300 mg) and male stock beagles (10-13 kg) were anesthetized with isoflurane 4%, and a midline abdominal incision was made exposing the bladder. The thin attachments of the perivesical fat were sharply freed from the bladder exposing the bladder neck. Suspensory sutures were then placed in the bladder muscle for ease of manipulation. The entire dorsal surface of the bladder, which corresponded to 40 - 70 % in the rat and 30 - 45 % in the dog, was resected and discarded. A patch of SIS graft, slightly larger than the defect and oriented with the stratum compactum surface facing the bladder lumen, was sewn into place with mild chromic suture in a water tight fashion. The average size patch used for augmentation was 1 cm^2 in the rat and 20 cm^2 in the dog. Four permanent marking sutures were placed at the 3, 6, 9, and 12 o'clock position around the SIS augmentation grafts, and the distance between sutures were measured, with the bladder in a partially distended state, for future reference. Finally, adjacent perivesical fat or a small piece of omentum was used to cover the graft, and the abdomen was then closed in layers with absorbable suture.

In Vitro Contractility Studies

To better characterize the regenerated smooth muscle cells, a series of similar *in vitro* contractility experiments were designed for both the rat and canine models. SIS regenerated muscle strips were obtained from within the area outlined by the permanent marking sutures in 11 month and 15 month post-augmentation rats and dogs, respectively. The full thickness longitudinal strips were then suspended from a platinum/iridium electrode in a 10 ml organ bath of modified Krebs solution and attached to a Grass FT .03 force transducer. Changes in isometric tension were recorded and analyzed with a M5000 Signal Processing Center using XYZ Realtime Software (Modular Instruments, Inc., Malvern, PA).

RESULTS

Initial Augmentation Study in Rats

The initial augmentation study was performed in 22 Sprague-Dawley rats which underwent the augmentation technique described above.[18] Urinary diversion or drains were not used in the immediate postoperative period for the rats nor did any of the rats receive antibiotics during the study period. The rats were sacrificed at periods ranging between 2 weeks and 11 months post-augmentation, and their bladders were studied grossly and with routine histology.

Macroscopically, the inflated bladder specimens were uniformly dilated without evidence of diverticular formation in the region of the graft. Grafts were barely discernible as a slightly whitened area compared to the surrounding native bladder. There were minimal adhesions to the surrounding perivesical fat or other intraabdominal structures. The permanent marking sutures left at the time of augmentation were easily identifiable, and grossly the SIS augmentation graft had not contracted. Free floating, ivory-tan calculi, not adherent to the graft, were noted in the bladders of three of the four rats harvested at three months, one out of two of the rats harvested at six months and three of the four animals harvested at eleven months. Routine stone analysis on the bladder calculi revealed 100 % hydroxyapatite. No other animals had any evidence of bladder stones. There was no macroscopic evidence of hydroureteronephrosis in any of the animals.

At two weeks, histologically, the SIS grafts were completely covered by apparently normal transitional epithelium. Grafts showed a prominent vascular proliferation at their edges and infiltration of their entire surface by fibroblasts and new capillaries. There was a moderately intense mononuclear inflammatory cell response in the periphery of the graft. Centrally, inflammation was less intense, and fibroblasts and small capillaries appeared to infiltrate the collagen fibers of the SIS.

By four weeks, the grafts were completely infiltrated by small well-formed blood vessels and fibroblasts. In the periphery there were scattered fibers with morphologic and staining characteristics consistent with smooth muscle. Mononuclear inflammatory cell infiltrates were minimal at the edges but moderately intense in the central region.

At three months, inflammatory activity throughout the graft had largely subsided, and small but well-formed blood vessels persisted. Collagen was organized into a linear array, and scattered fibroblasts were interspersed among the individual collagen fibers. Small fibers with morphologic and staining properties of smooth muscle were scattered throughout the graft. In addition, distinct bundles of smooth muscle now were evident in a small portions of the grafts.

In the six and eleven month animals, all three layers of the normal bladder (urothelium, smooth muscle and serosa) were present and grossly and microscopically indistinguishable from the normal rat bladder. The marking sutures were the only guide to where the SIS graft had been surgically implanted. Animals with bladder calculi present had a global urothelial inflammation in both the SIS regenerated and remaining normal bladder. Animals without calculi had no signs of inflammation.

From the above results it was evident that complete bladder regeneration was occurring. Histologically, large bundles of smooth muscle were present. To determine if the SIS regenerated muscle cells were 1) contractile, 2) express receptors for neurotransmitters, and 3) functionally innervated, a series of *in vitro* contractility experiments on SIS regenerated muscle strips were performed.[19]

In the rat, the results of the *in vitro* contractility show that both the SIS regenerated and control bladder strips demonstrated contractile responses to a muscarinic agonist (carbachol 1μm) and a purinergic agonist (a, ß-methylene ATP 1μm) and relaxation to a ß-adrenergic agonist (isoprenaline 1 μm). In addition, both tissues contracted in response to electrical field stimulation (EFS) (100V, 1 msec duration, 5 sec. trains) in a frequency-dependent manner. These contractions were antagonized in both control and SIS strips by atropine (1μm) and showed desensitization to a, ß-methylene ATP (100 μm). These results were consistent with what has been previous published on *in vitro* contractility in the bladder.[20-23] The only functional difference between the SIS regenerated bladder and control bladder was that the maximum contractile force generated by the SIS bladder was only 50% of control bladder contractile force.

Long Term Augmentation Study in Dogs

The encouraging results obtained from the pilot study in rat led to studying SIS bladder augmentations in a long term canine model.[24] Augmentation was performed in eight adult male beagles as described in detail above. Urinary diversion was instituted with a urethral catheter for the first 48 hours post-operatively, and thereafter the dogs were allowed to void spontaneously. All dogs received prophylactic antibiotic therapy for one week post-operatively. All dogs had extensive pre-operative and periodic post-operative evaluations with complete blood counts (CBC), serum chemistries, urine cultures, cystograms, intravenous urograms (IVU) and cystometrograms (CMG). The CMG protocol consisted of a set of three sequential CMG's performed at a filling rate of 5 ml/min. All CMG's were performed under light sedation with pentobarbital (20 mg/kg i.v.). Bladder capacity and peak maximal voiding pressure were defined at the volume at which first release of urine occurred resulting from either overflow incontinence (n=4) or spontaneous micturition contraction (n=4).

All eight dogs recovered from the initial surgery and survived the entire study period (15 months) without complication. All dogs were observed to void spontaneously without difficulties or evidence of retention. All eight dogs had normal blood work, negative urine cultures, normal IVU's and cystograms throughout the entire study period. Preoperative CMG's revealed a mean bladder capacity of 51 mls (range 39-66 mls) with a mean maximal voiding pressure of 52 cm H_20 (range 39-69 cm H_20). CMG's remained stable and similar throughout the study period and at 14 months post-augmentation the bladders remained compliant with a flat tonus limb, no evidence of unstable contractions, a mean bladder capacity of 55 mls (range 42-70 mls) with a mean maximal voiding pressure of 45 cm H_20 (range 28-52 cm H_20). For both preoperatively and postoperatively CMG's, equal proportion of dogs demonstrated either micturition contractions or overflow incontinence. At some point in the postoperative study period all dogs except one demonstrated micturition voiding contractions. Even preoperatively this same dog did not demonstrate a micturition voiding contraction. The inability to consistently demonstrate a micturition voiding contraction cystometrically is most likely a result of variable levels of barbiturate anesthesia which could not be avoided.

All eight dogs were sacrificed at 15 months post-augmentation and grossly the SIS regenerated bladder was remarkably similar to the remaining normal bladder. There were minimal adhesions and no evidence of bladder calculi in any of the animals. There was no significant change in the measured area of the permanent marking suture left around the original graft material, thereby excluding graft shrinkage. Histologically, the SIS regenerated bladder resembled normal canine bladder (e.g. all three layers; transitional mucosa, smooth muscle, and serosa were present and the thickness of the transitional cell layer was indistinguishable from the normal bladder). The only notable difference between SIS regenerated bladder and normal bladder was the quantity and organization of the smooth muscle fibers.(Fig 1) Specifically, the SIS regenerated bladder had a larger amount of collagen interspersed between what appeared to be a smaller number of smooth muscle bundles. (Computer assisted morphometric analysis are in progress.)

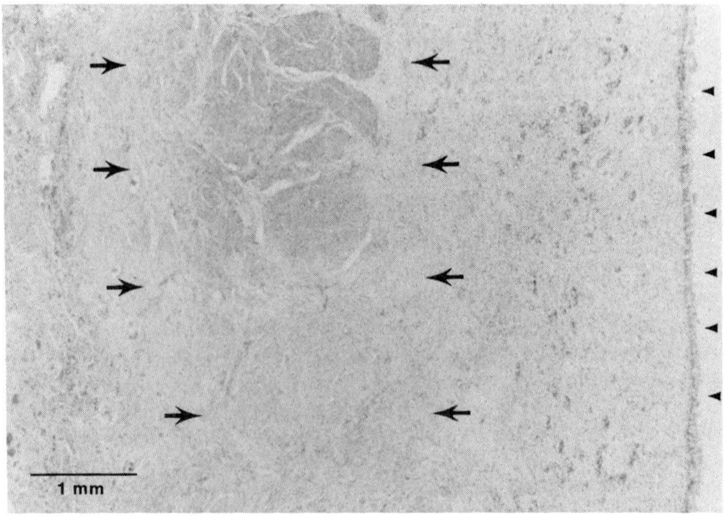

Figure 1. Photomicrograph of hematoxylin and eosin stained SIS regenerated dog bladder (15 month post-augmentation). Lumenal surface towards right. Arrowheads indicate full-thickness transitional cell lining. Arrows demarcate smooth muscle layer. Notice well-formed smooth muscle bundles.

In vitro contractility studies were also performed on the SIS regenerated dog bladder,[25] as described in the rat studies above, which confirmed 1) contractile activity of the smooth muscle cells, 2) expression of muscarinic, purinergic, and ß-adrenergic receptors, and 3) functional cholinergic and purinergic innervation (Fig. 2) that is similar to the normal canine bladder muscle. As with the rat the only difference between the SIS regenerated dog bladder and the control dog bladder was the maximal contractile response was decreased by 50%.

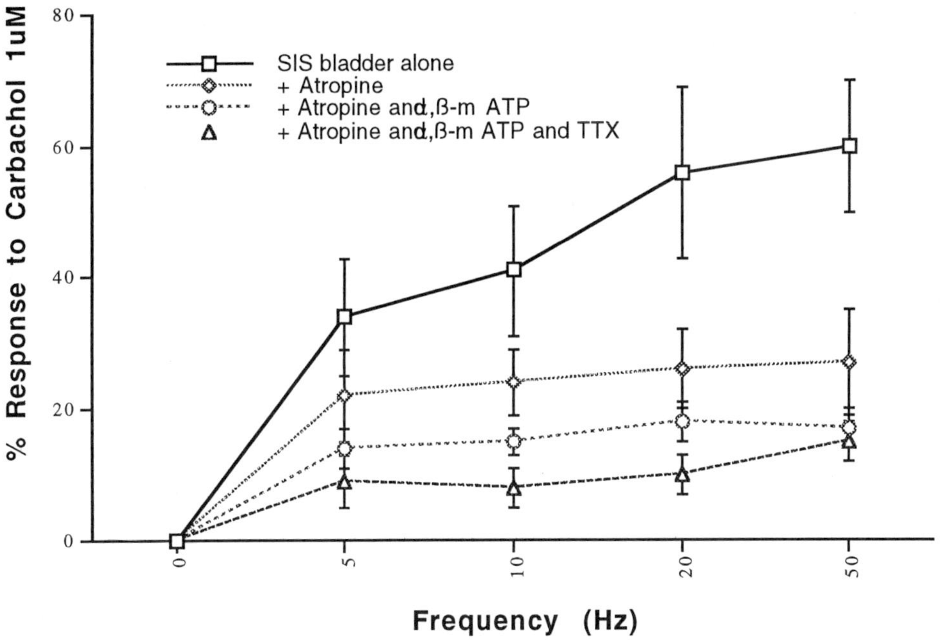

Figure 2. Frequency-response curves (100 V, 1.0 msec, 5 sec train) of SIS regenerated dog bladder strips (15 months post-augmentation). Note that neurogenic frequency-dependent contractions were mediated through muscarinic and purinergic mechanisms. (Doses: Atropine 1 μM, a,ß-methylene ATP 200 μM, Tetrodotoxin (TTX) 3.1 μM)

GENERAL DISCUSSION

Finding a suitable alloplastic or biodegradable material to function as a patch or bladder replacement, which would allow for bladder regeneration with low morbidity and mortality or other associated complications, has been a urologic dilemma for several decades. Previous research has shown that the collagen based biodegradable materials appear to give the most successful and reproducible results.[8,9,11,12] However to date, only pericardial tissue has been shown to produce long term acceptable results in a canine model. Histologically, there was a smooth epithelialized inner surface but absence of a smooth muscle layer in the pericardial graft.[10] Therefore, although pericardial tissue can be used to replace part of the bladder wall it does not allow for complete bladder regeneration.

Therefore, an initial rat bladder augmentation study was designed using small intestinal submucosa (SIS), a new biodegradable collagen-rich, xenogenic membrane harvested from porcine small intestine, which has been shown to allow tissue regeneration in multiple different sites. The conclusions obtained from the pilot study were that complete regeneration of all three layers (mucosa, smooth muscle, serosa) of the bladder will occur given the proper scaffold and that SIS was acting as a suitable scaffold without evidence of graft shrinkage or decreased bladder size. However, one major concern of this initial study was the presence of the bladder calculi in some of the long term animals. Stone analysis ruled out a infectious cause for the stone formation and review of previously published long term augmentation experiments in rats also revealed bladder calculi present in their control group, gastrocystoplasty group, ileocystoplasty group, and colocystoplasty group.[26] Although stone analysis was not reported in this series, it would appear that any type of surgery on the rat bladder predisposes it to the formation of calculi with long term follow up. Further support that the SIS alone was not the cause of the stone formation is derived from the fact that none of the long-term dogs developed bladder calculi.

Further conclusions from the long term canine study were: 1) SIS acts as a scaffold for bladder regeneration similar to that seen in the rat. 2) SIS bladder regeneration does not appear to be species specific. 3) The SIS regenerated bladder will allow continued normal filling and emptying without high pressures or deleterious renal or systemic side effects.

Finally, conclusions from the *in vitro* contractility experiments in the dog and rat indicate that SIS regenerated muscle 1) demonstrates contractile activity, 2) expresses muscarinic, purinergic, and ß-adrenergic receptors, and 3) exhibits functional cholinergic and purinergic innervation that is similar to the normal rat and canine bladder muscle. One explanation for the decrease in maximal contractile force seen in both animal models could be related to the subjective histological observation of decreased quantity of muscle in the regenerated bladder compared to normal bladder. Further studies will be required to completely understand this phenomena.

As with many of the biomaterials that have been used experimentally, short term results that confirm the regenerative capacity of the bladder are relatively easy to reproduce. The ultimate test for any biomaterial has to be in a long term, large animal study that includes evaluation of the upper tracts, preservation of bladder capacity and compliance, and no postoperative or systemic complication. It would seem to be an added bonus if the bladder that regenerated would function similarly to normal bladder. From this initial review of SIS regenerative bladder augmentations in two different species, it appears that a functionally normal bladder is regenerating using SIS as a scaffold for this regenerative process. The regenerated bladder is remaining compliant and allowing for normal micturition contraction with preservation of the upper tracts. Furthermore, the procedure is technically feasible and has had no morbidity or mortality. Therefore, SIS appears to be a viable alternative to the use of the gastrointestinal tract for bladder augmentations and urinary reconstruction.

ACKNOWLEDGEMENTS

The authors would like to acknowledge the contributions of Jeffery Vaught, MD., Marian Rippy, DVM., Barry Sawyer, John Hays, DVM., Bonita Shelton, LVT., Mary Katofiasc, LVT., Mark Muhlhauser, Harlan Shannon, PhD., Mark Adams, MD., Michael Keating, MD., and Richard Rink, MD.

REFERENCES

1. Gleeson, M. J. and Griffith, D. P.: The use of alloplastic biomaterials in bladder substitution. J Urol, 148: 1377, 1992.
2. Tizzoni, G. and Poggi, A.: Die Wiederherstellung der Harnblase: experimentelle Untersuchungen. Centralbl f Chir, 15: 921, 1888.
3. Bohne, A. W., Osborn, R. W. and Hettle, P. J.: Regeneration of the Urinary Bladder in the Dog, Following Total Cystectomy. Surg Gynecol Obstet, 100: 259, 1955.
4. Bohne, A. W. and Urwiller, K. L.: Experience with Urinary Bladder Regeneration. J Urol, 77: 725, 1957.
5. Kudish, H. G.: The use of Polyvinyl Sponge for Experimental Cystoplasty. J Urol, **78:** 232, 1957.
6. Swinney, J., Tomlinson, B. E. and Walder, D. N.: Urinary Tract Substitution. Brit Urol, 33: 414, 1961.
7. Ashkar, L. and Heller, E.: The silastic bladder patch. J Urol, 98: 679, 1967.
8. Agishi, T., Nakazono, M., Kiraly, R. J., Picha, G. and Nose, Y.: Biodegradable material for bladder reconstruction. J Biomed Mater Res, 9: 119, 1975.
9. Scott, R., Mohammed, R., Gorham, S. D., French, D. A., Monsour, M. J., Shivas, A. and Hyland, T.: The evolution of a biodegradable membrane for use in urological surgery. A summary of 109 in vivo experiments. Br J Urol, **62:** 26, 1988.
10. Kambic, H., Kay, R., Chen, J. F., Matsushita, M., Harasaki, H. and Zilber, S.: Biodegradable pericardial implants for bladder augmentation: a 2.5- year study in dogs. J Urol, 539, 1992.

11. Gorham, S., McCafferty, I., Baraza, R. and Scott, R.: Preliminary development of a collagen membrane for use in urological surgery. Urol Res, 12: 295, 1984.
12. Fishman, I. J., Flores, F. N., Scott, B., Spjut, H. J. and Morrow, B.: Use of Fresh Placental Membranes for Bladder Reconstruction. J Urol, 138: 1291, 1987.
13. Badylak, S. F.,diameter vascular graft in the dog. J Surg Res, 47: 74, 1989.
14. Lantz, G. C., Badylak, S. F., Coffey, A. C., Geddes, L. A. and Blevins, W. E.: Small intestinal submucosa as a small-diameter arterial graft in the dog. J Invest Surg, 3: 217, 1990.
15. Lantz, G. C., Badylak, S. F., Coffey, A. C., Geddes, L. A. and Sandusky, G. E.: Small intestinal submucosa as a superior vena cava graft in the dog. J Surg Res, 53: 175, 1992.
16. Knapp, P. M., Lingeman, J. E., Siegel, Y. I., Badylak, S. F. and Demeter, R. J.: Biocompatablility of Small-intestinal Submucosa in Urinary Tract as Augmentation Cystoplasty Graft and Injectable Suspension. J Endourol, 8: 125, 1994.
17. Badylak, S. F.: The Immunogenic Response of SIS. Personal Communication 1994.
18. Kropp, B. P., Eppley, B. L., Prevel, C. D., Harruff, R. C., Badylak, S. F., Adams, M.C., Rink, R. C. and Keating, M. A.: Experimental Assessment of Small Intestine Submucosa as a Bladder Wall Substitute. J Urol, 151: 501, 1994.
19. Vaught, J. D., Kropp, B. P., Sawyer, B., Shannon, H., Badylak, S. F. and Thor, K. B.: Detrusor Smooth Muscle Regeneration in the Rat Using Porcine Small Intestinal Submucosal Grafts: Functional Innervation and Receptor Expression. Submitted: American Urological Association Meeting Abstracts, 1995.
20. Maggi, C. A. and Meli, A.: Modulation by beta-adrenoreceptors of spontaneous contractions of rat urinary bladder. J Auton Pharmacol, 2: 255, 1982.
21. Kasakov, L. and Burnstock, G.: The use of the slowly degradable analog, a,ß-methylene ATP, to produce desensitization of the P2- purinoceptor: effect on non-adrenergic, non-cholinergic responses of the guinea-pig urinary bladder. Eur J Pharmacol, 86: 291, 1982.
22. Burnstock, G., Dumsday, B. and Smythe, A.: Atropine-resistant excitation of the urinary bladder: the possibility of transmission via nerves releasing a purine nucleotide. Br J Pharmacol, 44: 451, 1972.
23. Brading, A. F. and Williams, J. H.: Contractile responses of smooth muscle strips from rat and guinea-pig urinary bladder to transmural stimulation: effects of atropine and a,ß-methylene ATP. Br J Pharmacol, 99: 493, 1990.
24. Kropp, B. P., Rippy, M., Badylak, S. F., Adams, M. C., Keating, M. A., Rink, R. C. and Thor, K. B.: Small Intestinal Submucosa: Urodynamic and Histopathologic Evaluation in Long Term Canine Bladder Augmentation. Submitted: American Urological Association Meeting Abstracts, 1995.
25. Kropp, B. P., Sawyer, B. D., Shannon, H. E., Badylak, S. F. and Thor, K. B.: Regenerative Bladder Augmentation with Small Intestinal Submucosa: In Vitro Contractile Assessment in Canine Regenerated Bladder. In Preparation.
26. Little, J. S. J., Klee, L. W., Hoover, D. M. and Rink, R. C.: Long-term Histopathological Changes Observed in Rats Subjected to Augmentation Cystoplasty. J Urol, 152: 720, 1994.

THE EFFECTS OF HIGH URINARY POTASSIUM CONCENTRATION ON PELVIC NERVE MECHANORECEPTORS AND 'SILENT' AFFERENTS FROM THE RAT BLADDER

Jiang Wen and JFB Morrison

Department of Physiology
University of Leeds
Leeds LS2 9JT
United Kingdom

INTRODUCTION

It has been known for many years that the uroepithelium in the bladder and the region of the lamina propria are innervated by nerves which contain peptides such as Substance P, VIP and CGRP (see Torrens and Morrison, 1987). These fibres have free nerve endings and appear to be unmyelinated; these nerves are believed to be afferent because they are absent from rats treated neonatally with capsaicin. However there are few suggestions as to the physiological role of these nerves. It is known that some afferent fibres from the bladder are sensitive to acute administration of capsaicin and to irritant chemicals introduced into the bladder (McMahon and Abel, 1987; H_bler et al, 1988, 1990; Maggi, 1990), and it may be that these substances act on sensory endings in the epithelium. There have also been several reports recently that pelvic nerve afferents from the cat bladder include a population of fibres that do not respond to distension in the physiological and noxious ranges of pressure. These fibres are given the label 'silent' afferents. The present experiments have investigated the effects of intravesical potassium at concentrations near the maximum level observed in normal rat urine.

METHODS

Rats were anaesthetized with pentobarbitone sodium (60 mgs/Kg ip, with additional doses of 1.5mgs as required), and kept in good physiological condition. The bladder was cannulated, and stimulating electrodes were attached round the right pelvic nerve. Laminectomies were performed so as to expose the L6 and S1 dorsal roots, and filaments of dorsal roots were dissected down to filaments that contained 1-4 units which could be activated at constant latency by electrical stimulation of the pelvic nerve. The bladder was distended with saline or the experimental solutions at 37 degrees C at a rate of 180 microlitres/minute until the intravesical pressure exceeded 40 mmHg. Typically the duration of the slow infusion was about 4 minutes. Two types of solution were used: initially we used solutions with an osmolality of 2000 mosmols, with 300-500 mM KCl and 2M urea; later we used 200 or 300 mM KCl. Units were studied first with saline distensions, usually three in number, to ensure that repeated distension did not sensitise the afferent endings. Three or more distensions with the experimental solutions were then performed, and changes in the response to distension and the development of resting activity, when the bladder was empty, was observed. In some units, the responses to saline distensions were observed on several occasions following the experimental solutions.

Muscle, Matrix, and Bladder Function,
Edited by S. Zderic, Plenum Press, New York, 1995

RESULTS

Conduction Velocity

The conduction velocity of units was estimated from the latency and distance between the pelvic nerve stimulating electrodes and the dorsal root. Units with a conduction velocity of <1.3 m/sec were classified as C fibres (Waddell, Lawson and McCarthy, 1989).

Low and High threshold Mechanoreceptors

Nineteen out of thirty six fibres studied had mechanosensitive properties during control distensions with isotonic saline. Of these units one was classified as a high threshold unit, with a pressure threshold in excess of 40mm Hg, and was a C-fibre. The remaining 18 units had low thresholds, and twelve of these had A-delta conduction velocities. Repeated distension with solutions containing potassium caused either an increase in discharge rate, or a decrease (usually to zero), which was probably due to potassium block of the afferent endings. In fourteen units, the discharge rate increased, and in eleven this excitation was followed by a decrease, usually to near zero. Sensitization was often present following the potassium effect, irrespective of whether excitation or inhibition were seen during the application of potassium.

'Silent' Afferents

Seventeen fibres which did not respond to bladder distension, but were activated at constant latency by electrical stimulation of the pelvic nerve, were classified as 'silent' afferents; the majority were C-fibres. Almost all the 'silent' units responded to vesical distension with potassium solutions by developing an irregular discharge when the bladder was empty or partially filled, and this discharge outlasted the stimulus for at least two hours. Some, mainly C-fibres, also developed mechanosensitivity.

DISCUSSION

These studies show that mechanoreceptors arising from the bladder, and travelling in the pelvic nerves are present in the rat; some of these were C-fibres, and this is in contrast to the findings in the cat (Janig & Koltzenburg, 1990; Habler et al.1993; Bahns et al.,1987), where pelvic nerve C-fibre afferents were almost always insensitive to distension in the range 5-50 mm Hg. Unlike the pelvic nerve, the cat hypogastric afferent neurones form the bladder include C-fibre mechanoreceptors with low thresholds (Floyd, Hick and Morrison, 1976; Bahns, Ernsberger, J_nig and Nelke, (1986).

The 'silent' C-fibre afferents in the cat could occasionally be sensitized by noxious chemical stimulation, but in this study on the rat, the majority of C-fibre 'silent' afferents responded to distensions with potassium solutions. About half of the 'silent' C-fibres became mechanosensitive, and responded to slow filling of the bladder within the physiological range of pressures. In the cat, the mechanosensitivity that develops after noxious chemical stimulation, such as the application of turpentine or croton oil is generally high threshold sensitivity, i.e. mechanosensitivity in the supra-physiological range of bladder pressures. In the present experiments, the C-fibres that became mechanosensitive generally had thresholds in the same range as the C-fibres that showed mechanosensitivity at the start of the experiment (during distensions with saline).

The site of the sensory endings of the afferent fibres in these experiments is the subject for speculation: it is unlikely that potassium would reach the smooth muscle layers of the bladder wall in the in vivo situation. This is because the normal blood flow should have removed any potassium that passed across the epithelium, and there was no indication that either smooth muscle or the nerves that innervate it were excited during the distensions using potassium solutions in these experiments: there were no changes in the cystometrogram, nor were these spontaneous contractions or changes in tone of the smooth muscle, which might have been expected if potassium were reaching the smooth muscle layers. It is possible that the endings are actually in the epithelium of mucosa of the bladder. This would seem most appropriate for an action of urine on sensory nerve endings. In the gastrointestinal tract there is ample evidence for vagal nerve endings that are sensitive mechanoreceptors in the mucosa, and respond to concentrations of ions in the lumen.

It is concluded that low threshold mechanoreceptors and some of the 'silent' afferents in the rat bladder can be sensitised by concentrations of urinary potassium near the upper limit of normality, and that a proportion of the silent fibres become mechanosensitive following the chemical stimulus. It seems likely that some of these endings are in the mucosa or epithelium, and this is of particular interest in that it is possible for potassium to pass from normal urine into the mucosa in pathological situations when the permeability of the uroepithelium is increased.

ACKNOWLEDGEMENT

We would like to thank the Wellcome Trust and Pfizer plc for supporting this work.

REFERENCES

Bahns, E., Ernsberger, U., Janig, W. & Nelke, A., 1986, Functional characteristics of lumbar visceral afferent fibres from the urinary bladder and the urethra in the cat. Pflugers Archiv, European Journal of Physiology. 407:510-518.

Bahns, E., Halsband, U. & Janig, W., 1987, Responses of sacral visceral afferents from the lower urinary tract, colon, and anus to mechanical stimulation. Pflugers Archiv, European Journal of Physiology. 410:296-303.

Floyd, K., Hick, Verity E. and Morrison, J.F.B., 1976, Mechano-sensitive afferent units in the hypogastric nerve of the cat, J. Physiol. 259:457-471.

Habler, H.J., Janig, W., Koltzenburg, M., 1988, A novel type of unmyelinated chemospecific nociceptor in acutely inflamed urinary bladder, Agents and Actions. 25:219-221.

Habler, H.J., Janig, W., Koltzenburg, M., 1990, Activation of unmyelinated afferent fibres by mechanical stimuli and inflammation of the urinary bladder in the cat, Journal of Physiology. 425:545-562.

Habler, H. J., Janig, W. & Koltzenburg, M., 1993, Myelinated primary afferents of the sacral spinal cord responding to slow filling and distension of the cat urinary bladder, Journal of Physiology. 463:449-460.

Janig, W. & Koltzenburg, M., 1990, On the function of spinal primary afferent fibres supplying colon and urinary bladder, Journal of the Autonomic Nervous System. 30 Suppl:S89-S96.

Maggi C.A., 1990, The dual function of capsaicin-sensitive sensory nerves in the bladder and urethra, in 'Neurobiology of Incontinence' pp 77-83 Ciba Foundation Symposium.

McMahon S.B., and Abel, C., 1987, A model for the study of visceral pain states: chronic inflammation of the chronic decerebrate rat urinary bladder by irritant chemicals, Pain. 28:109-27.

Torrens, M. and Morrison, J.F.B., 1987, "The Physiology of the Lower Urinary Tract," Springer-Verlag, London.

Waddell, P.J., Lawson, S.N. and McCarthy, P.W., 1989, Conduction Velocity changes along the processes of rat primary sensory neurones, Neuroscience. 30:577-584

MUSCARINIC RECEPTOR SUBTYPES IN NORMAL, FETAL, AND GRAVID RABBIT BLADDER, HEART AND UTERUS

Steven B. Brandes MD#, Michael R. Ruggieri PhD*#

Department of Urology#
Department of Pharmacology*
Temple University School of Medicine
Philadelphia, PA

Supported by NIH grants #DK39086, DK43333 and DK42890

ABSTRACT

In the rabbit bladder, pregnancy has been shown to induce a significant decrease in both muscarinic receptor density and response to muscarinic stimulation. Neonatal rabbit bladders have a high muscarinic receptor density and contractile response to bethanechol stimulation. The bladders from 7 gravid rabbits, 7 age–matched virgin controls, and 32 fetal rabbits of 3 week gestation were studied. Compared to control tissue, filtration binding demonstrated receptor density to be 24.3% lower in gravid bladder dome, 41.2% lower in gravid bladder base, and 114.8% higher in fetal bladders. While total receptor density was not different from control in gravid heart, fetal hearts showed a 2.5 fold increased receptor density. There was also a 61% reduction in muscarinic receptor density in the gravid uterus.

Immunoprecipitation assays using muscarinic receptor subtype specific antisera were used to measure the relative levels of m1, m2, m3 and m4 receptors. The m2 receptor was the predominant subtype in the bladder and uterus, and the only subtype detected in rabbit heart. The m3 receptor protein was also present, but in lower levels in the bladder and uterus. The m1 and m4 receptors were not detected in any of the tissues studied. Furthermore, the relative percent of each receptor did **not** statistically change for the gravid or fetal rabbit bladder, uterus, or heart, when compared to its control. Differences in the contractile response to cholinergic stimulation of the gravid bladder and uterus, and of the fetal bladder then, can be attributed to changes in muscarinic receptor density and **not** to changes in receptor subtype.

INTRODUCTION

In the gravid rabbit, pregnancy has been shown to induce a significant (50 percent) decrease in both muscarinic receptor density and response to muscarinic stimulation (1). Neonatal rabbit bladders show a marked, yet stable muscarinic receptor density and contractile response to bethanechol and field stimulation (2). To understand the mechanism of these changes in muscarinic response, we studied the muscarinic receptor subtypes and receptor densities in the gravid, virgin, and fetal rabbit bladder. Studies were also carried out on the hearts of these animals since this organ has a similar muscarinic receptor subtype distribution as the bladder. The uterus was also studied since it is another smooth muscle organ that is dramatically affected by pregnancy.

Muscle, Matrix, and Bladder Function,
Edited by S. Zderic, Plenum Press, New York, 1995

Molecular cloning studies have identified five muscarinic receptor genes (m1–m5) that are expressed in multiple tissues (3). By using purified receptors from pig heart (m2), fusion proteins of the nonconserved segments of the third intracellular loop (m1), or c–terminal regions (m3 and m4) of these genes as antigens, subtype specific antisera have been developed, and were used here in immunoprecipitation assays (4,5). These assays provide a direct measure of the amount of the molecule directly involved in transducing the neurotransmitter signal (the receptor protein) and not merely the amount of mRNA for the receptor protein, which often does not correlate at all with amounts of receptor protein. In addition, these immunoprecipitation assays avoid reliance on muscarinic receptor subtype selective drugs which only provide a limited degree of subtype selectivity.

MATERIALS AND METHODS

All chemicals were of analytic grade. Tris (tris–hydroxymethylamine), EDTA (ethylenediaminetetra–acetic acid), atropine, fine grade Sephadex G–50, goat antimouse IgG_1–agarose, and sodium cholate were purchased from Sigma Chemical Company (St. Louis, MO). Pansorbin cells were obtained from Calbiochem (La Jolla, CA). Protease inhibitors, pepstatin, leupeptin, soybean and lima bean trypsin inhibitors, apoprotein, and alpha–2 macroglobulin were from Boehringer Mannheim Biochemicals (Indianapolis, IN). Digitonin was purchased from Gallard–Schlesinger Industries (Carle Place, NY). [^{3}H] Quinuclidinyl benzilate (QNB, 43Ci/mMol) was purchased from Dupont–New England Nuclear Research Products (Wilmington, DE). Number 30 glass fiber filters were from Schleicher & Schuell (Keene, NH). Biosafe II scintillation cocktail was from Fisher Scientific (Pittsburgh, PA). Antisera to the m1–i_3 loop, m2–i_3 loop, m3–c terminal, and m4–c terminal receptor subtypes have been previously described (4,5) and were a generous gift from Dr. Gary R. Luthin (Hahnemann University, Philadelphia, PA). Tissue from six month old age–matched virgin controls and three–week gravid New Zealand White rabbits (4 to 5 kg weight), as well as from their three–week fetuses were used (HRP, Denver, PA).

Tissue Preparation

The bladder body and base, heart ventricle, and uterine fundus of 7 gravid rabbits and 7 age–matched virgin controls (6 month–old), and the whole bladders and hearts of their 32 fetuses were studied by immunoprecipitation and radioligand filtration binding assays. Each rabbit was euthanized by cervical dislocation, followed by the removal of the urinary bladder, uterus, and heart. The bladder was removed in its entirety proximal to the urethra, and separated at the level of the ureteral orifices into a bladder base and dome. From each 21 day–old fetal rabbit the whole urinary bladder and heart were removed and used for study. Each tissue was then frozen on dry ice and stored at –80 ^{o}C.

Immunoprecipitation Assay

Immunoprecipitations were performed each time in triplicate, as previously described by Luthin et. al. (5). Each tissue was weighed, thawed and minced at 4 ^{o}C in 100mg/ml of TE buffer (1mM Tris, 1mM EDTA at pH 7.5) and 10ug/ml of each protease inhibitor (PIC; pepstatin, leupeptin, aprotinin, trypsin inhibitors from soy and lima bean, and alpha 2–macroglobulin). The tissue was homogenized with a Brinkman Polytron homogenizer by two–ten second bursts at setting 10, with 2 minutes of cooling between bursts. Membranes were incubated with ^{3}H–QNB (1.5nM) at 25 ^{o}C for 30 minutes and then centrifuged at 20,000 X g for 10 minutes at 4 ^{o}C. To solubilize the receptors, the pellet was resuspended in 1% TEDC (1mM Tris, 1mM EDTA, 1% digitonin, and 0.2% cholate) and PIC (100ug/ml) and incubated for 45 minutes at 4 ^{o}C. The homogenate was then centrifuged at 30,000 X g for 30 minutes at 4 ^{o}C and the supernate collected and transferred in 450ul volumes to microfuge tubes. 50 ul of muscarinic antisera (m1–m4) at their optimized concentration was added to the solubilized receptors. They were then incubated at 4 ^{0}C for 16–18 hours by continuous inversion on a Scientific Industries Rotator. Bound and free ligand were separated by desalting the 500ul samples over 3 ml Sephadex G–50 columns (0.7x13cm) that were equilibrated with 4ml of 0.1% TEDC buffer. Bound ligand was eluted using 1.5ml of buffer, and the receptor–antibody complexes were precipitated using pansorbin (m1, m3 and m4 receptors) or goat anti–mouse IgG_1 beads (m2 receptors). Each precipitate was then counted at 28% efficiency in 15ml of Bio–safe II scintillation cocktail, using a Beckman LS 5000TA scintillation counter. By using clonal cell lines transfected with each of the muscarinic receptor subtypes, the relative efficiencies of the m1, m2, m3 and m4 antibodies to precipitate receptors were determined to be 100%, 75%, 85%, and 100% respectively. All data were corrected accordingly for antibody efficiency.

Radioligand binding assay

Each frozen tissue was weighed, thawed and minced at 4 ^{0}C in 100mg/ml of TE buffer (pH 7.5) then homogenized using two–ten second bursts on a Brinkman Polytron homogenizer. Homogenates were centrifuged at 20,000 X g for 30 minutes and resuspended in 100mg/ml of TE buffer. For each binding assay performed in triplicate, 200ul of homogenate and 700ul of TE buffer were each added to six varying concentrations of ^{3}H–QNB (43 Ci/mMol) ranging from 31pM to 2nM. For each concentration of radioligand, atropine (10uM) was used to determine nonspecific binding. After 120 minutes of incubation at 32 ^{0}C, each sample was quenched with 5 ml of ice cold TEN (TE, 150mM NaCl at pH 7.5). The samples were filtered slowly under mild vacuum (1 ml/second) over glass fiber filters (#30, Schleicher & Schuell, Keene, NH) and washed twice with 5ml of TEN. Filters were counted at 28% efficiency in 4ml of Bio–safe II scintillation cocktail (Fisher Scientific). Protein concentrations of heart, bladder, and uterine membrane fractions were determined using the method of Lowry (6).

Analysis of data

The results shown in Figures 1–4 are representative triplicate determinations performed on 3 to 5 separate animals, except for the fetal tissues, where only one triplicate determination could be performed on the tissues pooled from the 32 animals. Saturation isotherms were transformed using the method of Scatchard (7). Kd and Bmax values were estimated by using unweighted linear regression analysis of the transformed data. Unless otherwise stated, all data were analyzed by analysis of variance or a two–tailed Student's t–test, where appropriate. P values < 0.05 were considered to be statistically significant.

RESULTS

No significant difference was noted in bladder dome wet weight between control bladders (1.57 ± 0.16gm) and gravid bladders (1.60 ± 0.40gm). Figure 1 presents filtration binding saturation curves that show gravid bladder dome maximal binding decreased by 24.3% ($P<0.001$) and increased in fetal bladder by 114.8% ($P<0.001$), with gravid Bmax at 38 ± 1.2 fMol/mg, fetal at 109 fMol/mg, and control at 56 ± 7.6 fMol/mg. Scatchard analysis is also presented, demonstrating essentially unchanged dissociation constants, with gravid bladder Kd at 0.09 ± 0.01 nM , fetal at 0.04nM , and control at 0.12 ± 0.07 nM. Figure 2 presents filtration binding for the bladder base, with 41.2% fewer muscarinic receptors in gravid bladder (33 ± 13.2 fMol/mg; P=0.005) as compared to control bladder (56 ± 7.6 fMol/mg). In the bladder base Kd was found to be statistically different (gravid at 0.19 ± 0.12 nM and control at 0.09 ± 0.01 nM; $P<0.001$).

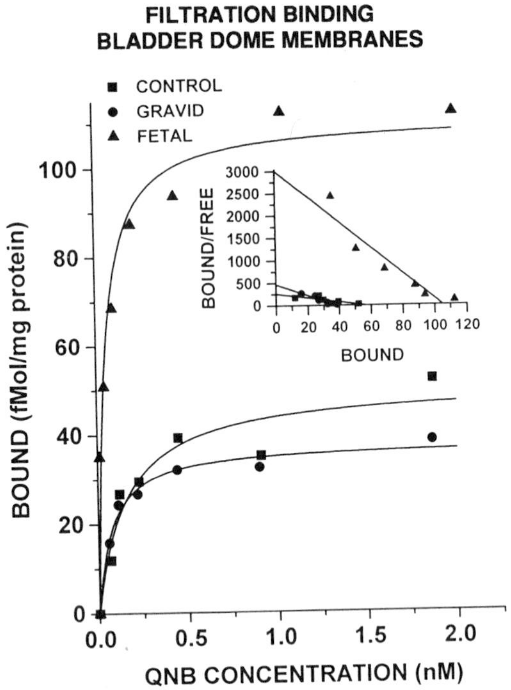

Figure 1. Representative saturation isotherms of specific [^{3}H] QNB binding to bladder dome membranes from gravid and virgin control rabbits and to whole bladder membranes from pooled rabbit fetuses. The Scatchard replot of these data is shown in the inset.

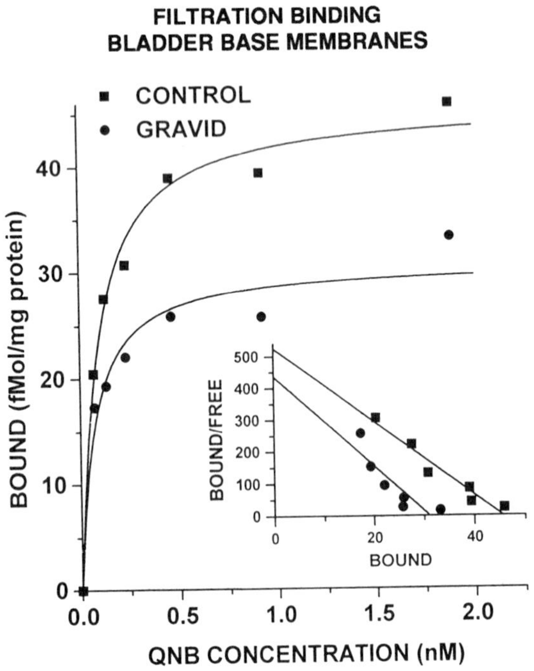

Figure 2. Representative saturation isotherms and Scatchard replots of specific [^{3}H] QNB binding to bladder base membranes from gravid and virgin control rabbits.

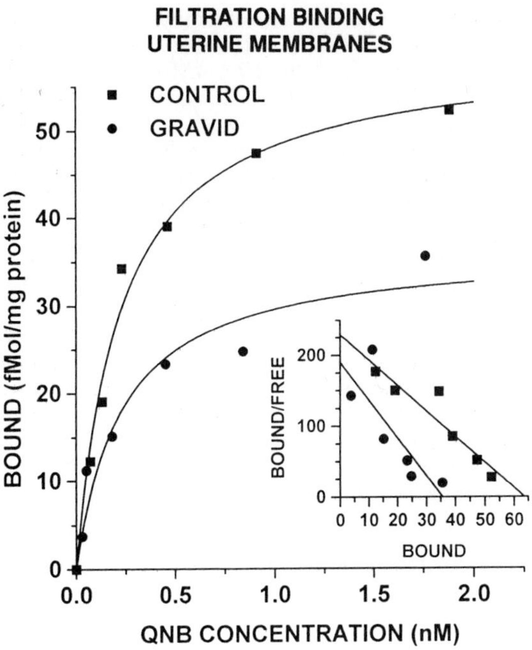

Figure 3. Representative saturation isotherms and Scatchard replots of specific [^{3}H] QNB binding to uterine fundus membranes from gravid and virgin control rabbits.

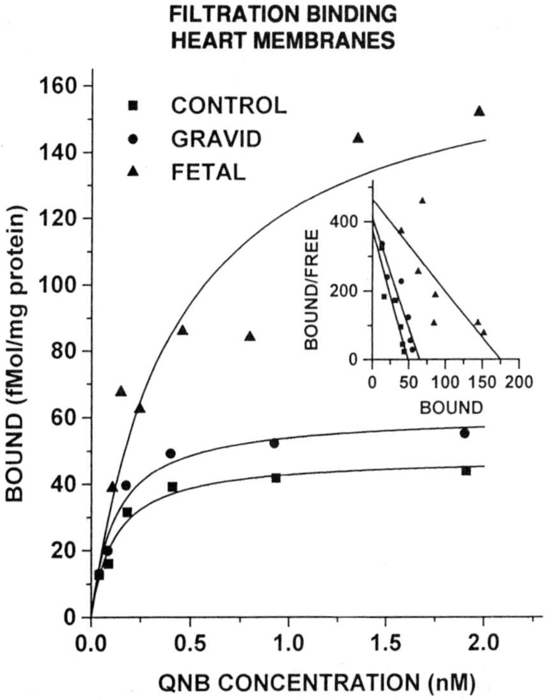

Figure 4. Representative saturation isotherms and Scatchard replots of specific [^{3}H] QNB binding to heart ventricle membranes from gravid and virgin control rabbits and to whole heart membranes from pooled rabbit fetuses.

Figure 3 demonstrates the effects of pregnancy on muscarinic receptor density in another smooth muscle model, the uterine fundus. A 61% reduction in density is demonstrated (15±11.3 fMol/mg vs. 39±19.6 fMol/mg; P=0.02). Pregnancy also significantly altered uterine muscarinic receptor Kd (control Kd at 0.04±0.01 nM and gravid Kd at 0.14±0.05 nM; P<0.001). Figure 4 shows that heart ventricle muscarinic receptor density remains unchanged by pregnancy, with gravid Bmax at 57±6.5 fMol/mg and control at 50±2.3 fMol/mg. Fetal heart receptor density, however, is 147% greater at 125 fMol/mg (P<0.001). The affinities, however, are not significantly different for the gravid (0.15±0.01 nM), fetal (0.21 nM) or control heart tissues (0.29±0.21 nM).

Table 1 represents the relative distribution of muscarinic receptor subtypes in the gravid and control virgin bladder dome and base, heart ventricle and uterine fundus, and in the whole fetal bladder and heart by fMol receptor precipitated per assay and by percent of total receptor determined by desalting over Sephadex G50 columns. The relative distribution of m2 and m3 receptors did **not** statistically change (p>0.05) for each of the gravid or fetal organ tissues, when compared to their respective controls. Thus, pregnancy induced changes in muscarinic receptor density in the bladder dome and base and the uterine fundus can be attributed to an overall and proportional down regulation of all muscarinic receptors detected, rather than of a specific subtype. Furthermore, differences in fetal muscarinic receptor density and responses to cholinergic agonists are then not attributable to a changes in receptor subtype.

TABLE 1. Distribution of Muscarinic Receptors Subtypes

Rabbit Tissue	Total Receptors (fMol/assay)	m1	m2	m3	m4
Bladder Dome					
–Control	73.69±20.43	—	52.91±19.13 (71.28±9.85)	9.35±4.77 (12.19±3.76)	—
–Gravid	66.77±18.88	—	44.44±16.31 (65.41±9.43)	9.77±3.29 (14.48±2.27)	—
Bladder Base					
–Control	66.8	—	37.4	4.7	—
–Gravid	55.8	—	32.6 (56.0)	6.1 (7.1)	—
Fetal Bladder	123.6*	—	93.5* (75.7)	20.3* (16.5)	—
Heart Ventricle					
–Control	138.2	—	122.6 (88.7)	1.2 (0.9)	—
–Gravid	79.1	—	78.7 (89.2)	0.4 (0.5)	—
Fetal Heart	199.1*	—	192.4* (96.7)	0.1 (0.1)	—
Uterus Fundus					
–Control	42.85±27.44	0.36 (0.7)	28.02±24.20 (60.82±21.22)	1.88±0.85 (5.28±2.06)	0.35 (0.7)
–Gravid	23.09+10.30	0.04 (0.2)	13.75±8.46 (58.37+14.92)	1.25±0.88 (5.28+2.06)	0.05 (0.2)

Muscarinic receptors were immunoprecipitated by subtype specific antisera. Shown are the means and standard deviations of all experiments in fMol/assay, each done in triplicate. The percentage of total receptors are shown in parentheses.

* statistically different from control, as determined by Student's t test.

DISCUSSION

High affinity estradiol and progesterone receptors have been clearly demonstrated in the female rabbit bladder, urethra, and uterus, with receptor density greatest in the uterus (8–10). To study the effect of these hormones on the autonomic function of the bladder, most prior studies have used exogenously administered estrogen, progesterone, or ovariectomy (11–15). The generated data has been conflicting, yet clearly shows an alteration in the physiology and pharmacology of the bladder (11–15). When given chronically, which more closely resembles a pregnancy model, when estradiol is administered by a subcutaneous osmotic pump to adult female rabbits over 3 weeks, there is a significant decrease in bladder body muscarinic receptor density, and no change in the bladder base or urethra (13). In contrast, when estradiol is given for 4 days, twice daily by injection at super physiological levels, dome and mid bladder muscarinic receptor density increases, while bladder base stays unchanged (12).

The hormonal changes of pregnancy induce changes in the physiology and pharmacology of the rabbit bladder. Despite no significant change in gravid bladder weight or in–vitro bladder capacity, pregnancy induced a 35% decrease in muscarinic receptor density, based on radioligand binding (34 fMol/mg vs 22), a 50% reduction in response to bethanechol, and a markedly increased response to ATP (16). Our data closely mirrors and reconfirms that pregnancy induces muscarinic receptor down regulation in the bladder, as well as in the uterus. These changes were again in the absence of change in bladder dome weight, as noted by others (13). This decrease in muscarinic receptor density can logically explain the functional decrease to bethanechol stimulation.

Fetal Bladder

1 day and 1 week old rabbit bladders compared to mature rabbit bladders have been shown to generate intravesical pressures 40% higher to bethanechol stimulation and 266% higher to ATP (17). With our 3 week gestation fetal whole bladders, we observed a 115% greater muscarinic receptor density. This increased density can explain the increased generated pressures. This has been confirmed by others, where muscarinic receptor density, response to cholinergic agonists, and bladder cholinergic innervation are present at high levels at birth (4 weeks gestation) and remain stable with increasing age, up to 15 weeks of life (18). In addition, neonatal bladder responses to ATP, serotonin, histamine and substance P were found to be 2.5 to 4.8 fold greater than in adult bladders (2). Nonadrenergic, noncholinergic neurotransmitters appear to play a greater role than once thought in the control of fetal and neonatal bladder function. Immunoprecipitation assays demonstrated m2 as the predominant receptor in fetal bladder, and to lesser degree m3. The distribution of fetal bladder subtypes did not differ with increased age of the control bladders. Based on prior studies in multiple species, only the m2 and m3 receptor subtypes have been isolated in the bladder, and we so limited our immunoprecipitation assays on bladder (19).

As others (4), the percent of precipitated receptors did not add up to 100 percent in all tissues studied. One possible explanation is that our data were not adjusted for nonspecific binding. Preliminary immunoprecipitation assays on rabbit bladders were performed in the presence of 1uM atropine, and could only account for between 1.2 to 4.5 percent nonspecific binding thus these nonspecific binding assays were subsequently not routinely performed. The antisera precipitation efficiencies for the muscarinic receptors subtypes were determined on transfected clonal cell lines which may not accurately reflect the precipitation efficiencies in rabbit tissue.

Gravid and Virgin Bladders

In general, bladder emptying is regulated pharmacologically primarily by muscarinic and purinergic mechanisms. Muscarinic cholinergic stimulation accounts for 60% of in–vitro whole bladder pressure generation in response to field stimulation and 40% by purineric stimulation. However, 90% of in–vitro bladder emptying to field stimulation is do to muscarinic cholinergic receptors. (20).

The density of most autonomic receptors including muscarinic receptors can be regulated by modifying the degree of receptor activation (21,22). Following prolonged administration of a muscarinic receptor agonist or increased neuronal stimulation, there is a decreased muscarinic receptor density, and thus decreased response to agonists. Conversely, following prolonged muscarinic receptor antagonist administration there is an associated increase in muscarinic receptor density (23–26). As has been previously described by our laboratory and others, m2 and m3 receptors are present in the urinary bladder in multiple species (19). In the pregnant human the bladder undergoes many functional changes, among them an increase in residual volume, decrease in bladder tone and increase in urinary stress incontinence. The decreases in muscarinic receptor density in both the rabbit bladder dome and base by 24% to 41%, and the previously detailed decreases in gravid bladder contractile response to bethanechol (16), help to explain these clinical findings.

Heart

Similar to the bladder, the fetal heart was found to contain significantly greater density of muscarinic receptors than either virgin controls or gravid adult rabbits. The physiologic significance of this increased receptor density in the heart, as in the bladder is not clear. Whether this increased fetal muscarinic receptor density is only found in tissues such as the heart and bladder that contain a predominance of the m2 subtype or is increased in all muscarinic receptor containing tissues awaits further investigation.

Uterus

There is some controversy as to which muscarinic receptors are present in the uterus. Dorje et. al. (4) demonstrated by immunoprecipitation assays that the rabbit uterus predominantly has m2 (65%), m4 (15%), and to a lesser degree m3 (4.5%) receptors. We also clearly demonstrated in our rabbits that the m2 receptor predominates (55 to 60%), with a small number of m3 receptors (6.6 to 7.9%); m1 and m4 receptors were not detected in our series. With total receptors precipitated totalling only 68%, it is possible that other receptor subtypes are present (e.g. m4 or m5), but were not detected. We found a 61% decreased muscarinic receptor density in gravid uterus. This may offer an explanation of how smooth muscle uterine contractions are suppressed until the time of labor, which is essential for a successful pregnancy. As in bladder tissue, the virgin and pregnant uterus retained the same relative distribution of m2 and m3 muscarinic receptor subtypes. It appears then, that pregnancy does not cause a selective down regulation of a specific muscarinic receptor in smooth muscle, but rather a global and proportional reduction in all muscarinic receptor subtypes.

ACKNOWLEDGEMENT

The authors wish to acknowledge Dr. Gary Luthin for gifts of subtype specific antisera as well as guidance in performing the immuniprecipitation assays.

BIBLIOGRAPHY

1. Zderic, S.A., Plazak, J.E., Duckett, J.W., Snyder, H.M., Wein, A.A., Levin, R.M.: Effect of pregnancy on rabbit urinary bladder physiology. Pharmacology, 41: 124, 1990.
2. Zderic, S.A., Duckett, J.W., Wein, A.A., Snyder, H.M., Levin, R.M.: Developmental factors in the contractile response of the rabbit bladder to both autonomic and nonautonomic agents. Pharmacology, 41: 119, 1990.
3. Bonner, T.I., Buckley, N.J., Young, A.C., Brann, M.R.: Identification of a family of muscarinic acetylcholine receptor genes. Science, **237**:527, 1987.
4. Dorje, F., Levey, A.F., Brann, M.R.: Immunological detection of muscarinic receptor subtype proteins (m1–m5) in rabbit peripheral tissues. Mol. Pharmacol., 40: 459, 1991.
5. Luthin, G.R., Harkness, J., Artymyshyn, R., Wolfe, B.: Antibodies to synthetic peptide can be used to distinguish between muscarinic acetylcholine receptor binding sites in brain and heart. Mol. Pharm., 34: 327, 1988.
6. Lowry, U.H., Roseborough, N., Farr, A.L., Randall, R.J.: Protein measurement with the folin reagent. J. Biol., 193: 265, 1951.

7. Scatchard, G.: The attraction of protein for small molecules and ions. Ann. N.Y. Acad. Sci., 51: 660, 1949.
8. Iosif, C.S., Batra, S.W.: Estrogen receptors in the human female lower urinary tract. Am. J. Obstet. Gynec., 141: 817, 1981
9. Batra,S.W., Iosif, C.S.: Female urethra: a target for estrogen action. J. Urol., 129: 418,1983.
10. Batra,S.W., Iosif, C.S.: Progesterone receptors in the female lower urinary tract. J. Urol., 138: 1301, 1987.
11. Hodgson, D.R., Heesch, C.M.: Effect of estrogen on sensitivity of rabbit bladder and urethra to phenylephrine. Invest. Urol.,16: 67, 1978.
12. Levin, R.M., Shofer,F.S., Wein, A.J.: Estrogen induced alterations in the autonomic responses of the rabbit urinary bladder. J Pharmacol. Exp. Ther., 215:614, 1980.
13. Shapiro, E.: Effect of estrogens on the weight and muscarinic cholinergic receptor density of the rabbit bladder and urethra. J. Urol., 135: 1084, 1986.
14. Batra, S., Andersson, K.E.: Oestrogen induced changes in muscarinic density and contractile responses in the female rabbit urinary bladder. Acta Physiol. Scand., 137: 135, 1989.
15. Levin, R.M, Jacobowitz, D., Wein, A.A.: Autonomic innervation of rabbit bladder following estrogen administration. Invest. Urol., 17: 449, 1981.
16. Levin, R.M., Zderic S.A., Ewalt D.H., Duckett, J.W., Wein, A.J.: Effects of pregnancy on muscarinic receptor density and function in the rabbit urinary bladder. Pharmacology, 43:69, 1991.
17. Keating, M.A., Duckett, J.W., Snyder, H.M., Wein, A.A., Potter, L., Levin, R.M.: Ontogeny of bladder function in the rabbit. J. Urol, 144:766, 1990.
18. Levin, R.M., Malkowicz, S.B., Jacobowitz, D.,Wein, A.J.: The ontogeny of the autonomoic innervation and contractile response of the rabbit urinary bladder. J. Pharmacol. Exp. Ther., 219:250, 1981.
19. Wang, P., Luthin, G.R., Ruggieri, M.R.: Muscarinic acetylcholine receptor subtypes mediating urinary bladder contractility and coupling to GTP binding proteins. J. Pharmacol. Exp. Ther., in press, 1994
20. Levin, R.M., Ruggieri, M.R., Wein, A.J.: Function effects of the purinergic innervation of the rabbit urinary bladder. J. Pharmacol. Exp. Ther., **236**:452. 1986.
21. Ruffolo, R.R, Kopia, G.A., Importance of receptor regulation in the pathophysiology and therapy of congestive hesrt failure. JAMA, 80 (suppl 2b): 67, 1986.
22. Axelrod, J. Regulation of circadian rhythms of indolamines in the pineal gland. In: The pineal gland and its endocrine role, ed. Axelrod, J.,Fraschini, F., and Velo, G.P., Plenum Press, New York, pp. 1–11, 1983.
23. Cawley, T.A., Shickley, T.J., Ruggieri, M.R. and Luthin, G.R. Effect of chronic neuroleptic treatment on central and peripheral muscarinic receptors. J Pharmacol Exp Ther, 267, 134– 139, 1993.
24. Levin, R.M., Chung, A.L., Aitada, S., Garber, B.B., Wein, A.J.: Effect of contractile activity on muscarinic receptor density and the response to muscarinic agonists. J. Pharmacol. Exp. Ther.,247:624, 1988.
25. Lee. W., Wolfe, B.B.: Regulation of muscarinic receptor subtypes and their responsiveness in rat brain following chronic atropine administration. Mol. Pharmacol., 36: 749, 1989.
26. Lee R.M., Malkowicz, S.B., Jacobowitz, D., Wein, A.J.: The ontogeny of the autonomic innervation and contractile response of the rabbit urinary bladder. J. Phamacol. Exp. Ther., 219: 250, 1981.

FETAL BOVINE BLADDER: PHYSIOLOGY AND PHARMACOLOGY

Harry P. Koo MD,[1] Edward J. Macarak, PhD,[2] John W. Duckett, MD,[1]
Howard M. Snyder, III, MD,[1] Alan J. Wein, M.D,[3] and
Robert M. Levin, PhD[3]

[1]Division of Urology
Children's Hospital of Philadelphia
[2]Department of Anatomy and Histology
University of Pennsylvania School of Dental Medicine
[3]Division of Urology, Hospital of University of Pennsylvania
Philadelphia, PA 19104

INTRODUCTION

The function of the lower urinary tract is to store urine at low intravesical pressure, and then to periodically expel the urine via a highly coordinated sustained contraction. To a large extent, urinary bladder function is dependent on the state of the neuronal innervation, the structure of the organ as a whole and the contractile response of smooth muscle elements[1,2].

Changes in bladder function may lead to significant morbidity in both pediatric and adult populations (incontinence, infection, renal dysfunction). Some causes of dysfunctional bladder may be the sequelae of postnatal events such as neurologic injury or outlet obstruction secondary to benign prostatic hyperplasia. However, children with posterior urethral valves or myelodysplasia are often born with poorly functioning, non-compliant bladders resulting from abnormal in utero development.

Although numerous experimental studies have addressed urinary bladder innervation, physiology and pharmacology, little information is available concerning the developmental aspects of bladder function[3-6]. This stems from reluctance to study normal infants, the difficulty with obtaining fresh fetal tissue, and the technical difficulty of performing studies on immature rodent and rabbit bladders. The fetal bovine bladders provide an excellent model to study the developmental physiology of the lower urinary tract. The cow's gestational period is approximately the same as man's. Virtually all gestational ages are available in unlimited quantities. All fetal bladders are obtained fresh, within minutes of maternal sacrifice at a local abattoir.

Bladder Compliance

One of the major aspects of bladder function that was correlated with gestational age was compliance. Compliance, which is determined by measuring the change in intravesical volume per change in intravesical pressure, is the property that allows the bladder to accommodate large volumes of urine while maintaining low intravesical pressure. By definition, "passive properties" refer to the compliance characteristics exhibited by the bladder in a calcium-free medium, whereas "resting properties" refer to the bladder's compliance characteristics in the presence of calcium. Compliance consists of both passive tension and an

Muscle, Matrix, and Bladder Function,
Edited by S. Zderic, Plenum Press, New York, 1995

active component ("tone"). Thus "passive properties" reflect passive tension only, while "resting properties" reflect both passive tension and "tone."

Coplen et al.[7] demonstrated using the fetal bovine model that compliance is low at early gestation, and increases subsequently. Their studies also indicated that a substantial component of the low compliance observed at early gestation was related to active tone, which could be eliminated by pre-incubation in calcium free buffer containing a calcium chelator, EGTA.

Active Properties of the Bladder

It is generally agreed that acetylcholine is the primary neurotransmitter controlling bladder contraction[1]. Muscarinic receptor density and response to muscarinic stimulation are greatest in the dome, and lowest in the base, thus making for efficient emptying[8]. Developmentally, cholinergic receptors (and responses to cholinergic stimulation) appear during early gestation whereas adrenergic alpha and beta receptors (and their stimulated response) appear later[6,9,10]. In a set of experiments utilizing isolated strips fetal bovine bladder, Lee et al.[11] demonstrated that the bladder's response to cholinergic stimulation develops prior to its response to field stimulation, and that its autonomic receptor density develops along a similar time-line as its response to specific autonomic agonists[12].

Although the studies by Lee et al.[11] correlate contractile response and autonomic receptor distribution to gestational age, there is presently no indication of how autonomic response development, as exhibited by the strips, correlates with the bladder's ability to generate pressure and empty. Studies correlating isolated strip techniques and whole-bladder methods have demonstrated that the linear isometric contraction of isolated strips does not always reflect the ability of the whole bladder to generate pressure or empty[1,13,14].

In Vitro Whole Bladder Technique

We have recently adapted the whole bladder techniques to fetal bovine bladders in order to provide a functional correlate to the previous muscle strip studies by Lee et al.[11] The isolated whole bladder preparation permits the investigator to have complete control of the internal and external environment and effectively evaluate bladder capacity, and functional responses of the bladder to electrical and pharmacological stimulations (pressure generation and ability to empty). Whole fetal bovine bladders were separated into three groups based on gestational age as determined by measurement of crown-rump length[15]: I. early gestation II. mid gestation III. term. Cystometrograms were performed in physiologic buffer. Bladder pressure response to field stimulation (in vitro representation of neuronal stimulation) and to specific pharmacological agonists (cholinergic, a-adrenergic, purinergic) were measured.

The pharmacological studies[16] demonstrated several unique responses of the fetal bladder: 1) The pressure response to field stimulation was approximately one-half that of bethanechol (muscarinic stimulation) for each age group. 2) There was no response to phenylephrine (a-adrenergic stimulation) or to ATP (purinergic). 3) Field stimulation emptied approximately 30% of the volume placed in the bladder. 4) Bethanechol stimulation emptied approximately 50% in the youngest fetal bladders and 100% in the two older groups. 5) There was a rapid phasic relaxation induced immediately upon the termination of field stimulation.

CONCLUSIONS

In conclusion, the fetal bovine bladder provides an excellent model for studying in utero bladder physiology and pharmacology. The studies using the developmental fetal bovine model demonstrate that bladder function in utero is both qualitatively and quantitatively different from postnatal bladder physiology and pharmacology. By understanding the steps in normal bladder development, it may be possible to gain insight in to the pathogenesis of bladder dysfunction and make it possible to design better pharmacological therapy.

REFERENCES

1. S.A. Zderic, R.M. Levin, A.J. Wein: Voiding function: relevant anatomy, physiology, pharmacology and molecular aspects, *in:* "Adult and Pediatric Urology," J.Y. Gillenwater, J.T. Grayhack, S.S. Howards, and J.W. Duckett, (eds.), 3rd Edition, Year Book Med. Pub. Inc., Chicago (1995).
2. A.R. Mundy, Structure and function of the lower urinary tract, *in:* "Scientific Basis of Urology," A.R. Mundy, ed., Churchill Livingstone, New York (1987).
3. C.A. Peters, S. Vasavada, D. Dator, M. Carr, E. Shapiro, H. Lepor, J.D. McConnell, A.B. Retik, and J. Mandell, The effect of obstruction on the developing bladder, *J. Urol.* 148:491 (1992).
4. M.A. Keating, J.W. Duckett, H.M. Snyder, A.J. Wein, L. Potter, and R.M. Levin, Ontogeny of bladder function in the rabbit, *J. Urol.* 144:766 (1990).
5. S.A. Zderic, J.W. Duckett, H.M. Snyder, A.J. Wein, and R.M. Levin, Ontogeny of bladder compliance, *Neurourol. Urodyn.* 9:595 (1990).
6. D. Mitolo-Chieppa, S. Schonauer, G. Grasso, E. Cicinelli, and M.R. Carratu, Ontogenesis of autonomic receptors in detrusor muscle and bladder sphincter of human fetus, *Urology* 25:599 (1983).
7. D.E. Coplen, E.J. Macarak, and R.M. Levin, Developmental changes in normal fetal bovine whole bladder physiology, *J. Urol.* 151:1391 (1994).
8. R.M. Levin, F. Shofer and A.J.Wein, Cholinergic, adrenergic, and purinergic response of sequential strips of rabbit urinary bladder *J. Pharmacol. Exp. Ther.* 212:536 (1980).
9. B.A. Kogan and H.S. Iwamoto, Lower urinary tract function in the sheep fetus: studies of autonomic control and pharmacologic response of the fetal bladder, *J. Urol.* 141:1019 (1989).
10. R.M. Levin, S.B. Malkowicz, D. Jacobowitz, and A.J. Wein, The ontogeny of the autonomic innervation and contractile response of the rabbit urinary bladder *J. Pharmacol. Exp. Ther.* 219:250 (1981).
11. J.G. Lee, D.E. Coplen, E.J. Macarak, A.J. Wein, and R.M. Levin, Comparative studies on the ontogeny and autonomic responses of the fetal calf bladder at different stages of development: involvement of the nitric oxide on field stimulated relaxation, *J. Urol.* 151:1096 (1994).
12. J.G. Lee, E.J. Macarak, D.E. Coplen, A.J. Wein, and R.M. Levin, Distribution and function of the adrenergic and cholinergic receptors in the fetal calf bladder during mid-gestional age, *Neurourol. Urodyn.* 12:599 (1993).
13. R.M. Levin, K. Brandler, and A.J. Wein: Comparative pharmacologic response of an in vitro whole bladder preparation (rabbit) with response to isolated smooth muscle strips, *J. Urol.* 130:377 (1983).
14. R.M. Levin, R.M. and A.J. Wein, Response of the in vitro whole bladder (rabbit) preparation to atutonomic agonists, *J. Urol.* 128:1087 (1982).
15. P. Eckstein and W.A. Kelly, Implantation and development of the conceptus, *in:* "Reproduction in Domestic Animals," H.H. Cole and P.T. Cupps, eds., Academic Press, New York (1977).
16. H.P. Koo, E.J. Macarak, J.W. Duckett, H.M. Snyder, A.J. Wein, and R.M. Levin, Developmental changes in pharmacologic and field stimulated response of the fetal bovine bladder, *J. Urol.*. 151:379A (1994).

TWO MATHEMATICAL MODELS PREDICT THE VARIATION IN CAPACITY AND COMPLIANCE OF HYPERTROPHIED BLADDERS

Margot S. Damaser[1] and Steven L. Lehman [1,2]

[1]Bioengineering Graduate Group
[2]Department of Human Biodynamics
University of California
Berkeley, CA 94720

Severe outlet obstruction causes acute bladder hypertrophy and overdistension. We have developed two mathematical models to investigate what effect hypertrophy, both with and without overdistension, has on filling mechanics. The first models hypertrophy due to a mild obstruction as an increase only in bladder mass. The second models hypertrophy and overdistension due to a severe obstruction as an increase in both bladder mass and rest volume.

METHODS

We took cystometrograms of normal catheterized, anesthetized dogs and found the equilibrium pressure at each volume using methods described previously.[1] After the dog was sacrificed, we measured bladder mass, V_w and measured wall thickness to deduce the rest volume (volume at zero pressure), V_o.

To model bladder hypertrophy due to mild outlet obstruction, we curve fit the thick walled sphere equilibrium equation for a nonlinear, isotropic, incompressible material[2] to a normal dog cystometrogram. We assumed that the stress-stretch relation remains the same in hypertrophy, and only the bladder mass increases. We simulated the cystometrogram of a hypertrophied bladder by computing the transmural pressure at 100 contained volumes for each of four values of bladder mass, V_w: two, three, four, and five times normal.

To model bladder hypertrophy due to severe outlet obstruction, we increased both V_w and V_o. Since we do not know the extent of bladder overdistension in outlet obstruction, we increased V_o to the same proportion as V_w. We curve fit the thin shell sphere approximation to a normal dog cystometrogram[1] and simulated the cystometrogram of a hypertrophied bladder by computing the bladder pressure at 100 contained volumes for each of four values of V_w and V_o: two, three, four, and five times normal.

RESULTS

Increasing V_w alone, as in the model of hypertrophy due to mild outlet obstruction, increases both plateau pressure and maximum pressure (figure 1). Increasing Vw also decreases compliance and likely decreases capacity since pressure rises at a lower volume. We define capacity as the volume at which the bladder reaches the maximum pressure in the normal cystometrogram (22 cm H_2O in figure 1). When V_w is increased to two, three, four or five times normal, capacity decreases to 89%, 79%, 66%, and 50% normal, respectively.

Muscle, Matrix, and Bladder Function,
Edited by S. Zderic, Plenum Press, New York, 1995

Increasing both V_w and V_0 proportionally as in the model of hypertrophy due to severe outlet obstruction, does not affect the maximum pressure but does increase the volume at which maximum pressure occurs, therefore likely increasing capacity (figure 2). Both capacity defined as above and compliance increase approximately linearly with V_0 in this model.

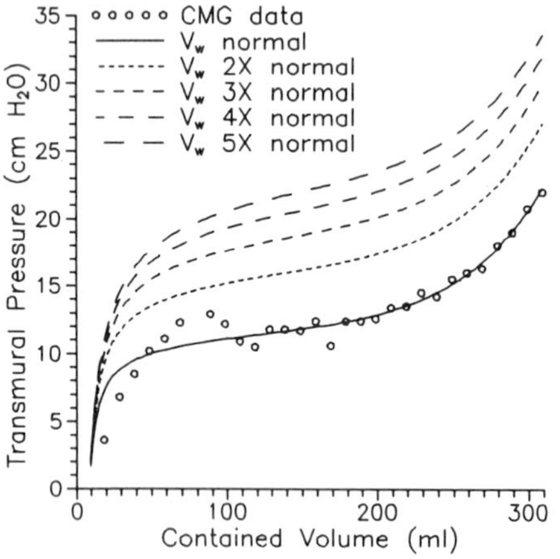

Figure 1. Simulated cystometrograms of hypertrophied bladders due to mild outlet obstruction show decreased capacity and decreased compliance

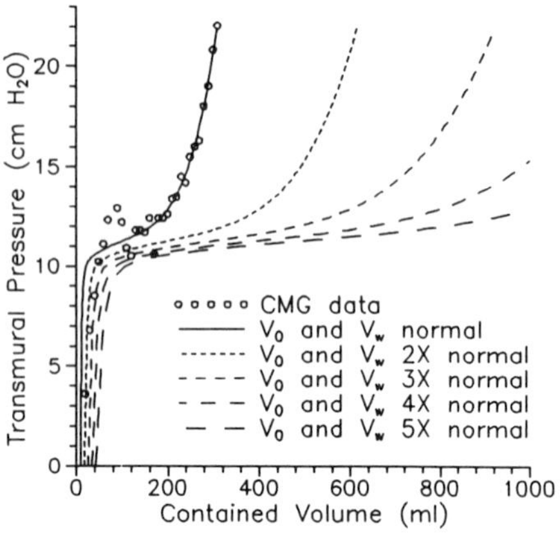

Figure 2. Simulated cystometrograms of hypertrophied bladders due to severe outlet obstruction show increased capacity and increased compliance

DISCUSSION

Our two models explain why patients with bladder hypertrophy can have a range of cystometrograms: decreased, normal, or increased capacity and compliance.[3,4] Hypertrophied bladders with decreased capacity and compliance are likely due to a mild or gradual outlet obstruction that did not cause acute bladder overdistension. Hypertrophied bladders with normal capacity and compliance are likely due to a moderate outlet obstruction that caused some overdistension. Hypertrophied bladders with increased capacity and compliance are likely due to a severe outlet obstruction which caused extensive acute overdistension.

REFERENCES

1. M.S. Damaser and S.L. Lehman, The effect of urinary bladder shape on its mechanics during filling. *J. Biomech.* in press.
2. R.F. Janz, B.R. Kubert, I. Mirsky, B. Korecky, and G.C. Taichman, Effect of age on passive elastic stiffness of rat heart muscle, *Biophys. J.* 16:281 (1976).
3. M.J. Speakman, K.K. Sethia, G.J. Fellows, and J.C. Smith, A study of the pathogenesis, urodynamics assessment and outcome of detrusor instability associated with bladder outflow obstruction, *Br. J. Urol.* 59:40, (1987).
4. J.G. Susset, Relationship between clinical urodynamics and pathologic findings in prostatic obstruction, *in:* "Benign Prostatic Hypertrophy," F. Hinman, Jr., ed., Sp[ringer-Verlag, New York (1983).

THE EFFECT OF CHOLINERGIC STIMULATION ON CULTURED SMOOTH MUSCLE CELLS

S.Bruce Malkowicz, Juanita Thomas, and Robert M. Levin

Division of Urology
University of Pennsylvania Medical Center and
Philadelphia Veterans Administration Medical Center
Philadelphia, PA 19104

ABSTRACT

Objective - To assess the role of cholinergic angonists and antagonists on the growth control of cultured rabbit detrusor smooth muscle cells.

Design - Dosage dependent growth curve analysis of smooth muscle cells in standard media to bethanechol and atropine at concentrations from 1 x10^{-5} to 1 x 10^{-9} mMol.

Results- Exogenous exposure to bethanechol results in a significant dose dependent increase in total cell number over a 10 day period. At a concentration of 1x10-6 and 1 x10-5 mMol a 20% to 50% growth increase is noted. Exogenous atropine at all concentrations decreased growth by approximately five fold.

Conclusions - The exogenous application of muscarinic cholinergic agonists and antagonists can significantly effect bladder muscle cell growth in vitro. The mechanism of growth control through these pharmacological receptors remains to be elucidated.

INTRODUCTION

The role of cholinergic transmission in the normal and pathologic function of the lower urinary tract has been a subject of intense investigation over the past two decades. The predominant role of the muscarinic cholinergic transmission in the contractile function of detrusor smooth muscle has been firmly established, yet its contribution to other aspects of cell viability have been explored to a lesser degree. Generally, cell growth and maintenance is presumed to be under the control peptide growth factors whose effects are mediated predominantly through second messenger systems. Similarly, muscarinic cholinergic transmission is mediated through secondary message systems, yet the effect of muscarinic activation beyond contractile function has not been carefully investigated in this organ system. Therefore the object of this study was to investigate the trophic effects of exogenous cholinergic stimulation and inhibition on rabbit bladder smooth muscle in a dose dependent fashion.

MATERIAL AND METHODS

Cell culture

Rabbit bladder smooth muscle was isolated from anesthetized 6-8 pound New Zealand rabbits in standard fashion and then cultured. After removal of the epithelial cells, the bladder walls were incubated in HBSS at 37° C in a 1.0 mM calcium and 0.05 mM magnesium solution with 1.0 mg collagenase/ml (type 3 sigma) for 15 minutes. The lumenal side was rescraped and the exterior surface dissected to remove the residual bladder adventitia. The remaining muscle

Muscle, Matrix, and Bladder Function,
Edited by S. Zderic, Plenum Press, New York, 1995

cells were then centrifuged at 1500 rpm for 5 minutes. The pellet was resuspended in M199 culture media with 10% fetal calf serum (FCS). The tissue suspension was added to 6 well plates pre-coated with 1% gelatin. A majority of the culture media was then removed leaving approximately 0.3 ml per well. After incubation in a 5% CO_2 incubator at 37° overnight, an additional 2 ml of culture media was added to each well. Media was changed twice weekly. Cells migrated form the explants to the substrate and proliferated. The primary passage was trypsinized with 0.0025 % trypsin/EDTA solution when the outgrowth covered 80% of the culture well. Cells were initially harvested from the tissue/cell suspension by filtration through a sterile nylon mesh. The harvested smooth muscle cells were seeded at a density of 1×10^4 cells /cm^2 in plastic culture flasks and subcultured at 8-9 day intervals. Cell cultures were incubated at 37° C in a humidified atmosphere of CO_2/95% air with media changes twice per week.

Cell growth assays

Smooth muscle cells were seed (triplicate) in 12 well plates at a density of 1×10^4 cells/cm^2. The cells were subsequently trypsinized and counted using a cell counter [Coulter Electronics Inc.) on days 1,3,5,8,10. Only cell passages 2-4 were employed in any assay. M199 culture media supplemented with 10% FCS was used for all experiments. The cultures were refed with fresh media and pharmacologic agent [1×10^{-5} mM to 1×10^{-9} mM] every three days to the completion of the experiment. Growth curves were statistically evaluated by analysis of variance at a $p < 0.05$ level. Atropine and bethanechol were obtained from standard commercial sources.

RESULTS

Immunohistochemical staining for alpha actin was performed on each batch of smooth muscle cells for phenotypic verification. All cells displayed strong positive staining (data not shown).

Smooth muscle response to exogenous bethanechol stimulation is demonstrated in figure one. Relative to control growth, a 20 to 50% increase in total cell numbers is seen after 10 days by those cells co-incubated with bethanechol.These changes are significant at the $p<0.01$ level. Increased growth is noted in a dose dependent fashion. While growth of the control muscle cells appears somewhat linear, a definite sigmoid curve is demonstrated by the cells chronically exposed to bethanechol.

Smooth muscle response to exogenous atropine administration is depicted in figure two. Relative to control growth, a 4-5 fold level of growth inhibition is seen at all concentrations of co-incubation with atropine. These changes are significant at the $p<0.001$ level.

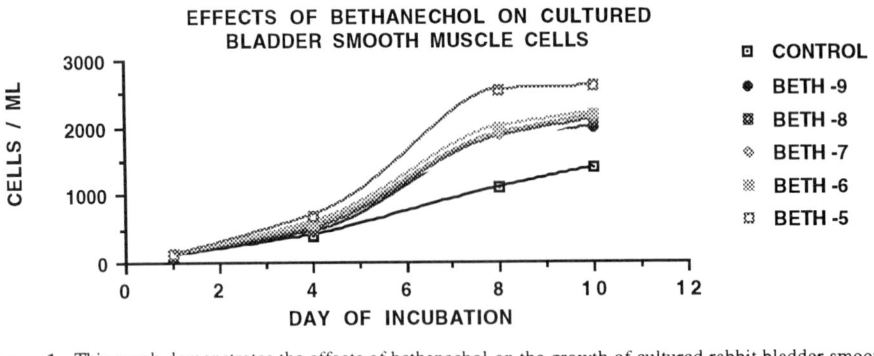

Figure 1 - This graph demonstrates the effects of bethanechol on the growth of cultured rabbit bladder smooth muscle cells. The range of concentrations is shown in a logarithmic notation. The growth rate is markedly increased in the cells grown in the presence of bethanechol at the 10th day, and the differences from control are highly significant ($P < 0.01$).

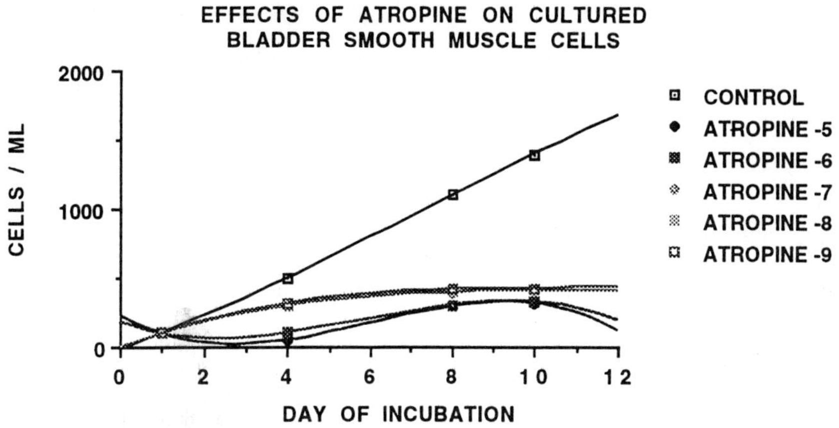

Figure 2 - This graph demonstrates the inhibitory effects of atropine on the growth of cultured rabbit bladder smooth muscle cells. The range of concentrations is shown in a logarithmic notation. The growth rates are remarkably inhibited at all atropine concentrations and are significantly different from control values at all time points. By days 8 and 10 $P < 0.01$ for all atropine concentrations tested.

DISCUSSION

Functional studies and direct bindings assays have demonstrated muscarinic cholinergic receptors in the bladder body. The muscarinic acetylcholine receptors exist as multiple subtypes with up to five demonstrated at this time (1). The M3 and M2 receptor have been described in several smooth muscle preparations and M3 receptors have be noted predominately in the rabbit destrusor. The M3 receptor activation is couple to the G protein Gq which initiates the formation of inositol triphosphate [IP3] and diacylglycerol. Both protein phosphorylation and intracellular calcium release are mediated by these substances. Data also exists to suggest a contributory role of M2 receptors to contractile activity through the Gi protein. This G protein acts negatively on andenylate cyclase and thus inhibits relaxation caused by beta- adrenergic stimulation.

Recent evidence exists however which suggests that muscarinic receptor activation may play a role other than mediation of contractractile response (2,3). Transfection of the muscarinic subtypes in parental NIH 3T3 cells created no change in the cell phenotype in the absence of a cholinergic agonist. In the presence of carbachol however M1,3,5 transfectants displayed foci of transformation while M2 and4 transfectants did not. Furthermore the antiproliferative compound L651582 has demonstrated the ability to inhibit muscarinic receptor stimulated calcium flux and arachidonic acid release (4). In PC12 cells which express M1 receptors, neurite outgrowth can be stimulated by the application of muscarinic agonists and be completely inhibited by the application of atropine (3). The exact secondary messenger pathways which effect these growth related changes are not clearly deliniated.

Pharmacologic receptors other than the muscarinic receptor family may similarly mediate growth effects. The activation of adrenergic receptors may play a role in the hypertrophic response of cardiac muscle (5). It is felt that this may be directed through the activation of protein kinase C.

The inhibitory effect of atropine is clearly more profound than the growth stimulation induced by bethanechol. The reasons for this are as yet unclear. The potential toxic effects of atropine must be considered, yet virtually the same level of growth inhibition is noted at extremely low doses of this agent. It is also possible that other growth pathways are being inhibited by atropine. The stimulatory and inhibitory effects of muscarinic receptor activity need to be investigated under different growth conditions since peptide growth factors may augment growth or abrogate inhibition elicited by cholinergic stimulation. To fully explore these issues however a better understanding of secondary messenger system interactions will be necessary if trans activation of growth factor pathways are to be evoked a possible explanation of this phenomenon.

In summary it appears that the muscarinic cholinergic system can play a role in the growth control of several in vitro systems. These initial data also display the significant growth stimulatory and inhibitory role of muscarinic activation/inhibition in in vitro on bladder muscle cells. It is possible that the cholinergic system may play a greater role in the growth response of urinary bladder smooth muscle then previously thought.

ACKNOWLEDGEMENT

Supported by the Veterans Administration, and American Cancer Society CDA 92-51.

REFERENCES

1. Eglen RM, Reddy H, Watson N, and Challis RA. Muscarinic acetylcholine receptor subtypes in smooth muscle. Trends in Pharm Sci. 15:114-119, 1994.

2. Gutkind JS, Novotny EN, Brann MR, and Robbins KC. Muscarinic acetylcholine receptor subtypes as agonist-dependent oncogenes. Proc. Natl. Acad. Sci. 88:4703-4707, 1991.

3. Pinkas-Kramarski Ronit, Stein R, Lindenboim L, and Sokolovsky M. Growth factor-like effects mediated by muscarinic receptors in PC12M1 cells. J. Neurochemistry 59:2158-2166, 1992.

4. Felder CC, Ma AL, Liotta LA and Kohn EC. The antiproliferative and antimetastatic compound L651582 inhibits muscarinic acetylcholine receptor-stimulated calcium influx and arachiodonic acid release. J. Pharm. Ex. Ther. 257:967- 971, 1991.

5. Chien KR, Knowlton KU, Zhu H, and Chien S. Regulation of cardiac gene expression during myocardial growth and hypertrophy:molecular studies of an adaptive physiologic response. FASEB Journ. 5:3037-3046, 1991.

AUTOLOGOUS PERIURETHRAL FAT INJECTIONS FOR THE CREATION OF BLADDER OUTLET OBSTRUCTION

Douglas A. Canning

Division of Urology
University of Pennsylvania
School of Medicine
Children's Hospital of Philadelphia
Philadelphia Veterans Administration Medical Center
Philadelphia, PA. 19104 (USA)

INTRODUCTION

The enormous social, emotional, and financial impact of urinary incontinence in the United States is most apparent when one considers two major facts: 1) urinary incontinence is the primary reason many elderly Americans require institutional care, and 2) the total cost of incontinence to our society is estimated to exceed 10 billion dollars in 1987 (1). Over the years a number of surgical procedures were developed to provide an anatomic solution to this problem, and while they have enjoyed much success they are not applicable in every instance. Furthermore, there is controversy regarding the long term outcomes of continence surgery. For years urologists have searched for an ideal substance that could be endoscopically injected alongside the urethra to enhance outlet resistance and improve urinary incontinence. When one considers the cost of open surgical procedures with the resulting admission and/or loss of time from work, any effective endoscopic procedure would be a major advance.

The problem has been the identification of the ideal injectable agent because the currently available substances have been criticized as dangerous or transient. Polytef Paste® has the benefits of being easily injectable and permanent and has been used with success in several clinical applications such as urinary incontinence (2) and reflux (3). It's use was curtailed in the United States by reports associating Teflon® long term local and remote inflammatory responses (4). Organic xenografts such as bovine collagen break down with long term follow-up and require reinjections (5) Chondrocytes in alginate suspension appear promising in subdermal injection (6) and have been recently tested in the genitourinary tract (7) but long term results are not yet available. We have been experimenting with autologous fat as a possible alternative to these substances in clinical practice.

These techniques also have application in the experimental study of bladder outlet obstruction. Most available models produce an acute outlet obstruction (8), yet this is rare in clinical practice. Most infravesical obstructions are chronic and incomplete. The model most commonly employed at the University of Pennsylvania calls for placing a silk ligature around the urethra in which an 8 french catheter has been placed. Usually only one person is responsible for these obstructions to try and produce uniformity. Despite this the obstruction that result are variable, and the onset is acute. Some bladders compensate following this initial insult, while others remain decompensated with poor contractile performance (8). Mostwin (9) has suggested the use of a jewelers jump ring placed around the urethra of a neonatal guinea pig which allows the animal to slowly "grow into the obstruction" as an alternative method. The goal of this review is to highlight some of our previous work in the use of autologous fat injections, and it particular

Muscle, Matrix, and Bladder Function,
Edited by S. Zderic, Plenum Press, New York, 1995

to see if by serial injections it is possible to induce changes in bladder function that are consistent with outlet obstruction.

MATERIALS AND METHODS

Female New Zealand White rabbits were divided into two groups: 8 received fat injections and 6 were given equal volumes of saline injections (control group). Frequency and micturition volumes were determined in metabolic cages for each rabbit prior to surgery, and these urodynamic patterns were compared to those seen after fat injection. Ketamine/Xylazine mixture (29.2 mg/ml Ketamine, 8.3 mg/ml Xylazine) was used for sedation, and adequate anesthesia was maintained with intravenous Nembutal. Perivesical space fat was minced into pieces with a scalpel to a size allowing for injection via a 20 gauge needle. Fat was injected via a small anterior cystotomy incision just superior to the bladder neck. Aliquots of 0.5 cc were injected below the urothelium of the bladder neck at the 3, 6 and 9 o' clock positions approximately 1 cm inferior to the trigone. Six weeks later a second set of identical fat injections were carried out in the experimental group. Controls received an equivalent volume of saline at identical positions. Two weeks after the second injection, we transferred all rabbits to the metabolic cages and measured micturition frequency. Seventy-two hours later, we cystoscoped each rabbit and removed the bladder including the bladder neck and proximal urethra.

All bladders were weighed and three longitudinal muscle strips 10 x 2 mm (mean weight 58.1 mg: no significant difference between the experimental and control group) were taken from the bladder body in each rabbit. Muscle strip studies were carried out in a standard fashion using well described techniques from our laboratory (10). The length of optimal force generation (Lo) was determined by stretching the strips and testing their response to field stimulation with a Grass S-88 field stimulator. Active contractile responses to electrical stimulation and to cholinergic stimulation with Bethanechol was determined. All data was expressed as mean $\pm$ standard error of the mean (SEM) and the contractile responses were normalized to gram tension/100 mg tissue. Statistical analysis was performed with the unpaired Student's T test with a $P < 0.05$ considered statistically significant.

RESULTS

Rabbits voided more frequently and with small volumes following fat injection. The average 24 hour frequency of micturition increased from 2.6 ± 1.5 in the control to 5.7 ± 0.8 in the obstructed group ($P<0.05$). Cystoscopy revealed excellent preservation of the fat implants and the bladder necks were soft and supple; the control animals all appeared endoscopically normal. Bladder weights were increased in the animals with autologous fat injections (4.38 ± 0.2 g.) as compared with controls (3.6 ± 0.2 g.) ($P < 0.05$) which is evidence in support of hypertrophy. Muscle strips from the fat-injected rabbits generated increased tension (21 ± 2 g/100 mg tissue) (in response to 32 Hz field stimulation) when compared to saline injected controls (13 ± 2 g/100 mg tissue). The strips from the fat injected group also generated greater tension in response to bethanechol stimulation than did those strips from saline injected controls (22 ± 2 versus 16 ± 2) ($P < 0.05$). Length tension curves showed the strips from fat injected animals were less compliant than their saline injected counterparts.

DISCUSSION

Our previous work has shown very clearly that in a rat model, small volumes of fat, taken from the perivesical area and transplanted to the bladder dome persist for as long as 6 months (11). Transplanted fat cells initially rely on diffusion to survive the initial injection trauma. Long term survival requires in growth of peripheral vessels to the center of the implant (12). If too large an initial volume is injected, the center of the injected graft may undergo ischemic necrosis. Very little is known about the effectiveness of fat achieving urethral obstruction in the rabbit. The extensive experience of the University of Pennsylvania urology lab in using the rabbit model of acute outlet obstruction led us to consider using this approach to create a model of outlet obstruction. This model, with the potential for slowly increasing the degree of obstruction, could provide a better alternative to the acute models used currently.

It is clear even from this small scale study that adipose injected below the urothelium creates enough of a pressure gradient to induce hypotrophy of the urinary bladder smooth muscle (13). This was apparent because of the increased bladder mass and increased contractile response to electrical field stimulation and bethanechol. That these effects were significant in vivo was also supported by the findings of increased micturition frequency with diminished volumes. Kato made similar observations in rabbits with more traditional models of obstruction producing acute onset mild obstructions. The two-fold increase in micturition frequency in our study is less than the four fold increase observed in rabbits obstructed with a silk ligature (14). This data supports our contention that the obstructions produced by fat injections are milder. There is reason to suspect that in acutely induced moderate obstructions there is an initial phase of decompensation associated with ischemia. As this resolves, the bladder may then undergo compensatory changes. Our data suggests that the degree of obstruction in rabbits obstructed with fat is mild, resulting in no initial period of overdistension, ischemia and decompensation.

The ability to create mild bladder outlet obstruction which consistently stimulates increased detrusor contractility and bladder hypertrophy provides a new research tool in the study of bladder outlet obstruction. The next step in this work is to perfect the technique of endoscopic injection of these fat deposits such that several small injections could be carried out over several months to produce a slow occlusion of the outlet. The study of which fat types (visceral versus subcutaneous) work best is also essential if the work is to be applied to human use. An operational rabbit model would be ideal to test these options, and also offer the urologic scientist access to a model of long term mild chronic obstruction.

REFERENCES

(1) Hu, T. (1990: Impact of incontinence on health care costs. J Am Geriatric Soc 38: 292-295.

(2) Politano, V. A. (1978): Periurethral Teflon injection for urinary incontinence. Urol Clin No Amer 5: 415-422.

(3) Canning, D. A. (1991): New implants for endoscopic treatment of reflux. Dialogues in Ped Urol 14: 608.

(4) Malizia, A.; Reiman, H.; Myers, R.; Sande, J.; Barham, S.; Benson, R.; Dewanjee, M.; Utz, W. (1984): Migration and granulomatous reaction after periurethral injection of Polytef (Teflon). JAMA 251: 3277-3281.

(5) Shortliffe, L.; Freiha, F.; Kessler, R.; Stamey, T.; Constantinou, C. (1989): Treatment of urinary incontinence by the periurethral implantation of glutaraldehyde cross-linked collagen. J Urol 141: 538-541.

(6) Vacanti, C. A.; Cima, L. G.; Woo Seob, K.; Atala, A. (1993): Injectable chondrocyte alginate suspension as a potential treatment for incontinence. J Urol 149: 291-A.

(7) Atala, A.; Kim, W.; Paige, K.; Vacanti, J.; Vacanti, C. A.; Retik, A. (1993): In Abstract #65, American Academy of Pediatrics Annual Meeting; Washington DC.

(8) Zderic, S.A., Wein, A.J., Monson, F.C., Levin, R.M. (1991): The effects of infravesical obstruction on the bladder. Problems in Urology 5:357-368.

(9) Mostwin, J.; Karim, O.; Van Koeveringe, G.; Brooks, E. (1991): The guinea pig as a model of gradual urethral obstruction. J Urol 145: 854-858.

(10) Zderic, S.A., Liu, G., Plzak, J., Monson, F., Gong,C., Levin, R.M. (1994): What is the most accurate way to study the active properties of bladder smooth muscle? Pharmacology 48:380.

(11) Matthews, R. D.; Christensen, J. P.; Canning, D. A. (1994): Persistence of autologous free fat transplant in bladder submocosa of rats. J Urol 152: 819-821.

(12) Billings, E.; May, J. (1989): Historical review and present status of free fat graft autotransplantation in plastic and reconstructive surgery. Plastic and Reconstructive Surgery 83: 368-381.

(13) Canning, D.A., Seibold, J., Daito, M., Zderic, S.A., Snyder, H.M., Duckett, J.W., Levin, R.M. (1995): Use of injectable fat to obstruct the urethra. Neurourol. & Urodynamics (in press).

(14) Levin, R.; Longhurst, P.; Monson, F.; Kato, K. (1990): Effect of bladder outlet obstruction on the morphology, physiology, and pharmacology of the bladder.The Prostate Supplement 3: 9-26.

PARTICITANTS AND FACULTY (*)

Anthony Atala
Children's Hospital
Boston, MA 02115

Ebrahim Azizi
Department of Pharmacy, PCPS
Philadelphia, PA 19104

*Laurence Baskin
University of California, San Francisco
Department of Urology
San Francisco, CA 94143-0738

Stuart Bauer
Children's Hospital
Urology Department
Boston, MA 02115

J. Thomas Benson
1633 N. Capitol Ave. 436
Indianapolis, IN 46202

Linda Brubaker
1725 W. Harrison Ste 818
Chicago, IL 60612

Ralph Buttyan
630 W. 108th St.
New York, NY 10027

Fortune Campbell
1 Hospital Drive
Columbia, MO 65212

Douglas Canning
Children's Hospital of Philadelphia
Division of Urology
34th St. & Civic Center Blvd.
Philadelphia, PA 19104

Michael Carr
4800 Sand Point Way NE
PO Box 5371
Seattle, WA 98105

Marc Cendron
Children's Hospital at Dartmouth
Department of Pediatrics
Lebanon, NH 03756-0001

*Samuel Chacko
University of Pennsylvania
School of Veterinary Medicine
Department of Pathobiology
Philadelphia, PA 19104

Min-Wei Chen
630 W. 108th St.
New York, NY 10032

Serban Constantinescu
Children's Hospital of Philadelphia
Division of Urology
Philadelphia, PA 19104

*Douglas E. Coplen
Washington University School of Med.
Urology Research Laboratories
St. Louis, MO 63110

Sandra Culbertson
Magee Womens Hospital
300 Halket St.
Pittsburgh, PA 15213

Margot Damaser
University of California
Bioengineering Graduate Group
Berkeley, CA 94720

Ross M. Decter
Hershey Medical Center
Department of Urology
P.O. Box 850
Hershey, PA 17033

*William C. Degroat
University of Pittsburgh
School of Medicine
Department of Pharmacology
Pittsburgh, PA 15261

Steven G. Docimo
The Johns Hopkins Hospital
Baltimore, MD 21287

*John W. Duckett
Children's Hospital of Philadelphia
Division of Urology
Philadelphia, PA 19104

*H. Paul Ehrlich
Milton S. Hershey Medical Center
Department of Surgery
Hershey, PA 17033-0850

*David Ewalt
Children's Medical Center
Department of Renal Urol, 5th floor
1935 Motor St.
Dallas, TX 75235

Gamal M. Ghoniem
1430 Tulane Ave.
New Orleans, LA 70112

Chao-Liang Gong
University of Pennsylvania
School of Medicine
Department of Urology
Philadelphia, PA 19104

John Gearhart
The Johns Hopkins Hospital
Division of Urology
Baltimore, MD 21287-2101

Marilyn E. Hess
University of Pennsylvania
36 John Morgan Bldg, 36th St.
Philadelphia, PA 19104

*Pamela Howard
Children's Hospital of Philadelphia
Division of Urology, Connective
Tissue Research Institute
34th St. & Civic Center Blvd.
Philadelphia, PA 19104

Hueih-Shing Hsu
Veterans General Hospital-Taipei
Department of Surgery, Div. of Urology
No. 201, Sec. Z, Shih-Pai Road
Taipei, Taiwan

Radim Kocvara
Children's Hospital of Philadelphia
Division of Urology
34th St. & Civic Center Blvd.
Philadelphia, PA 19104

*Barry Kogan
University of California,
School of Medicine
Department of Urology
San Francisco, CA 94143

*Harry P. Koo
Children's Hospital of Philadelphia
Division of Urology
34th St. & Civic Center Blvd.
Philadelphia, PA 19104

R. Lawrence Kroovand
Medical Center Blvd.
Winston-Salem, NC 27157

Chung L. Lau
University of Pennsylvania
School of Veterinary Medicine
Department of Pathobiology
Philadelphia, PA 19104

*Robert Levin
University of Pennsylvania
School of Medicine
Department of Urology
Philadelphia, PA 19104

*Bertram J. Levy
P.O. Box 218
Port Townsend, WA 98368

Victor Lin
University of Texas Southwestern
Medican Center,
5323 Hines
Dallas, TX 75235-9110

My Lindh
1800 Concord Pike
Wilmington, DE 19897

Penelope A. Longhurst
University of Pennsylvania
School of Medicine
Division of Urology
Philadelphia, PA 19104

*Edward J. Macarek
Children's Hospital of Philadelphia
Division of Urology, Connective
Tissue Research Institute
Philadelphia, PA 19104

John D. McConnell
University of Texas School of
Medicine
Division of Urology
Dallas, TX 75235-9110

Michael E. Mitchell
4800 Sand Point Way NE,
PO Box 5371
Seattle, WA 98105

*Jacek L. Mostwin
The Johns Hopkins Medical Center
Brady Urology Institute
600 N. Wolfe St.
Baltimore, MD 21287-2411

*Lee N. Nyberg, Jr.
National Institute of Diabetes and
Digestive & Kidney Diseases
National Institute of Health
Westwood Bldg., Ste 3A-05
Bethesda, MD 20892

Robert J, Padley
One Abbott Park Rd.
Abbott Park, IL 60064-3500
*Craig A. Peters
Children's Hospital
Harvard Medical School
Boston, MA

Lisa Potter-Lee
1800 Concord Pike
Wilmington, DE 19897

Mr. F. Quinn
Senior Registrar
Department of Urology
Great Ormond Street Hospital
London, England WC1N 3JH

Sidney Radomski
Toronto Western Hospital
Toronto, Ontario MST 258
Canada

*Timothy Ratliff
Washington University School of
Medicine
Urology Research Laboratories
St. Louis, MO 63110

Philip A. Rawsley
29 Orde Hall St.
London, England WClN 35L

*Joel Rosenbloom
University of Pennsylvania
School of Dental Medicine
Department of Anatomy & Histology
Philadelphia, PA 19104

*Ellen Shapiro
New York University
School of Medicine
Department of Urology
New York, NY 10016

Grahame H. Smith
Children's Hospital of Philadelphia
Division of Urology
Philadelphia, PA 19104

J. Sosinski
University of Pennsylvania
School of Veterinary Medicine
Department of Pathobiology
Philadelphia, PA 19104

*William Steers
University of Virginia Health Science Center
Department of Urology
Box 422
Charlottesville, VA 22908

*William S. Stevenson
National Cancer Institute
Laboratory of Pathology
Bldg. 10, Room 2A-33
Bethesda, MD 20892

*Gary Striker
National Institutes of Digestive &
Kidney Diseases
National Institutes of Health
Bethesda, MD 20892

*Bengt Uvelius
Lund University Hospital, S-221
Department of Urology
85 Lund, Sweden

Frank Wang
University of Pennsylvania
School of Veterinary Medicine
Department of Pathobiology
Philadelphia, PA 19104

Dana J. Weaver
1 Hospital Drive
Columbia, MO 65212

*Alan Wein
University of Pennsylvania
School of Medicine
Division of Urology
3400 Spruce St.
Philadelphia, PA 19104

Andrea Westernhausen-Larson
Magee Womens Hospital
300 Halket St.
Pittsburgh, PA 15213

Gregg D. Yasay
1800 Concorn Pike
Wilmington, DE 19897

*Stephen A. Zderic
Children's Hospital of Philadelphia
Division of Urology
34th St. & Civic Center Blvd.
Philadelphia, PA 19104

INDEX